Rare Metal Technology

2015

TMS2015

144th Annual Meeting & Exhibition

March 15-19, 2015 • Walt Disney World • Orlando, Florida, USA

Rare Metal Technology 2015

Proceedings of a symposium sponsored by

The Minerals, Metals & Materials Society (TMS)

held during

TMS2015
144th Annual Meeting & Exhibition

March 15-19, 2015
Walt Disney World • Orlando, Florida, USA

Edited by:

Neale R. Neelameggham

Shafiq Alam | Harald Oosterhof

Animesh Jha | David Dreisinger

Shijie Wang

Editors
Neale R. Neelameggham
Shafiq Alam
Harald Oosterhof
Animesh Jha
David Dreisinger
Shijie Wang

ISBN 978-3-319-48606-2 ISBN 978-3-319-48188-3 (eBook)
DOI 10.1007/978-3-319-48188-3

Chemistry and Materials Science: Professional

Published by Springer International Publishers, Switzerland, 2016
Reprint of the original edition published by John Wiley & Sons, Inc., 2015, 978-1-119-07830-2

Printed on acid-free paper

This Springer imprint is published by Springer Nature
The registered company is Springer International Publishing AG
The registered company address is: Gewerbestrasse 11, 6330 Cham, Switzerland

TABLE OF CONTENTS
Rare Metal Technology 2015

Preface ix
About the Editors xi
Session Chairs xvii

Rare Metal Processes

Industrial Practice of Biohydrometallurgy in Zambia 3
J. Wang, H. Zhao, W. Qin, X. Liu, and G. Qiu

Industrial Oxygen and Its Advanced Application Technology for Hydrometallurgy Process 11
W. Zhongling, Z. Deliang, L. Yongguo, T. Le, Y. Qing, and T. Niehoff

Electrochemical Removal Impurity of NaCl from LiCl-KCl Melts 19
B. Li, M. Shen, Y. Chen, and Q. Wang

Modern Beryllium Extraction: A State-of-the-Art Kroll Reduction Plant 27
E. Vidal, J. Yurko, and K. Smith

How to Recover Minor Rare Metals from E-Scrap 37
T. Nakamura

Solvent Extraction of Cu^{2+} with Laminar Flow of Microreactor from Leachant Containing Cu and Fe 45
J. Feng, L. Chuanhua, P. Jinhui, Z. Libo, and J. Shaohua

Precious Metals

Extraction of Gold from a Low-Grade Double Refractory Gold Ore Using Flotation-Preoxidation-Leaching Process 55
Y. Yang, S. Liu, B. Xu, Q. Li, T. Jiang, and P. Lv

Gold Extraction from a High Carbon Low-Grade Refractory Gold Ore by Flotation-Roasting-Leaching Process 63
Y. Yang, Z. Xie, B. Xu, Q. Li, and T. Jiang

Gold Leaching from a Refractory Gold Concentrate by the Method of Liquid Chlorination 71
C. Li, H. Li, X. Yang, S. Wang, and L. Zhang

The Effects of Common Associated Sulfide Minerals on Thiosulfate Leaching of Gold 79
Y. Yang, X. Zhang, B. Xu, Q. Li, T. Jiang, and Y. Wang

Hydrometallurgical Extraction of Precious, Rare and Base Metals Using an Oxidizing Acid Chloride Heap Leach 87
D. Dreisinger, N. Verbaan, C. Forstner, and R. Fitch

Recovery of Platinum Group Metals Using Perovskite-Type Oxide 101
K. Nagai, H. Kumakura, S. Yanai, and T. Nagai

Research on Process of Hydrometallurgical Extracting Au, Ag, and Pd from Decopperized Anode Slime 107
Y. Yang, W. Yin, T. Jiang, B. Xu, and Q. Li

Rare Earth Metals

Status of Separation and Purification of Rare Earth Elements from Korean Ore 117
J. Kim, H. Kim, M. Kim, J. Lee, and J. Kumar

Optimization of Rare Earth Leaching 127
G. Wallace, S. Dudley, W. Gleason, C. Young, L. Twidwell, J. Downey, H. Huang, R. James, and E. Rosenberg

Numerical Simulation of the Mass Transfer for Rare-Earth Concentrate in Leaching Process 135
T. Liu, Y. Sheng, T. Yang, B. Wang, L. Han, and Q. Liu

Apatite Concentrate, A Potential New Source of Rare Earth Elements 145
T. Sun, M. Kennedy, G. Tranell, and R. Aune

Rare Earth Elements Gallium and Yttrium Recovery from (KC) Korean Red Mud Samples by Solvent Extraction and Heavy Metals Removal/Stabilization by Carbonation 157
T. Thriveni, J. Kumar, C. Ramakrishna, Y. Jegal, and J. Ahn

Rare Earth Element Recovery and Resulting Modification of Resin Structure 169
S. Dudley, M. Chorney, W. Gleason, E. Rosenberg, L. Twidwell, and C. Young

Ultra High Temperature Rare Earth Metal Extraction by Electrolysis 177
B. Nakanishi, G. Lambotte, and A. Allanore

Vanadium-Molybdenum-Tungsten

A Novel Technology of Vanadium Extraction from Stone Coal 187
M. Wang, X. Wang, and B. Li

Mechanical Activation of Processing of Egyptian Wolframite 193
A. Abdel-Rehim and M. Bakr

Leaching of Vanadium from the Roasted Vanadium Slag with High Calcium Content by Direct Roasting and Soda Leaching 209
X. Yan, B. Xie, L. Jiang, H. Guo, and H. Li

Solvent Extraction of Vanadium from Converter Slag Leach Solution by P204 Reagent 217
T. Zhang, Y. Zhang, G. Lv, Y. Liu, G. Zhang, and Z. Liu

Effect of Solution Compositions on Optimum Redox Potential in Bioleaching of Chalcopyrite by Moderately Thermophilic Bacteria 225
H. Zhao, J. Wang, W. Qin, and G. Qiu

Poster Session

Research on Quality Improvement of Titanium Sponge by Process Optimization 231
L. Li, K. Li, Q. Miao, and C. Wang

Recovery of Rare Earth Elements from NdFeB Magnet Scraps by Pyrometallurgical Processes 239
Y. Bian, S. Guo, K. Tang, L. Jiang, C. Lu, X. Lu, and W. Ding

Study on Electrolysis for Neodymium Metal Production 249
G. Lee, S. Jo, C. Lee, H. Ryu, and J. Lee

Experimental Investigation of Recycling Rare Earth Metals from Waste Fluorescent Lamp Phosphors 253
P. Eduafo, M. Strauss, and B. Mishra

Author Index 261

Subject Index 263

PREFACE

Rare Metal Technology 2015 is the second proceedings of the symposium on Rare Metal Extraction & Processing initiated in 2014 and sponsored by the Hydrometallurgy and Electrometallurgy Committee of the TMS Extraction and Processing Division. The symposium has been organized to encompass the extraction of rare metals as well as rare extraction processing techniques used in metal production.

The intent of the symposium is to cover discussions on the extraction of rare metals, that is, less common metals or minor metals not covered by other TMS symposia. The elements considered include antimony, bismuth, barium, beryllium, boron, calcium, chromium, gallium, germanium, hafnium, indium, manganese, molybdenum, platinum group metals, rare earth metals, rhenium, scandium, selenium, sodium, strontium, tantalum, tellurium, tungsten, etc. These are rare metals of low-tonnage sales compared to high-tonnage metals such as iron, copper, nickel, lead, tin, zinc, or light metals such as aluminum, magnesium, or titanium and electronic metalloid silicon. Rare processing included biometallurgy, hydrometallurgy, electrometallurgy, as well as extraction of values from EAF dusts, and less common waste streams not discussed in recycling symposia. Rare high temperature processes to be included microwave heating, solar-thermal reaction synthesis, and cold crucible synthesis of the rare metals and the design of extraction equipment used in these processes as well as laboratory and pilot plant studies.

This proceedings volume covers about 15 rare metal elements and 15 rare earth elements in 29 papers. The symposium is organized into four sessions encompassing (1) rare metal process, (2) precious metals, (3) rare earth metals, and (4) V–Mo–W (vanadium–molybdenum–tungsten). There is also a poster session. In addition, most of the poster papers have the manuscripts included in the proceedings.

We acknowledge the efforts by the organizing and editing team consisting of Neale R. Neelameggham, Shafiq Alam, Harald Oosterhof, Animesh Jha, David Dreisinger, and Shijie Wang. We thank Bing Li who helped in the review process. In addition to above six, we thank Takashi Nakamura, Tohoku University, Japan and Joon Soo Kim, Chonnam National University, Gwangju, Korea who will form the session co-chairs. Our thanks to Trudi Dunlap and Patricia Warren of TMS in assembling the proceedings, and Matt Baker for facilitating the proceedings. We appreciate all the authors who contributed to this proceedings as well as column reformatting which was requested during the review.

Neale Neelameggham

EDITORS

Lead Editor: Neale R. Neelameggham

Neale R. Neelameggham is 'The Guru' [Professor] at IND LLC, involved in Technology marketing and international consulting in the field of light metals and associated chemicals (boron, magnesium, titanium, lithium, and alkali metals), rare earth elements, battery and energy technologies, etc. He has been a senior consultant for global engineering companies on various metal and chemical processes. In 2012 he was a visiting expert at Beihang University of Aeronautics and Astronautics, Beijing, China and was a plenary speaker at the recent Light Metals 2014 Conference in Pilanesberg, South Africa. Neale has been an advisor for several other metal production and energy process and engineering firms.

Dr. Neelameggham has over 38 years of expertise in magnesium production technology from the Great Salt Lake Brine in Utah, involved in process development of its startup company NL Magnesium through to the present US Magnesium LLC from where he retired in 2011. He was instrumental in most of the process and equipment development of all areas of the plant from the raw material source – the Great Salt Lake Brine to metal. In addition, Dr. Neelameggham's expertise includes an in-depth and detailed knowledge of all competing technologies worldwide of magnesium production, both electrolytic and thermal processes, as well as alloy development. He also has extensive knowledge of titanium chemicals and metal production. He and his partners are promoting a thiometallurgical process—a new concept of using sulfur as the reductant and or fuel—for magnesium and titanium production with the least greenhouse gas emissions. His expertise in process development extends to the development of a convective model for the global anthropogenic warming from thermal emissions which is based on all energy conversions irrespective of whether it is combustion conversion or from renewables.

Dr. Neelameggham holds 13 patents and patent applications on sustainable light metal production. He has several technical papers to his credit. As a member of TMS, AIChE, and a former member of American Ceramics Society he is well versed in energy engineering, bio-fuels, rare-earth minerals and metal processing and related processes. He has served in the Magnesium Committee of the TMS Light Metals Division (LMD) since its inception in 2000, chaired it in 2005, and in 2007 he was made a permanent co-organizer for the Magnesium Technology Symposium. He has been a member of the Reactive Metals Committee, Recycling Committee, and Titanium Committee, and a Programming Committee Representative for LMD. In 2008, LMD and the Extraction and Processing Division created the Energy Committee following the CO_2 Reduction Metallurgy Symposium that Dr. Neelameggham initiated, and he was the first chair of that committee. He has been a co-organizer of the Energy Technology symposium each year since 2008.

Dr. Neelameggham is currently the chair of the Hydrometallurgy and Electrometallurgy Committee, and he organized the Rare Metal Technology 2014 and 2015 symposia. He is the co-editor of the *Essential Readings in Magnesium Technology* compendium of TMS papers published in 2014. He is a co-editor of 2015 symposium on *Drying, Roasting and Calcining*.

Dr. Neelameggham received the LMD Distinguished Service Award in 2010. He holds a doctorate in extractive metallurgy from the University of Utah.

Shafiq Alam

Shafiq Alam is an associate professor at the University of Saskatchewan, Canada. In 1998, he received his Ph.D. degree in chemical engineering from Saga University, Japan. From 1999 to 2001 he was appointed as a post-doctoral research fellow at the University of British Columbia and the University of Toronto, Canada.

Dr. Alam has extensive experience in industrial operations, management, engineering, design, consulting, teaching, research, and professional services. Before joining academia in 2006, he worked with many different companies, such as Shell, Process Research ORTECH Inc., Fluor Canada Ltd., and the National Institute of Advanced Industrial Science and Technology (AIST), Japan. Dr. Alam is highly experienced in the area of mineral processing and extractive metallurgy. He possesses two patents and has over 120 publications in the area of extractive metallurgy. He is the co-editor of two books and an associate editor of the *International Journal of Mining, Materials and Metallurgical Engineering* (*IJMMME*).

Dr. Alam is a registered professional engineer and has worked on projects with many different mining industries including Xstrata, Phelps Dodge, INCO, Barrick Gold Corporation, Rambler Metals, and Anaconda Mining. He is an Executive Committee Member of the Hydrometallurgy Section of the Metallurgy and Materials Society (MetSoc) of CIM and currently he holds the office of secretary with MetSoc (2013-2015). Dr. Alam is also the Vice-Chair of the Hydrometallurgy and Electrometallurgy Committee of the Extraction and Processing Division (EPD) of TMS (2013-2015). He is actively involved in organizing different international conferences, such as, the Conference of Metallurgists (COM) in Canada; Ni-Co Symposium at TMS 2013 in

Texas, USA; the Copper 2013 Conference in Chile; and the Rare Metal Extraction & Processing Symposium at TMS 2014 and TMS 2015 in San Diego and Orlando, USA.

Harald Oosterhof

Harald Oosterhof graduated as a chemical engineer from Twente University in The Netherlands in 1994. In the same year, he assumed a position as researcher at TU Delft where he worked in the laboratory for process equipment. His research on anti-solvent crystallization of well-soluble salts was rewarded with two patents and a dozen publications. After receiving his Ph.D. from Delft University in 1999, he assumed the position of project manager at Umicore, a global materials and technology group that is based in Belgium. During his first assignment as Project Leader – Hydrometallurgy, he focused on the refining of cobalt, nickel, and germanium. Since 2011, Dr. Oosterhof has worked as scientist in the Recycling and Extraction Technology group at Umicore's Central R&D department. His main competence areas are special metals hydrometallurgy, recycling and refining of rare earth metals, base metal hydrometallurgy, and recycling of spent rechargeable batteries. In his current job, Dr. Oosterhof is frequently involved in business development of scarce metals recycling and he is heading a team of hydrometallurgical specialists.

Animesh Jha

Animesh Jha has been a professor of materials science at the University of Leeds in the United Kingdom since August 2000. He obtained his Bachelor of Engineering (BE) degree in metallurgy in June 1979 from the University of Roorkee (India); his Master of Engineering (ME) degree in metallurgical engineering in July 1981 from the Indian Institute of Science, Bangalore in India; and Ph.D. and DIC from Imperial College (London) in October 1984 in the area of chemical and process metallurgy. After earning his Ph.D., he did a short postdoctoral research engagement at Purdue University (US) in 1985-86, before returning to Sheffield, UK in December 1986 where he started research on phase transformations in novel salt-based and metallic glass systems. In May 1989, he took his academic position at Brunel University, Uxbridge, UK where he developed his academic career in the areas of IR glasses and mineral processing. He has since 1989 continued research in these two areas and produced over 24 Ph.D. theses from 1992 to 2014.

He has published over 400 research papers in peer-reviewed journals and reviewed conference proceedings. He has been a member of TMS since 1992, a fellow of the Institute of Physics in London, and member of the Optical Society of America and IEEE. He serves as an external reviewer for overseas research agencies including US, Canada, and EU countries. He is also actively involved in translational research, which allows the lab work to reach industry.

David Dreisinger

David Dreisinger holds the position of Professor and Industrial Research Chair in Hydrometallurgy at the University of British Columbia (UBC). Dr. Dreisinger received B.Sc. and Ph.D. degrees in metallurgical engineering from Queen's University in Kingston before beginning his career at UBC in 1984.

At UBC, Dr. Dreisinger supervises a wide ranging program of research and development in pressure leaching of ores and concentrates, solution purification, and use of electrochemical methods for metal recovery. Dr. Dreisinger has published extensively and has received (with co-workers) 17 U.S. patents for work in areas such as pressure leaching, ion exchange removal of impurities from process solutions and use of thiosulfate as an alternative to cyanide in gold leaching.

Dr. Dreisinger has received a number of awards including the Sherritt Hydrometallurgy Award (1993), the Metal Chemistry Award from CMSC (2001), the Extraction and Processing Division Science Award from the TMS (2005), and the INCO Medal (2008). Dr. Dreisinger has been named a Fellow of the Canadian Academy of Engineering and the Canadian Institute of Mining, Metallurgy and Petroleum. Dr. Dreisinger has been actively organizing conferences and teaching short courses around the world through the Industrial Research Chair in Hydrometallurgy. The Chair is sponsored by 18 international companies with an interest in Hydrometallurgy.

Shijie Wang

Shijie Wang received his B.Sc. in Mineral Processing from China and his Masters and Ph.D. in Metallurgical Engineering from the University of Nevada at Reno. Dr. Wang has experience working at the Beijing General Research Institute for Non-Ferrous Metals, ASARCO Inc., and Phelps Dodge Corporation. He is currently Superintendent of Precious Metals/Rare Metals Operations at Rio Tinto Kennecott Utah Copper. Dr. Wang has been active in extractive metallurgy and has experience in metallurgical process development, existing operation optimization, and troubleshooting. His work interests include metal recovery, operational efficiency, and profitability as well as process safety management. Dr. Wang holds three U.S. patents and has published 20 journal papers including non-ferrous metals', precious metals', rare metals', and rare earth metals' resourcefulness and recovery. He is a copper, gold, silver, selenium and tellurium refining subject matter expert. Dr. Wang has been a TMS member since 1991 and is former chair of the Hydrometallurgy and Electrometallurgy Committee of TMS from 2011 to 2013.

SESSION CHAIRS

Rare Metal Processes
Shafiq Alam
Takashi Nakamura

Precious Metals
David Dreisinger
Shijie Wang

Rare Earth Metals
Neale R. Neelameggham
Joon Soo Kim

Vanadium-Molybdenum-Tungsten
Harald Oosterhof
Bing Li

Rare Metal Technology 2015

Rare Metal Processes

Rare Metal Technology 2015
Edited by: Neale R. Neelameggham, Shafiq Alam, Harald Oosterhof, Animesh Jha, David Dreisinger, and Shijie Wang
TMS (The Minerals, Metals & Materials Society), 2015

INDUSTRIAL PRACTICE OF BIOHYDROMETALLURGY IN ZAMBIA

Jun Wang[1,2*], Hongbo Zhao[1,2*], Wenqing Qin[1,2], Xueduan Liu[1,2], Guanzhou Qiu[1,2]
([1]School of Minerals Processing & Bioengineering, Central South University, Changsha 410083, Hunan, PR China;
[2]Key Lab of Biohydrometallurgy of Ministry of Education, Changsha 410083, Hunan, PR China)

Keywords: Bio-hydrometallurgy; Industrial practice; Zambia

Abstract

Bio-hydrometallurgy technology was applied for the extraction of copper from the raw ores of Chambishi Mine in Zambia. A copper extraction of 93.29% was obtained for small scale column bioleaching within 63 days, while a copper extraction of 89.05% was achieved for large scale column bioleaching in 90 days, thus confirming the amenability of the raw ores for effective extraction via bioleaching. The bacteria were cultured in a 6-stage enlarge cultivation. The bacterial cultures with cell concentration of more than 1×10^8 cells/mL were added into a spray pond of sulfuric acid to be applied in the dump leaching, and a copper extraction of about 50% was achieved within 2 months. The production report revealed that the copper extraction increased by approximate 20%, and the acid consumption was reduced to around 35% as a consequence of adding the bacteria. The industrial demonstration of the bio-hydrometallurgy technique is now well established in Zambia, and further applications of bio-hydrometallurgy in both Zambia and Congo are currently in progress.

1 Introduction

As the world demand for copper has been increasing consistently, the metal holds as an important role in the modern industry. Currently, as high grade copper resources have or are largely exploited, low grade copper ores remain. However, those copper ores of low grade are difficult and costly processed by traditional technologies [1]. Bio-hydrometallurgy, as a simple, efficient, low-cost and eco-friendly technology, has been widely applied in the processing of low grade ores containing secondary oxides and sulfides [2-6]. Heap leaching offers the advantages of simple equipment, low investment and low operation cost [7, 8]. Currently, more than 20% of copper is produced globally is achieved using hydrometallurgy technology. The Kennecott copper company started the application of heap bioleaching of copper sulfides in the 1950s, and obtained the first patent for bioleaching. The technology of heap bioleaching-solvent extraction-electrowinning (SX-EW) was commercially applied in the year of 1970, and the total copper production in Chile and the USA using SX-EW reached greater than 2.1 Mt in 2001[9-12]. In China, the majority of the copper resources are of low grade, which are difficult and costly to be processed by traditional technologies. The first commercial heap bioleaching plant started commercial operation in the Dexing Copper Mine by the end of 1997, which was followed by the establishment of heap bioleaching plant in the Zijin Mine in 2001[13]. Zambia, especially the Chambishi district has abundant copper resources. The reserve of copper is more than 7 million tons. Therefore, the application of

* **Corresponding author1**: Jun Wang, associated professor. Email: wjwq2000@126.com, Tel: 86-731-88876557
* **Corresponding author2**: Hongbo Zhao, PhD. Email: alexandercsu@126.com, Tel: 86-731-88876557

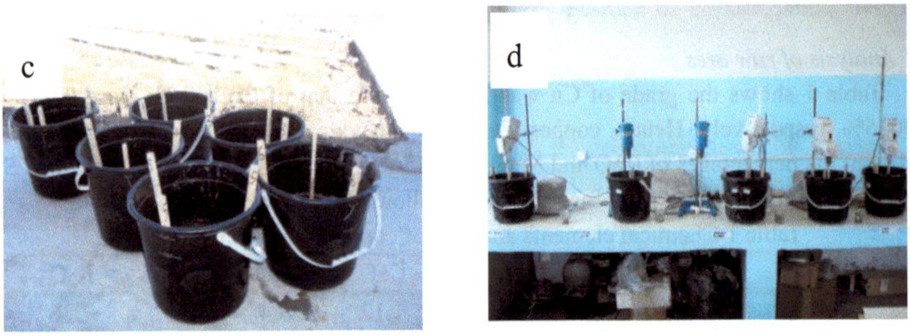

Fig. 3 Self-made stirred tanks: Stirring paddle (a), vertical view (b), front view (c), and operation (d).

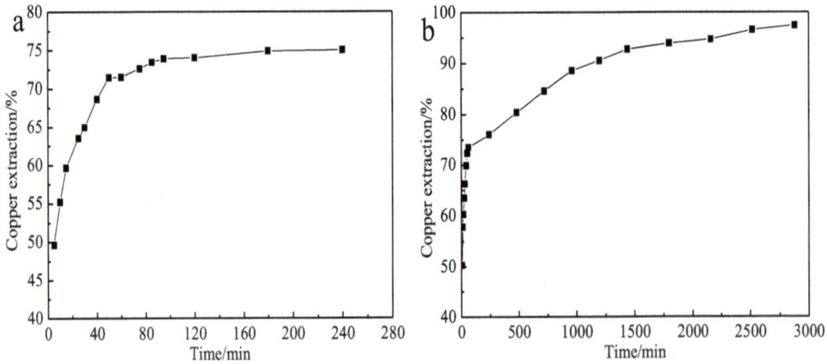

Fig. 4 Copper extractions from the leaching of raw ores: Acid leaching (a) and bacterial leaching (b).

3.3 Column bioleaching of raw ores

Column bioleaching experiments were carried out in the equipment shown in Fig. 1. The results obtained from the small scale column bioleaching and large scale column bioleaching are shown in Table 3 and Fig. 5, respectively. A copper extraction of 93.29% was achieved after small scale bioleaching for 63 days, and 89.05% was obtained after large scale bioleaching for 90 days. Therefore, copper can be effectively extracted from the raw ores by column bioleaching, and the industrial practice of heap bioleaching can most likely be further implemented.

Table 3 Chemical elements analysis of the residues from the small scale column bioleaching.

m (raw ores)/kg	m (residues)/kg	Grade (raw ores)/%	Grade (residues)/%	Results
		Cu	Cu	Cu extraction/%
	18.56	2.07	0.139	93.29

Rare Metal Technology 2015
Edited by: Neale R. Neelameggham, Shafiq Alam, Harald Oosterhof, Animesh Jha, David Dreisinger, and Shijie Wang
TMS (The Minerals, Metals & Materials Society), 2015

INDUSTRIAL PRACTICE OF BIOHYDROMETALLURGY IN ZAMBIA

Jun Wang[1,2*], Hongbo Zhao[1,2*], Wenqing Qin[1,2], Xueduan Liu[1,2], Guanzhou Qiu[1,2]
([1]School of Minerals Processing & Bioengineering, Central South University, Changsha 410083, Hunan, PR China;
[2]Key Lab of Biohydrometallurgy of Ministry of Education, Changsha 410083, Hunan, PR China)

Keywords: Bio-hydrometallurgy; Industrial practice; Zambia

Abstract

Bio-hydrometallurgy technology was applied for the extraction of copper from the raw ores of Chambishi Mine in Zambia. A copper extraction of 93.29% was obtained for small scale column bioleaching within 63 days, while a copper extraction of 89.05% was achieved for large scale column bioleaching in 90 days, thus confirming the amenability of the raw ores for effective extraction via bioleaching. The bacteria were cultured in a 6-stage enlarge cultivation. The bacterial cultures with cell concentration of more than 1×10^8 cells/mL were added into a spray pond of sulfuric acid to be applied in the dump leaching, and a copper extraction of about 50% was achieved within 2 months. The production report revealed that the copper extraction increased by approximate 20%, and the acid consumption was reduced to around 35% as a consequence of adding the bacteria. The industrial demonstration of the bio-hydrometallurgy technique is now well established in Zambia, and further applications of bio-hydrometallurgy in both Zambia and Congo are currently in progress.

1 Introduction

As the world demand for copper has been increasing consistently, the metal holds as an important role in the modern industry. Currently, as high grade copper resources have or are largely exploited, low grade copper ores remain. However, those copper ores of low grade are difficult and costly processed by traditional technologies [1]. Bio-hydrometallurgy, as a simple, efficient, low-cost and eco-friendly technology, has been widely applied in the processing of low grade ores containing secondary oxides and sulfides [2-6]. Heap leaching offers the advantages of simple equipment, low investment and low operation cost [7, 8]. Currently, more than 20% of copper is produced globally is achieved using hydrometallurgy technology. The Kennecott copper company started the application of heap bioleaching of copper sulfides in the 1950s, and obtained the first patent for bioleaching. The technology of heap bioleaching-solvent extraction-electrowinning (SX-EW) was commercially applied in the year of 1970, and the total copper production in Chile and the USA using SX-EW reached greater than 2.1 Mt in 2001[9-12]. In China, the majority of the copper resources are of low grade, which are difficult and costly to be processed by traditional technologies. The first commercial heap bioleaching plant started commercial operation in the Dexing Copper Mine by the end of 1997, which was followed by the establishment of heap bioleaching plant in the Zijin Mine in 2001[13]. Zambia, especially the Chambishi district has abundant copper resources. The reserve of copper is more than 7 million tons. Therefore, the application of

* **Corresponding author1**: Jun Wang, associated professor. Email: wjwq2000@126.com, Tel: 86-731-88876557
* **Corresponding author2**: Hongbo Zhao, PhD. Email: alexandercsu@126.com, Tel: 86-731-88876557

bio-hydrometallurgy in Zambia can contribute to the increase of copper production and efficient utilization of copper resources.

2 Materials and methods

2.1 Bacteria and culture conditions

Bacteria were obtained from the Key Lab of Biohydrometallurgy of Ministry of Education, Central South University, Changsha, China. Bacteria were cultured in 250 mL shake flasks in an orbital incubator with a stirring speed of 170 r/min and temperature of 30℃. The 9K medium used for cell cultivation consisted of the following compositions: $(NH_4)_2SO_4$ (3.0 g/L), $MgSO_4 \cdot 7H_2O$ (0.5 g/L), K_2HPO_4 (0.5 g/L), KCl (0.1 g/L) and $Ca(NO_3)_2$ (0.01 g/L). All the bacterial cultures were sub-cultured into a basal salts medium supplemented with ferrous sulfate ($FeSO_4$) and raw ores as the energy source. The resulting culture was used as inoculums for the enlarge cultivation.

2.2 Bioleaching experiments

Small scale bioleaching experiments were carried out in 250 mL shake flasks in an orbital incubator with a stirring speed of 170 r/min and temperature of 30°C. The equipment used in column bioleaching is shown in Fig. 1, and the technological process of heap bioleaching is shown in Fig. 2. The concentrations of metal ions were analyzed by inductively coupled plasma-atomic emission spectrometer (ICP-AES) (America Baird Co. PS-6).The pH values were measured with a pH meter (PHSJ-4A).

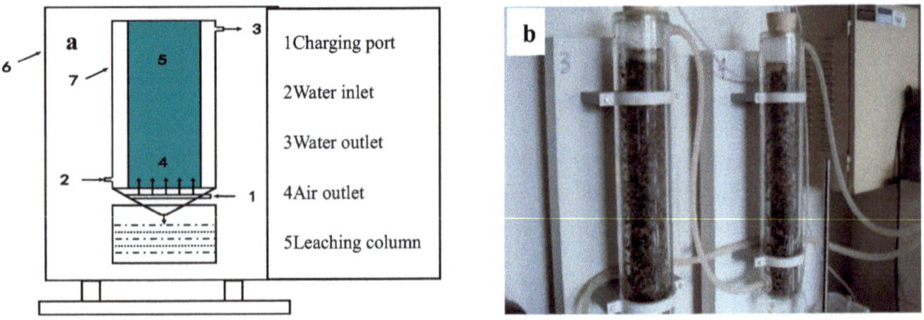

Fig. 1 Diagrammatic drawing of the small size column bioleaching system (a) and Photo of the small size column bioleaching system (b).

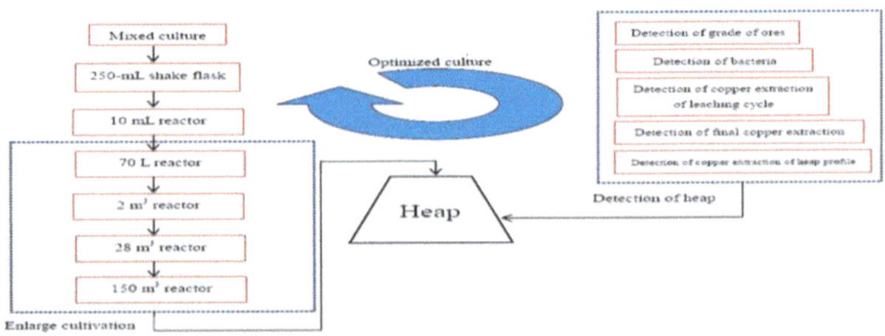

Fig.2 Diagrammatic drawing for the technological process of heap bioleaching.

3 Results and discussions

3.1 Analysis of raw ores

Table 1 shows the grade of Cu was 2.07%, while that of Co and Ni were 0.03% and 0.0035%, respectively. Hence, copper was the main valuable element of the raw ores. In addition, the percentages of CaO and MgO were low, and the main gangue mineral was SiO_2, with a content of 58.32%. Therefore, the sample should be amenable to bioleaching.

Table 1 Chemical elements analysis of copper ores of Chambishi/%.

Fe	FeO	SiO_2	Al_2O_3	CaO	MgO	MnO	K_2O
4.07	1.08	58.32	9.21	1.12	3.44	0.09	7.36
Na_2O	Cu	Co	Ni	S	Pb	Zn	Na_2O
0.20	2.07	0.03	0.0035	0.40	0.0010	0.0085	0.20

The compositions of the various copper phases shown in Table 2 indicate that 50% of the copper was distributed in free copper oxides, combined copper oxides account for 23.67%, and copper sulfides account for about 25.12%, of which were mainly chalcopyrite and bornite.

Table 2 Copper phase analysis of copper ores from Chambishi/%.

phase	AsCu	HFCu	Primary	Secondary	Total
Grade	1.06	0.49	0.475	0.045	2.07
Distribution	51.21	23.67	22.95	2.17	100

*AsCu is free copper oxide, HFCu is combined copper oxide.

3.2 Leachability experiments of raw ores

Leachability experiments were conducted in the self-made stirred tanks as shown in Fig. 3, and the results are shown in Fig.4. It illustrates that the copper extraction of 75.02% can be obtained by acid leaching, and 97.43% was achieved by using bioleaching after 2 days, thus confirming that bioleaching can be used to effectively recover copper from the raw ores.

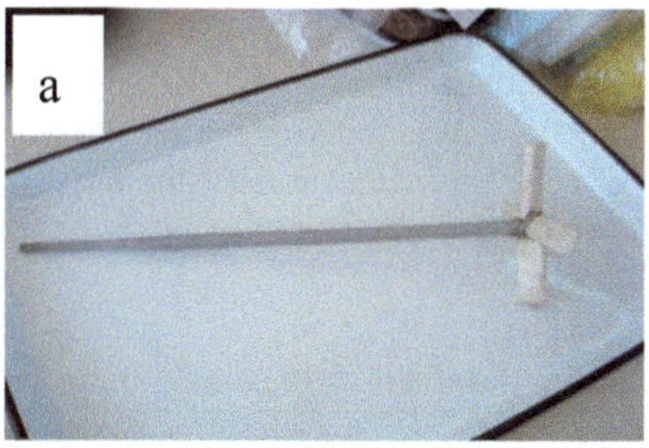

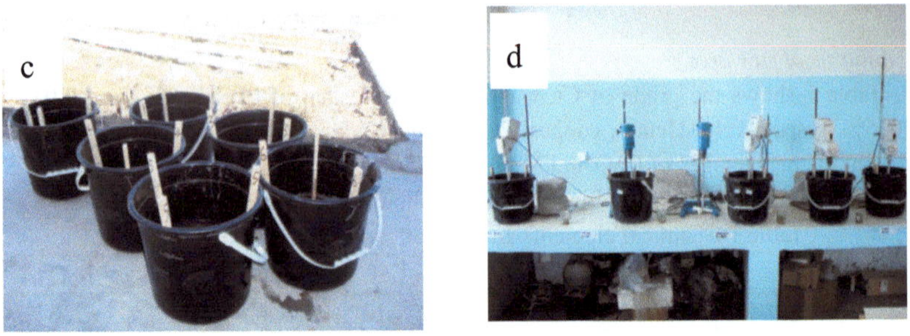

Fig. 3 Self-made stirred tanks: Stirring paddle (a), vertical view (b), front view (c), and operation (d).

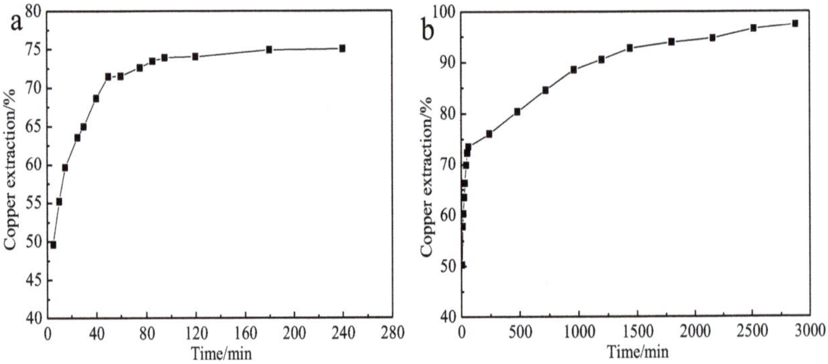

Fig. 4 Copper extractions from the leaching of raw ores: Acid leaching (a) and bacterial leaching (b).

3.3 Column bioleaching of raw ores

Column bioleaching experiments were carried out in the equipment shown in Fig. 1. The results obtained from the small scale column bioleaching and large scale column bioleaching are shown in Table 3 and Fig. 5, respectively. A copper extraction of 93.29% was achieved after small scale bioleaching for 63 days, and 89.05% was obtained after large scale bioleaching for 90 days. Therefore, copper can be effectively extracted from the raw ores by column bioleaching, and the industrial practice of heap bioleaching can most likely be further implemented.

Table 3 Chemical elements analysis of the residues from the small scale column bioleaching.

m (raw ores)/kg	m (residues)/kg	Grade (raw ores)/%	Grade (residues)/%	Results
		Cu	Cu	Cu extraction/%
	18.56	2.07	0.139	93.29

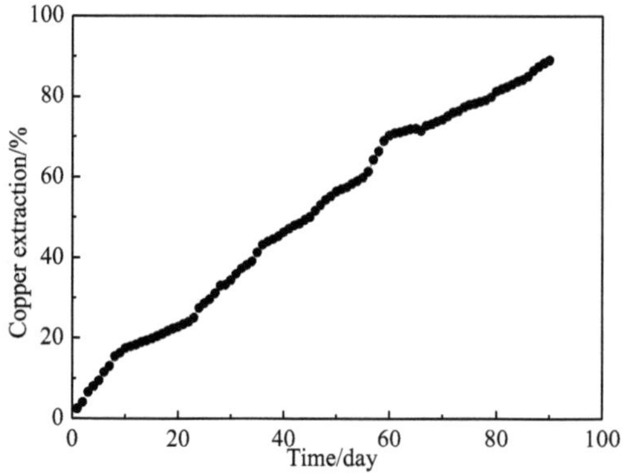

Fig. 5 Copper extraction for large scale bioleaching of raw ores.

3.4 Heap bioleaching

The bacteria were cultured in 6-stage enlarge cultivation using 250 mL-shake flasks, a 10 L-agitator, a 70 L-agitator, a 2 m^3-stirred tank, a 28 m^3-stirred tank and a 150 m^3-stirred tank. The corresponding enlarge cultivation process is shown in Fig. 6, and the corresponding cell concentrations of each stage are shown in Table 4. The bacterial cultures with cell concentrations of more than 1×10^8 cell/mL were incorporated into the spray pond of sulfuric acid to be applied in the dump leaching.

Fig. 6 6-stage enlarge cultivation of bioleaching bacteria in Chambishi.

Table 4 The cell concentration of each stage during the 6-stage enlarge cultivation.

Stage	1	2	3	4	5	6
concentrations	5×10^8	5×10^8	2×10^8	2×10^8	1×10^8	1×10^8

The 200 t and 600,000 t heap bioleaching experiments are shown in Fig. 7. A copper extraction of about 50% was achieved in 2 months. Information from the production report shown in Table 5 reveals that the copper extraction increased to about 20%, and the acid consumption reduced to approximate 35% as a result of the addition of bacteria.

Based on the successful industrial practice of bio-hydrometallurgy in Chambishi, the industrialization demonstration base of bio-hydrometallurgy was established in Zambia, and further applications of bio-hydrometallurgy in Luansha and Congo are currently in progress.

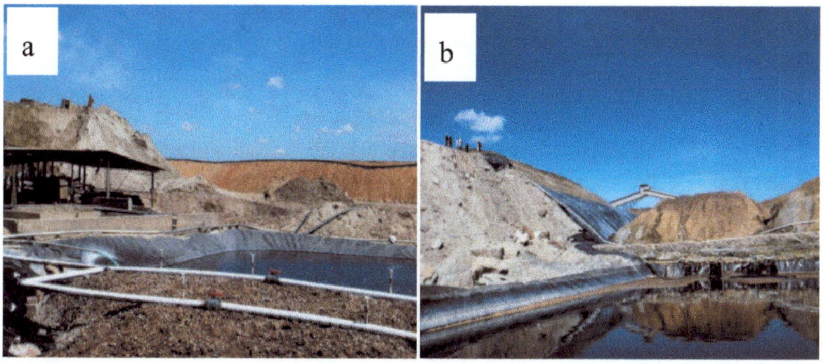

Fig. 7 Heap bioleaching in Chambishi: 200 t heap bioleaching (a) and 600,000 t heap bioleaching (b).

Table 5 Comparison of sulfur acid leaching and bioleaching processes.

Process parameters	Sulfur acid leaching	Bioleaching
Grade / %	≥1.6%	≥0.5%
Leaching Time	4-8 h	120 d
Acid concentration (g/L)	30-40	10
Bacteria	——	Mixed culture
Cu^{2+} concentration/(g/L)	4-8	4-6
Cu extraction/%	65-70%	85-90%
Acid consumption	4.88 t	2.18 t
pH of Leaching solution	1.4-1.6	1.8-2.1
Production	6000 t	>10000 t
Production cost	3344 dollars	2150dollars

4 Conclusions

Bio-hydrometallurgy technology was applied in the extraction of copper from the raw ores of Chambishi in Zambia. It was proven that the raw ores could be effectively extracted by bioleaching in the lab. The bacteria were cultured in a 6-stage enlarge cultivation through a 250 mL-shake flasks, 10 L-agitator, 70 L-agitator, 2 M^3-stirred tank, 28 M^3-stirred tank and a 150 M^3-stirred tank. The final bacterial cultures were added into a spray pond of sulfuric acid to be applied in the dump leaching, and a copper extraction of about 50% was achieved in 2 months. The obtained copper extraction increased by approximate 20%, and the acid consumption reduced to about 35% due to the addition of bacteria. Based on the successful industrial practice of bio-hydrometallurgy in Chambishi, the industrialization demonstration base for bio-hydrometallurgy was established in Zambia, and further applications of bio-hydrometallurgy in Zambia and Congo are currently in progress.

Acknowledgments

This work was supported by the National Natural Science Foundation of China (project No. 51374248), Program for New Century Excellent Talents in University (project No. NCET-13-0595), Hunan Provincial Innovation Foundation for Postgraduate (project No. CX2014B091) and the China Postdoctoral Science Foundation (project No. 2014T70692).

References

[1] S.E. Kesler et al, Earth's copper resources estimated from tectonic diffusion of porphyry copper deposits, *Geology*, 36 (2008) 255-258.

[2] G. Olson et al, Bioleaching review part B, *Applied Microbiology and Biotechnology*, 63 (2003) 249-257.

[3] T.G. T ROHWERDER et al, Bioleaching review (part A): Progress in bioleaching: fundamentals and mechanisms of bacterial metal sulfide oxidation [J], *Applied Microbiology and Biotechnology*, 63 (2003) 239-248.

[4] F. Acevedo, Present and future of bioleaching in developing countries, *Electronic journal of Biotechnology*, 5 (2002) 18-19.

[5] C. Brierley, How will biomining be applied in future?, *Transactions of Nonferrous Metals Society of China*, 18 (2008) 1302-1310.

[6] H.L. Ehrlich, Past, present and future of biohydrometallurgy, *Hydrometallurgy*, 59 (2001) 127-134.

[7] N. Pradhan et al, Heap bioleaching of chalcopyrite: A review, *Minerals Engineering*, 21 (2008) 355-365.

[8] S. Panda et al, Insights into heap bioleaching of low grade chalcopyrite ores-A pilot scale study, *Hydrometallurgy*, 125-126 (2012) 157-165.

[9] H. Watling, The bioleaching of sulphide minerals with emphasis on copper sulphides—a review, *Hydrometallurgy*, 84 (2006) 81-108.

[10] H. Watling, Chalcopyrite hydrometallurgy at atmospheric pressure: 1. Review of acidic sulfate, sulfate–chloride and sulfate–nitrate process options, *Hydrometallurgy*, 140 (2013) 163-180.

[11] W.G.L. Davenport, M. King, M. Schlesinger, A.K. Biswas, Extractive metallurgy of copper, Elsevier, 2002.

[12] H.A. Schnell, Bioleaching of copper, in: Biomining, Springer, 1997, pp. 21-43.

[13] Y. Songrong et al, Research and application of bioleaching and biooxidation technologies in China, *Minerals Engineering*, 15 (2002) 361-363.

Rare Metal Technology 2015
Edited by: Neale R. Neelameggham, Shafiq Alam, Harald Oosterhof, Animesh Jha, David Dreisinger, and Shijie Wang
TMS (The Minerals, Metals & Materials Society), 2015

INDUSTRIAL OXYGEN AND ITS ADVANCED APPLICATION TECHNOLOGY FOR HYDROMETALLURGY PROCESS

Wei Zhongling[1], Zhang Deliang[2], Liu Yongguo[2], Tang Le[2], Yin Qing[2] Thomas Niehoff[3]

1 Linde China Technolgy Center, Linde Gas, Shanghai, 201206, China;

2 Shandong Gold Mining (Xin-hui) Co., ltd, Shandong, 266715, China;

3 Linde Gas, Unterschleissheim, Munich, 85716, Germany

E-mail: rocky.wei@linde.com

Abstract: The gold ore of Shandong Gold Mining (Xinhui) was a refractory ore, which contained multi metallic sulfides (ZnS and PbS etc.) and carbonaceous. The gold ore concentrate came into leaching process and thickener and zinc replacement to recover gold. In order to improve gold recovery, the GOLDOX™ hydrometallurgy process was studied in Shandong Gold (Xinhui) by bench scale and 6 month plant scale test supported by Linde Gas. It was shown that the process promoted the leaching kinetics and improved gold recovery. The residue grade reduced 0.22 g/t in average after oxygen was applied. This indicates an operating profit increase 850,000 yuan/a in the plant. It was setting a good example for similar gold mine plant and would promote the oxygen application in Chinese gold mining industry.

Keywords: industrial oxygen, leaching kinetics, gold recovery, refractory ore

1. Introduction

The chemistry of gold cyanidation was studied in 1846 by Elsner. Almost 100 years after Elsner's recognition of the usefulness of oxygen in the cyanidation of gold, Boonstra discovered that the dissolution of gold in cyanide solutions was similar to the process of metal corrosion, where dissolved oxygen was reduced to H_2O_2 or OH^-. It was evident that for the reaction to proceed, both cyanide and oxygen must be present. The dissolved oxygen level was critical for the gold leaching process.

Some variables were affected by the addition of oxygen, for example, the required leaching time and cyanide consumption. But, some variables in the gold leaching process were unaffected by the addition of oxygen. For example, the pH value and relative density of pulp and the particle size distribution. However, when it comes to the chemistry of gold dissolution, oxygen comes into its own. Oxygen enhances the chemistry of gold dissolution, resulting in reduced reagent consumption, improved residence time, greater throughput and ultimately higher gold recoveries. In the gold mining context, oxygen is a relatively inexpensive commodity with untold potential. Its successful application in the process can hold considerable potential for the plant.

Linde Gases developed the GOLDOX™ hydrometallurgy process which involves the application of pure oxygen to stoichiometrically complete the gold leaching reaction. The process promotes kinetics of gold leaching, reduces cyanide consumption and improves recovery. Linde Gas developed technology which can help determine oxygen rates and injection methods to optimize gold recovery. GOLDOX has come about as a result of an industry requirement to improve gold production efficiency in metallurgical plants, thereby reducing production costs. Linde Gas South Africa, also known as AFROX, pioneered the application of adding pure oxygen to Rand Mine Crown leaching circuits in 1984. The GOLDOX™ process has since then been accepted worldwide. GOLDOX™ hydrometallurgy process is widely used by gold mines in Australia, South Africa and Malaysia.

The gold ore from Shandong Gold Mining (Xin-hui) was of a refractory ore which contained multi metallic sulfides (such as ZnS and PbS etc.) and carbonaceous. The gold ore concentrate goes into the leaching process followed by the thickener and zinc replacement stage to recovery gold. The deleterious effects of carbonaceous materials preg-robbing the dissolved gold in the pulp are well known. Carbonaceous material can reduce the gold recovery by restricting the release of gold from the carbonaceous matrix, or by absorbing dissolved gold from the leaching liquor. The treatment of these ores involves roasting, the addition of kerosene or heavy oils, floatation and aqueous chlorination. In the past years, a few methods were applied to improve the gold recovery, such as agent addition, increased pulp R.D. etc. In order to further improve gold recovery, GOLDOX™ hydrometallurgy process was studied at Shandong Gold (Xinhui) by bench scale and plant scale testing supported by Linde Gas in 2013.

2. Bench scale test result

2.1 Material and methods

Elemental analysis result of gold ore concentrate was listed in Table 1. In order to compare the gold recovery by normal air with oxygen in leaching process, bench scale test was done in lab of Shandong Gold (Xinhui) in September 2012. The sample was gold ore concentrate after floatation in the operating line of the plant. Every sample was 500 g gold ore concentrate, which was ball mill ground, 8 minute before bottle rolling.

Table 1: Elemental analysis of different samples of gold ore concentrate

samples	Au(g/t)	Zn（%）	Pb（%）	Fe（%）
1	31.96	12.27	11.45	0.91
2	32.65	12.53	12.05	1.02
3	39.60	16.64	14.66	0.83

Cyanide consumption was controlled at $5.0 \sim 6.0 \times 10^{-3}$, pH value control at 11 ~ 13, respectively, and leaching time was chosen 4 h, 12 h, 24 h. Each experiment was repeated twice, and the experimental results were averaged. Oxygen was supplied by oxygen cylinder.

2.2 Result and discussions

Figure 1 and Table 2 show the comparison of gold recovery by oxygen and air used in leaching process. It was found that dissolved oxygen (DO) level was 0.45 ~ 1.72 mg/l in the air bottle, much lower than saturation concentration of dissolved oxygen 8.2 mg/l. This may be due to the associated gold concentrate iron, zinc, lead sulfide ore, and lead to consume a lot of oxygen in the leaching bottle. DO level was greater than 20 mg/l in the bottle with oxygen injection. There was only a small amount of foam during the test when oxygen was injected into the bottle and DO reached greater than 20 mg/l. Each experiment was repeated twice, each sample was analyzed more than 3 times. The results were averaged.

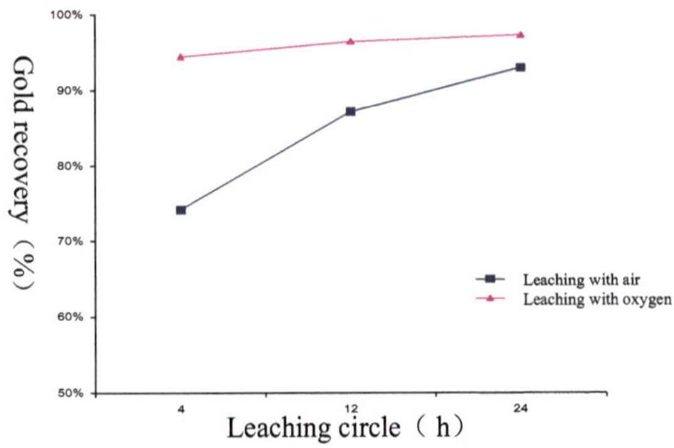

Fig.1: Compared oxygen with air in bottle roll leaching

Table 2: Result of application of oxygen and air in bottle roll leaching

Sample No.	Circle time (h)	Ore grade (Au,g/t)	Residue grade (Au,g/t)	Gold reocvery (%)
1-1	4 (oxygen)	31.96	1.74	94.56
1-2	4 (air)	32.65	8.40	74.27
2-1	12 (oxygen)	34.04	1.20	96.47
2-2	12 (air)	31.26	4.00	87.20
3-1	24 (oxygen)	42.10	1.15	97.27
3-2	24 (air)	39.60	2.77	93.01

From the result, three conclusions can be drawn. Oxygen can significantly promote gold leaching kinetics, shorten the leaching time. Oxygen injection can improve the leaching rate of gold in the same period. There was no big foam issue, which was a serious issue when air is injected in to the pulp.

The results showed that GOLDOX™ hydrometallurgical process is feasible for Shandong Gold (Xinhui), which can significantly improve the leaching rate of gold. Plant residue bottle roll test results also show that there was a certain amount of cyanide gold in the cyanide residue. It was planned to start up GOLDOX ™ oxygen leaching plant scale tests in Shandong Gold (Xinhui).

3. Plant scale test result

3.1 Material and methods

Oxygen was supplied as liquid oxygen (LOX). Linde was responsible to build a liquid oxygen station near the leaching workshop in Shandong Gold (Xinhui). It was stored in a liquid oxygen tank, whose size was 30 m^3, with a set of vaporizers. LOX was transported and charged by liquid oxygen truck. Oxygen was fed to the leaching tank via newly built stainless steel piping and flow control panel. The oxygen application system was shown in Fig. 2. The air compressor was stopped operating and all air piping was removed form leaching tanks.

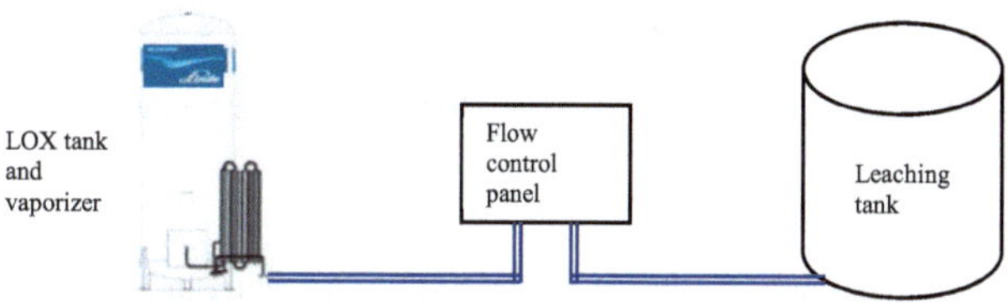

Fig.2: GOLDOX hydrometallurgy process supply system

3.2 Results and discussions

Half year trial was started in July 2013. The results are listed in Table 3, after GOLDOX™ hydrometallurgy process was implemented. It was compared with the same period in 2012 by leaching with normal air injection.

Table 3: Trail result in second half of 2013 compared with same period in the past

Month in the year	Residue grade (Au, %)	
	2012	2013
July	1.66	1.29
August	1.91	1.59
September	1.99	2.08
October	1.84	1.73
November	1.85	1.75
December	2.19	1.70
In average	1.91	1.69

It was found that the residue grade was decreased by 0.22 g/t from July to December 2013 in average compared to the same period in 2012. Plant produced residue of 24,000 t/a, gold price was chosen to 240 yuan/g as calculated, gold recovery was 99% in the follow-up process. The profit increment due to gold recovery improvement by GOLDOX™ application was about 1.25 million yuan (24,000 t * 0.22 g / t * 99% * 240 yuan/g = 125 million). The oxygen related cost was about 400,000 yuan/a. So, annual profit was increased 850,000 yuan in Shandong Gold (Xinhui) by GOLDOX™ hydrometallurgy process application.

4. Conclusions

Oxygen was a highly reactive industrial gas that promotes the leaching of gold from ores by enhancing the cyanide reactions with the slurry. Therefore, oxygen could be used to improve the leaching efficiency and increase gold recovery.

The GOLDOX™ hydrometallurgy process was successfully launched in Shandong Gold Mining (Xinhui) Co. Ltd., after a bench scale test and a 6-month plant scale test. It was setting a good example for similar gold mine plant and would promote the oxygen application in Chinese gold mining industry.

References:

1. Sir T.K. Rose, W.A.C. 1937,The Metallurgy of Gold (7th ed.)[M].Newman, USA.
2. B.A. wills ed., 2005, Advances in Gold Ore Process[M], Elsevier, Australia
3. John O. Marsden, C. Lain House, 2005, The Chemistry of Gold Extraction (2nd ed.)[M], SME, USA
4. Zhang De-liang etc., 2014, research on plant scale test of improved gold recovery by pure oxygen hydrometallurgy process application, Gold science and technology[J], submited (in Chinese)
5. Zhang De-linag, etc., 2013, reduce the gold grade for leaching residue with carbon inhibito, Gold scenece and technolgy[J], 21(3):91-94 (in Chinese)

6. Liu Yong-guo, Hao Jian-zhen, Yin Qing, 2003, Gold concentrate cyanidation practice [J]. Gold. 23(7) 41-43 (in Chinese)
7. Yin Ching,1997, Application of efficient thickener [J]. Gold. 17(7) 45-47(in Chinese)

Rare Metal Technology 2015
Edited by: Neale R. Neelameggham, Shafiq Alam, Harald Oosterhof, Animesh Jha, David Dreisinger, and Shijie Wang
TMS (The Minerals, Metals & Materials Society), 2015

ELECTROCHEMICAL REMOVAL IMPURITY OF NaCl FROM LiCl-KCl MELTS

Bing Li[1], Miao Shen[2], Yuqi Chen[1], Qiang Wang[1]
[1] East China University of Science and Technology, 130 Meilong Road, Shanghai, 200237, China
[2]Shanghai Institute of Applied Physics, Chinese Academy of Sciences, 2019 JialuoRoad, Shanghai ,201800,China

Keywords: electrochemical; impurity; NaCl; LiCl-KCl; revomal

Abstract

Electrochemical methods including cyclic voltammograms, square wave voltammograms, chronopotentiometry are applied to characterize the reduction potentials of Na^+ and Li^+ ions in LiCl-KCl melts. According to theoretical calculation, the reduction potential for Na^+ is slightly more negative than Li^+ in NaCl-LiCl-KCl melts at 450℃. However, by electrochemical measurements, the reduction potential of Na^+ is slightly more positive than Li^+ in the above melt and the potential difference between Na^+ and Li^+ reaches 0.16-0.2V on an inert W electrode in 1.0-2.0 wt% NaCl-LiCl-KCl melts. An active Pb electrode is used to improve the removal rate of Na^+ from LiCl-KCl melts.

The reduction potentials for both Na^+ ions and Li^+ ions on a Pb electrode shift to more positive values and the reduction currents become larger. After 6h electrolysis on a Pb electrode at -3.0V, the practical removal efficiency of Na^+ ions reaches 90.12% and the reduced products contain metallic Na and Li with a Na/Li mass ratio of 4.25/1.

1. Introduction

Currently, metallic lithium produced from LiCl-KCl molten salts electrolysis usually possesses a purity of about 98-99 wt% with about 0.1-0.8 wt% metallic Na, 1.0 wt% metallic K, 0.03 wt% metallic Al and 0.01-0.05 wt% metallic Ca [1]. These impurities mainly come from NaCl, KCl, $AlCl_3$ and $CaCl_2$ existing in the LiCl and KCl raw materials. In industry, traditional vacuum-distillation methods are applied in order to improve the purity of metallic lithium from 98.5 wt% to 99.9 wt%. This process is rather expensive due to both high energy consumption and serious corrosion of the reactor [2-4].

Electrochemical methods are relatively simple and can be carried out prior to the electrowinning of Li from LiCl-KCl melts. In our previous researches [5-6], an electrochemical method was employed to investigate the possibility of removing Ca^{2+} and Al^{3+} ions from $CaCl_2$-LiCl-KCl melts and $AlCl_3$-LiCl-KCl melts, respectively prior to Li^+ ions reduction.

The potential difference between the impurity ions and Li^+ ions has a significant effect on the capability of removing of the impurity ions from LiCl-KCl melts upon application of the

electrochemical method. According to Bard [7], at least a 0.2 V potential difference between the two species is required to achieve a total electroseparation.
According to table 1, only the theoretical potential difference between Al^{3+} and Li^{+} ions is about 1.7V-1.8V larger than 0.2V, while the other theoretical potential difference between K^{+}, or Na^{+} or Ca^{2+} ions and Li^{+} ions in chloride melts are less than ±0.2V in the temperature range of 400℃-800℃.

Table 1 Theoretical decomposition voltage

Chlorides	400℃(V)	500℃(V)	600℃(V)	800℃(V)
KCl	3.854	3.755	3.658	3.441
LiCl	3.722	3.646	3.571	3.457
NaCl	3.615	3.519	3.424	3.240
$CaCl_2$	3.607	3.534	3.462	3.323
$AlCl_3$	1.91	1.86	1.81	1.71

Liquid Zn electrodes can effectively remove impurity Al^{3+} from KCl-LiCl-$AlCl_3$ melts. After 10 hours of electrolysis, the removal efficiency of Al^{3+} on a liquid Zn cathode reaches 99.90 wt %. The cathode product is composed of Zn 67.80 wt% and Al 32.20 wt% without metallic Li [6].
In comparison to the tungsten electrode, liquid tin (Sn) electrodes have significantly increased the removal rate and removal efficiency of the Ca^{2+} ions in $CaCl_2$ (2.0wt%)-LiCl-KCl melts. Under stirring, 94.66 wt% Ca^{2+} ions are electrochemically removed at the Sn electrode for 6.0 h electrolysis, and the reduced product on the Sn electrode contains metallic Li and Ca with a Li/Ca mass ratio of 1/6. Obviously, during electrochemically removing Ca^{2+} ions, a certain amount of Li^{+} ions are inevitably reduced with Ca^{2+} together. Of course, it is acceptable that a certain amount of Li^{+} ions are consumed together with impurities with less than 2 wt% concentration.
According to table1, the reduction potential for Na^{+} ions is more close to that of Li^{+} ions, the reduction potential difference between them is less than 0.2V. So in this paper, an active electrode Pb is selected as the working electrode to explore the electrochemically removal efficiency of Na^{+} ions from LiCl-KCl melts with 1wt% to 2wt% NaCl.

2. Experimental

The chemicals used in the experiments include LiCl, KCl (both from Alfa Aesar Co. Ltd., AR,≥99 wt%) and NaCl (AR, ≥99 wt%). The procedures for preparation and purification of the melts are referred to our previous work [5]. The mixture of LiCl-KCl was dried under vacuum for 4h at 523K to remove water, then melted in an alumina crucible placed in a stainless steel cell located in an electric furnace. The temperature was measured with a nickel-chromium thermocouple sheathed by an alumina tube. Na^+ ions are introduced into the bath in the form of dehydrated NaCl powder.

The electrochemical measurement method, the working electrode, counter electrode and reference electrode are the same as that described as our previous researches [5]. In the electrolysis, liquid Pb is employed as the cathode and a pure graphite rod as the anode, a silver wire dipped into LiCl-KCl-AgCl melts contained in a porcelain tube was used as the reference electrode. All the potentials discussed in this paper will be measured with respect to Cl^-/Cl_2. The analysis method for the cathode products and the melts remain the same as stated in the literature [5].After electrolysis, the cathode products were analyzed by energy dispersive spectrometer and the content of Na^+ and Li^+ ions in the melts and in the cathodic products were analyzed by inductively coupled plasma-atomic emission spectrometry.

3. Results and discussion

3.1 The potential difference between Na^+ and Li^+ ions

According to Eq.(1) to Eq.(5), the theoretical potential difference between Na^+ and Li^+ ions is -0.05V at 723K and -0.095V at 953K, implying a more negative reduction potential for Na^+ ions than Li^+ ions. Therefore, based on the results, it is impossible to remove Na^+ ions prior to Li^+ ions reduction using electrochemical method.

$$Li^+ + Na \rightarrow Li + Na^+ \quad (1)$$

$$\Delta G_T = \Delta G_T^0 + \frac{RT}{ZF}\ln\frac{a_{LiCl}}{a_{NaCl}} \quad (2)$$

$$\Delta G^0 = nF\Delta E \quad (3)$$

$$\Delta E_{723} = -0.050V \quad (4)$$

$$\Delta E_{953} = -0.095V \quad (5)$$

In fact, as the NaCl concentration increases to the same as LiCl in KCl-LiCl melts, such as 18mol%, the potential difference between Na^+ and Li^+ ions at 450℃ is 0.3V according to the cyclic voltammograms (CV) results on a W electrode , as shown in Fig.1. Obviously, this value deviates from the results of previous calculation possibly as a result of ionic interactions. Therefore, according to Bard [7], the potential difference larger than 0.2V has provided the possibility of complete separation of Na^+ and Li^+ by the electrochemical method.

But as the NaCl concentration decreases below 2.0wt% (far less than LiCl concentration in KCl-LiCl melts), it becomes necessary to obtain the reduction potential difference between Na^+ and Li^+. Fig.2 shows the CVs recorded on a W electrode in LiCl-KCl-NaCl (1.0wt %) melts. The reduction peaks I_d and I_a are attributed to Na^+ ions and Li^+ ions reduction, respectively. So according to CV results, the reduction potential difference between Na^+ and Li^+ ions is about 0.085-0.096V. The chronopotentiograms at different current densities on a W electrode in LiCl-KCl-NaCl (2.0 wt%) melts at 450℃ are shown in Fig.3. As the current density is 0.09A cm^{-2}, a potential plateau at about -3.34V in LiCl-KCl-NaCl (2.0 wt %) melts is caused by Na^+ ions reduction. As the current density increases to -0.12A cm^{-2}, two potential plateaus occur. When the current density is further increased until to approximately 0.61 Acm^{-2}, two potential plateaus at about -3.40V and -3.60V are evident, which correspond to Na^+ and Li^+ ions reduction. Though the transition time for the plateaus changes with the applied current density, the potential difference between Na^+ and Li^+ ions almost remains unchanged at the value of about 0.16-0.2V.

As liquid Pb is used as the working electrode, a pronounced reduction current A starting at about -2.4V and reduction current D starting at about -3.0V are associated with Na^+ ions and Li^+ ions reduction in Fig.4. The reduction potentials for both of Na^+ ions and Li^+ ions on the Pb electrode shift to increasingly more positive values and the reduction currents become larger. But it is hard to distinguish the starting reduction current for Na^+ ions and Li^+ ions as well as their reduction potential difference.

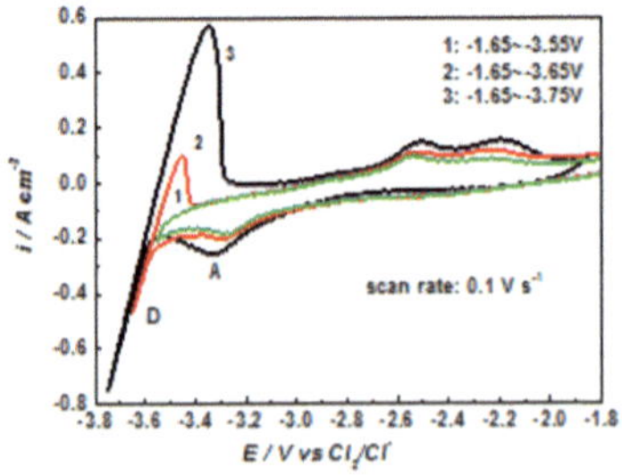

Fig.1 Typical cyclic voltammograms recorded on a tungsten electrode in KCl (64 mol%)-LiCl (18 mol%)-NaCl (18 mol%) with the scan rate $0.1Vs^{-1}$ at 450℃.WE:$0.165cm^2$; CE: Graphite; RE: Ag/AgCl

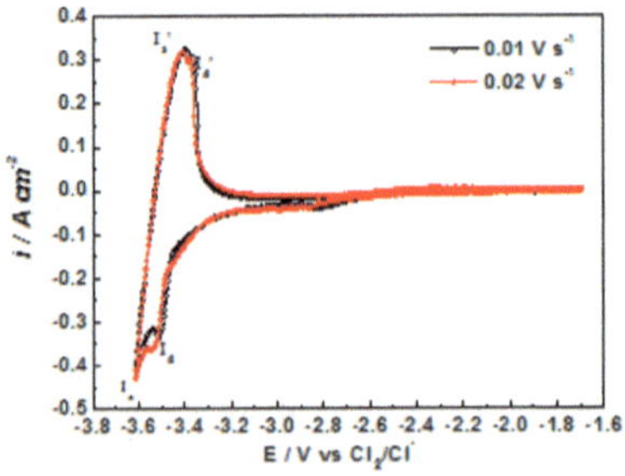

Fig.2 Cyclic voltammograms on tungsten electrode in KCl-LiCl-NaCl (1.0 wt%) melts at 450℃ with various scan rates. WE: $0.165cm^2$; CE: Graphite; RE: Ag/AgCl

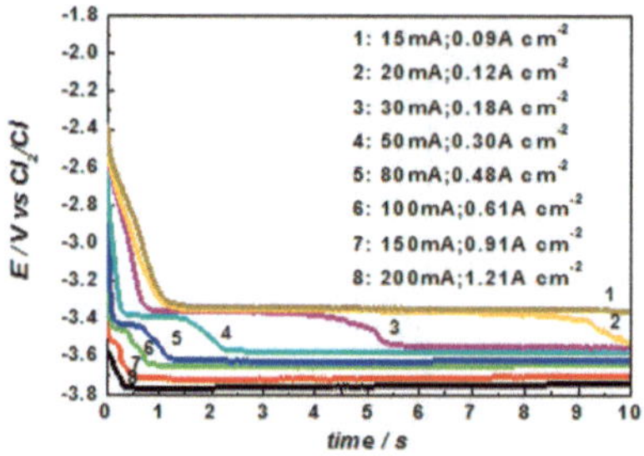

Fig.3 Chronopotentiograms obtained at different current densities on a tungsten electrode in the LiCl-KCl-NaCl(2.0 wt%) melts at 450℃.WE: 0.165cm^2; CE: Graphite; RE: Ag/AgCl.

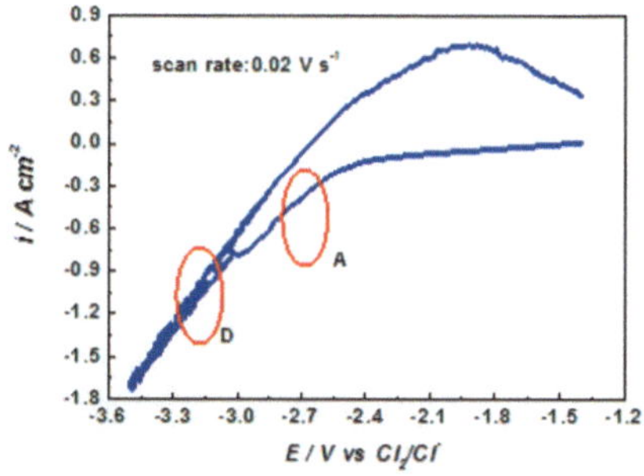

Fig.4 CVs recorded on a Pb electrode in LiCl-KCl-NaCl (2.0 wt%) melts at 953K ,scan rate: 50mV/s, WE:0.283cm^2; CE: Graphite; RE: Ag/AgCl

3.2 Potentiostatic electrolysis

Potentiostatic electrolysis on a liquid Pb electrode at various potentials has been carried out in order to characterize the extent of removal for Na^+ ions from LiCl-KCl-NaCl (2.0wt%) melts, as shown in Fig.5. Clearly, as the applied potential changes from -2.6V to -3.0V, the consumed electrical charge increases from 2571C to 3740C. According to the consumed electrical charge, the theoretical removal efficiencies for Na^+ ions after 3h potentiostatic electrolysis are 69.55wt% (-2.6V),79.99 wt% (-2.8V) and 101.17 wt% (-3.0V), respectively. But the practical removal efficencies for the Na^+ ions based on the cathode products analyzed by ICP-AES are 47.67wt% (-2.6V), 68.32wt% (-2.8V),75.71wt% (-3.0V), respectively. The mass ratios of reduced metallic Na and Li in the cathode products are 78.5/1(-2.6), 62.5/1(-2.8V), 9.74/1(-3.0V), respectively in Fig.5. The results imply that the main reduction species is Na^+ ions at -2.6V, and more Li^+ ions are reduced in the Pb electrode at the potential of -3.0V.

Further extending the electrolysis time to 6h on the Pb electrode at -3.0V, the theoretical removal efficiency for Na^+ ions increases from 101.17wt% to 160.18wt% and the practical removal efficiency increases from 75.71wt% to 90.12wt%. The Na/Li mass ratio in the cathode product is reduced to 4.25/1.

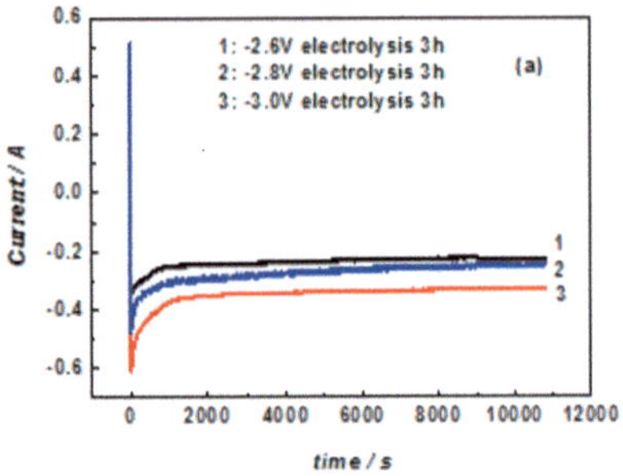

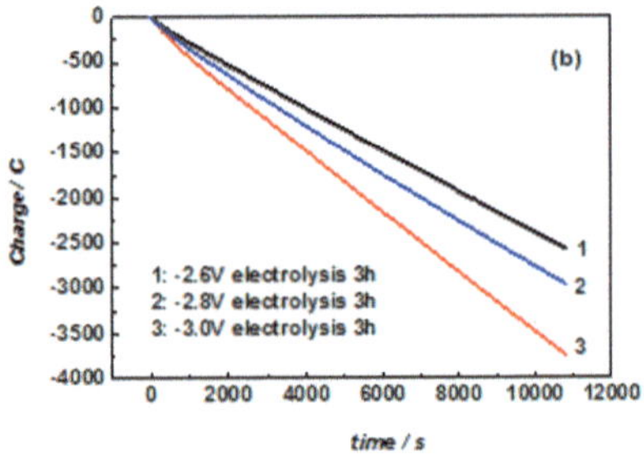

Fig.5 The relation of (a) current; (b) total electricity charge with time at 450℃ on a Pb electrode in LiCl-KCl-NaCl (2.0wt%) melts. WE: Pb（1.52cm^2）;CE: graphite; RE: Ag/AgCl

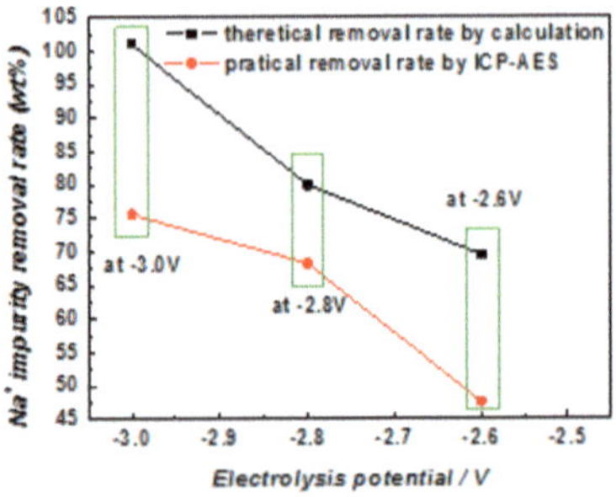

Fig.6 The relation of Na^+ removal efficiency with electrolysis potential in LiCl-KCl-NaCl(2.0wt%) melts after 3h electrolysis on the Pb electrode at 450℃. WE:1.52cm^2;CE:graphite； RE: Ag/AgCl

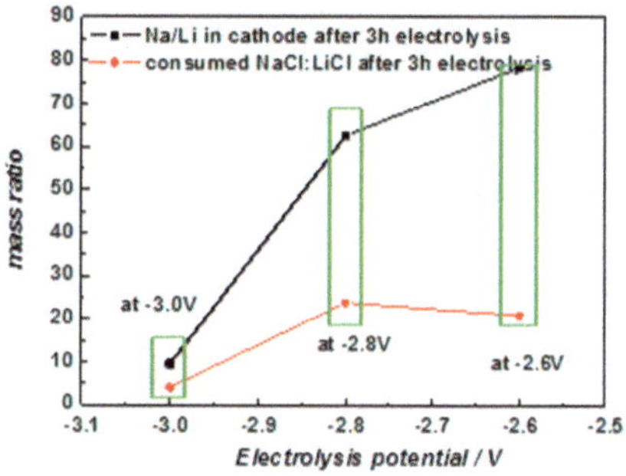

Fig.7 The mass ratios of Na/Li in cathode product and of the removed NaCl and the consumed LiCl in LiCl-KCl- NaCl (2.0wt%) melts on the Pb cathode at different potential at 680℃ .WE:1.52cm^2; CE: graphite; RE: Ag/AgCl

4. Conclusions

(1) According to the theoretical calculations, the reduction potential for the Na^+ is slightly more negative than Li^+ in NaCl-LiCl-KCl melts at 450℃. However, using electrochemical measurements, the potential difference between Na^+ and Li^+ ions at 450℃ is 0.3V in KCl-LiCl (18mol%)-NaCl (18mol%) melts, and the reduction potential of Na^+ is slightly more positive than Li^+ and the potential difference between Na^+ and Li^+ reaches 0.16-0.2V in 1.0-2.0wt% NaCl-LiCl-KCl melts.

(2) The reduction potentials for Na^+ ions and Li^+ ions on a Pb electrode both shift to more positive values and the reduction currents become increased. The practical removal efficiency for Na^+ ions after 3h potentiostatic electrolysis in LiCl-KCl-NaCl(2.0wt%) melts increases from 47.67wt% to 75.71wt% as the applied potential shifts from -2.6V to -3.0V. The Li/Na/mass ratio in the cathode product has increased from1 /78.5/to 1/9.74/.

(3) After 6h of electrolysis on a Pb electrode at -3.0V, the practical removal extent of Na^+ ions is 90.12wt% with a Li/Na mass ratio of 1/ 4.25.

Acknowledgements

This work was financially supported by the National Natural Science Foundation of China (No.51274102).

References

1.W.A. Averill, D.L. Olson, Energy 3(1978)305.
2.W.L.Chen,L.Y.Chai,X.B.Yang,Y.N.Dai,X.Yu,C.F.Zhang,Trans.Nonferr.Met.Soc.China 11(2001)937.

3.W.L.Chen,L.Y.Chai,X.B.Yang,Y.N.Dai,X.Yu,C.F.Zhang,Trans.Nonferr.Met.Soc.China 12(2002)152.
4.H. Lan, W. Zhao, Xinjiang Youse Jinshu 8(1996)55.
5.MiaoShen, Bing Li, Jianguo Yu, Investigation on electrochemical removal of $CaCl_2$ from LiCl–KCl melts, Electrochimica Acta, Volume 62, 15 February 2012, Pages 153-157.
6.M.SHEN, B.LI, S.Z.LI, and J.G. YU, Electrochemical Removal of $AlCl_3$ from LiCl-KCl Melts, METALLURGICAL AND MATERIALS TRANSACTIONS A,VOLUME 43A, MAY 2012,page 1662-1669.
7.A.J.Bard,L.R.Faulkner, Electrochemical Methods: Fundamentals and Applications,2nd ed., John Wiley & Sons, New York, 1980

Rare Metal Technology 2015
Edited by: Neale R. Neelameggham, Shafiq Alam, Harald Oosterhof, Animesh Jha, David Dreisinger, and Shijie Wang
TMS (The Minerals, Metals & Materials Society), 2015

MODERN BERYLLIUM EXTRACTION: A STATE-OF-THE-ART KROLL REDUCTION PLANT

Edgar E. Vidal, James A. Yurko and Keith Smith
Materion Beryllium & Composites; 14710 W. Portage River So. Rd.; Elmore, Ohio 43416, USA

Keywords: Beryllium, Metallurgy, Extractive, Kroll Process

Abstract

Beryllium, like many other reactive metals, is extracted from a halide (generally BeF_2) using magnesiothermic reduction, also known as the Kroll process. The beryllium is produced in a molten state, which when solidified resemble "pebbles". Although beryllium is used in limited quantities compared to other metals (< 100 Mt/annual), it is a critical material for the nuclear, defense/aerospace, and large science markets because of its unique nuclear and mechanical properties. Beryllium processing has unique environmental, health, and safety concerns which must be handled appropriately in the 21st century from a worker and environmental safety perspective, and must be designed with this first in mind.

In 2011, Materion began operations of its new Kroll plant – informally known as the "Pebbles Plant" - in Elmore, Ohio. The plant was majority funded by a U.S. Department of Defense Production Act (DPA) Title III grant, which is a unique government/industry partnership program to, "create assured, affordable, and commercially viable production capabilities and capacities for items essential for national defense." After a twelve year hiatus from primary beryllium production, Materion has reentered its extraction business and has relearned how to make pebbles. The purpose of this presentation is to give an overview on beryllium extraction and discuss some of the unique aspects of building and operating a safe beryllium facility during the production scale-up phase.

Introduction

Beryllium is one of the lightest metals used in engineered applications. Being the 4th element in the periodic table it has an atomic weight of 9.01 g and a density of only 1.82 g/cm^3, yet its melting point is 1287°C. Thanks to beryllium's high Young's Modulus – specific stiffness 6X of steel and 7X titanium – it has been used in electro-optical systems such as gimbal yokes, mirrors, benches and other components. Beryllium also has very unique nuclear properties that make it an ideal thermal neutron moderator and reflector. As such, this metal is used extensively in materials [nuclear] test reactors and radioisotope reactors. Beryllium is very unique in terms of thermal conductivity and heat capacity – both are very high when compared to other elements and materials. Because of this, it has found use in applications where heat dissipation is a must combined with low weight (i.e. aerospace and avionics systems). Beryllium is transparent to x-rays and thus it is the material of choice for use as windows in x-ray sources such as the ones used in the medical industry. More recently, beryllium is playing a key role in p-type semiconductors by outperforming carbon doped equivalents. These semiconductors that utilize beryllium are typically produced by molecular beam epitaxy or MBE. High-end acoustic speakers used by professional studios, bands and theaters also benefit from the properties of beryllium domes where very crisp sounds are achieved in a broad frequency range [1].

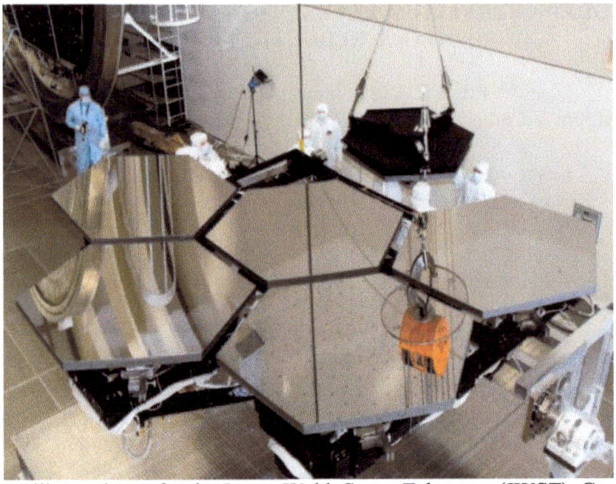

Figure 1 – Beryllium mirrors for the James Webb Space Telescope (JWST). Credits: NASA.

Compounds and alloys of beryllium have also found advance uses. For instance, aluminum-beryllium was developed to enhance the stiffness of aluminum while reducing its weight, yet diminishing the brittleness which is characteristic of pure beryllium. This metal-metal composite material has typically been fabricated either by powder metallurgy techniques or by casting. Beryllium is also found in many different "bulk metallic glasses" or BMGs, where it plays a critical role in maintaining the amorphous state of the alloy even when the cooling rate from the liquid state is not extreme. For example, Vitreloy 1b utilizes 3.8 wt% Be making it one of the most robust Zr-based BMGs available today [1].

Figure 2 – Common electro-optical components made of beryllium. Credits: Materion

Despite the controversy that exists around the safe use of beryllium metal and its compounds, these materials have made and continue to make possible the development of numerous advanced

technologies. Beryllium made it possible for humans to reach space, to develop nuclear energy, produce most of the medical isotopes used for cancer treatment, and develop alloys for the 21st century. Giant leaps have been made in the safe handling of beryllium to ensure the health and safety of workers that process this material.

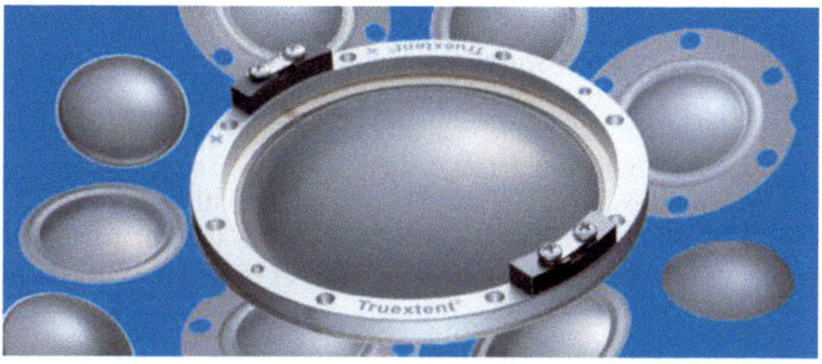

Figure 3 – Beryllium speaker domes. Credits: Materion

Early Extraction Methods

Beryllium – once known as "glucinium" – was first extracted into the form of an oxide from beryl minerals in 1798 by Vauquelin. Beryllium metal was later extracted from its chloride using potassium by Wohler and Bussy in 1828. Many attempts were made to extract beryllium electrolytically, being Lebeau's experiments the most notable. In 1898 he prepared an electrolyte of beryllium fluoride and sodium fluoride, and proved that beryllium metal could in fact be extracted. It was not until 1913 that Flitcher obtained a significantly purer beryllium and in larger quantities using Lebeau's process [1].

Many methods have been developed to produce metallic beryllium, but most required either non-existing engineered materials or were simply not economically feasible. The most logical approach used early on was to start with beryllium oxide (BeO) and perform metallothermic reductions. Because of the thermodynamic stability of BeO, not many candidates exist that can reduce it. For instance, zirconium and yttrium could in theory be used for the reduction of BeO, but the economics and the formation of beryllides do not allow their use. Using carbon as a reducing agent is also a logical approach, but thermodynamically the formation of beryllium carbide (Be_2C) is favored. Despite this, the use of carbon along with copper metal enhances the back reaction of the carbide phase into beryllium which dissolves in the copper [2].

Early on, calciothermic reduction of beryllium oxide was heavily studied. It was not only a non-economical approach to use calcium metal, but also the formation of calcium beryllides seemed to upset the process. In the same way, the Hunter process that utilizes gaseous sodium metal has been tried, but is not economical in large quantities.

The Kroll Process

William Justin Kroll was a metallurgist from Luxemburg who invented a process to extract titanium metal from ores using an intermediate halide phase and magnesium metal as the reducing agent. He later applied the same method to the extraction of zirconium. The "Kroll Process" as it is known is the main method used around the world for the production of titanium, zirconium and some rare earth metals.

In the Kroll Process, the metal halide – typically a chloride – is produced from the titanium or zirconium oxide by carbothermic reduction and in the presence of the halide gas, and purified by distillation. The metal halide gas is then passed through molten magnesium metal which then produces a mixture of solid metal (titanium or zirconium) and liquid magnesium chloride. The resulting "sponge" titanium or zirconium has entrapped magnesium chloride that is subsequently removed by a combination of leaching and vacuum distillation. The sponge metal is crushed and compacted into large electrodes which are vacuum arc remelted (VAR) producing solid slabs or rods of the pure metal.

Figure 4 – Examples of aluminum-beryllium castings showing the intricate shapes and features possible. Credits: Materion.

The Modified Kroll Process for Beryllium Extraction

As mentioned earlier, the beryllium oxide itself is troublesome to reduce into a pure metal. Using the same principles as in the Kroll Process, a reduction of beryllium halide with magnesium should be possible. In fact, as shown in figure 5, the thermodynamic stability of beryllium fluoride and chloride is lower than that of the magnesium halide.

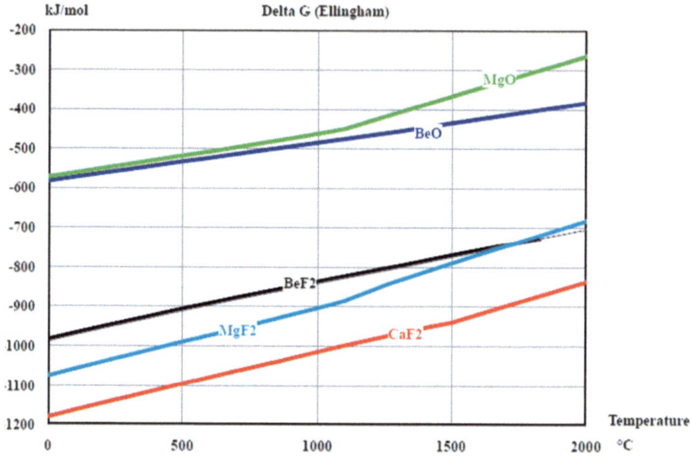

Figure 5 – Ellingham diagram of beryllium fluoride and oxide as it compares to other metals like calcium and magnesium.

A technique that derived from this modified Kroll Process, that does not use magnesium for the reduction but does start with a beryllium chloride, is known as the Pechiney Process. In this case, a beryllium chloride ($BeCl_2$) is produced by mixing beryllium hydroxide with a carbonaceous material and forming briquettes which are then treated in a chlorine gas atmosphere at elevated temperatures (700 to 800°C). This beryllium chloride is distilled and purified which becomes the input material for an electrolysis process. An electrolyte is made by adding sodium chloride to the beryllium chloride and the electrolysis is performed at around 350°C. The end product is a "flaky" beryllium that is intimately mixed with the electrolyte, which requires subsequent leaching and washing steps to obtain the pure beryllium. This technique was used commercially for many years by the Pechiney Co. of France [3].

An alternative to the beryllium chloride is the beryllium fluoride (BeF_2). Like in the Kroll Process, beryllium fluoride is reduced with magnesium metal producing beryllium metal and magnesium fluoride as a by-product. Contrary to the actual Kroll Process, the beryllium fluoride is not a gaseous state but rather co-melted with magnesium. For this reason, the beryllium produced in this process has the characteristic "pebble" like morphology which is indicative of surface tension differences with the magnesium fluoride.

The Materion Beryllium Extraction Process

For more than 70 years Materion (known then as the Brush Beryllium Company) has used the modified Kroll Process for the purification and extraction of beryllium. The process starts with beryllium hydroxide ($Be(OH)_2$) which is produced by the company in the mining facility in Delta, Utah. This beryllium hydroxide is produced with bertrandite ore from the mine and beryl ores that are brought in from different sources. This hydroxide undergoes three major steps [3]:

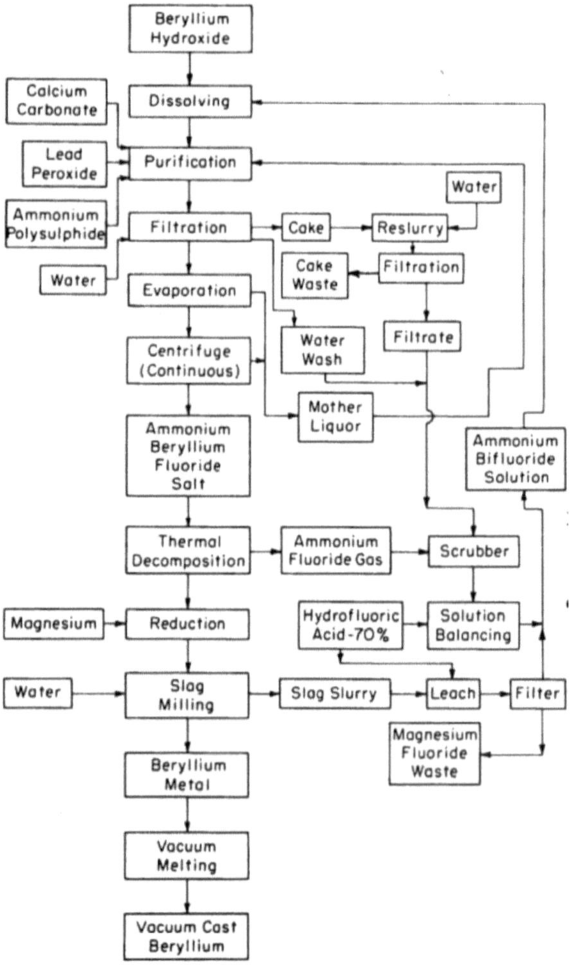

Figure 6 – Flow chart of the beryllium extraction process using the modified Kroll Process [2].

- Conversion of hydroxide to ammonium fluoroberyllate solution (ABF);
- Thermal decomposition of ABF salt into beryllium fluoride; and,
- Magnesiothermic reduction of beryllium fluoride into beryllium

To produce the ABFsolution, the hydroxide is dissolved with ammonium bifluoride solution according to the reaction:

$$Be(OH)_2 + 2NH_4.HF_2 \Rightarrow (NH_4)_2BeF_4 + H_2O \tag{1}$$

In this process the pH of the solution is carefully controlled so the precipitation of impurities found in solution is not allowed. Part of the purification process is attained by utilizing lead oxide as an oxidant to precipitate some elements like manganese and chromium. Other metals found in solution are removed by using ammonium polysulfide which also precipitate in the form of sulfides. The filtrate produced is essentially free of contaminants and ready for crystallization which is a critical step to obtain the appropriate concentration of ABF crystals or commonly referred to as salt. As with any other crystallization process, the concentration of "seed" is important to ensure accelerated growth of crystals.

Once the ABF salt is produced and collected it goes to the next step of the process which involves the thermal decomposition into beryllium fluoride:

$$(NH_4)_2BeF_4 \overset{\Delta}{\Rightarrow} BeF_2 + 2NH_4F \tag{2}$$

This thermal decomposition is carried out in an induction heated graphite retort at approximately 1000°C. As the ABF salts are added to the retort they decompose forming a molten phase of beryllium fluoride which is cast into discrete pellets. During the decomposition process, ammonium bifluoride gas is produced and collected and recycled back into the process.

The final step in the extraction process involves the actual modified Kroll Process where beryllium fluoride and magnesium are reacted together:

$$BeF_2 + Mg \Rightarrow Be_{(l)} + MgF_{2(l)} \tag{3}$$

Based on the stoichiometric reaction one mol of magnesium is needed for each mol of beryllium fluoride. The reaction is highly exothermic and proceeds to near completion. But if the stoichiometric proportions are used, a high purity magnesium fluoride is produced which is difficult to remove form the beryllium itself even when leaching with acid. For this reason, a sub-stoichiometric amount of magnesium metal is used (between 70 and 75 wt%). The charge is loaded into a graphite crucible and temperature increased until it is fully molten (above 1300°C). As the reaction proceeds, small droplets of beryllium metal are formed and coalesce into the larger common "pebbles." [2] Once the reaction is complete the product is discharged and fed into a ball mill where the magnesium/beryllium fluoride by product is partially dissolved in the milling solution, and the rest of the by-products are physically removed from the beryllium pebbles.

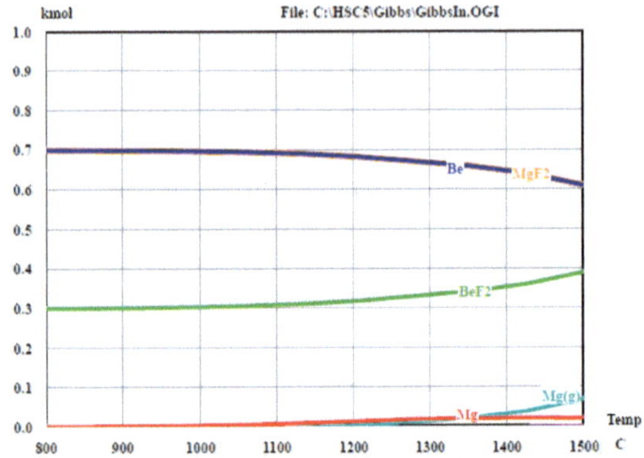

Figure 7 – Thermodynamic stability of the reaction between beryllium fluoride (stoichiometrically deficient) and magnesium metal.

Figure 7 shows a thermodynamic stability diagram showing that when less than stoichiometric amounts of magnesium is used, the end product will have some beryllium fluoride and free magnesium metal present. The discharge from the ball mill is separated into a beryllium pebbles stream and the magnesium fluoride slurry. The beryllium pebbles continue their processing by going through magnetic separators to remove any steel introduced in the process, and a final acid wash prior to sorting and classification. Up to 97% recovery of beryllium is attained in the extraction process from the hydroxide. The magnesium fluoride slurry is further treated with more acid solution to recover as much beryllium fluoride as possible then filtered into a relatively dry sand (15 wt% moisture). This magnesium fluoride is relatively pure and is commercialized as a by-product of the process.

Once the pebbles have been sorted, they are melted and vacuum refined where any magnesium left from the extraction process can be lowered to an acceptable amount. The molten beryllium is cast into billets which go into the final process of making powder for the powder metallurgy (PM) fabrication process.

Figure 8 – Typical beryllium pebbles produced in the Materion Pebbles Plant.

Classic Extraction – New Technology, Methods and Plant

In the year 2000, Materion closed its primary beryllium extraction operations at Elmore, leaving the country without a source of fresh beryllium supply. The original extraction facility was aging and there was enough beryllium available in the National Defense Stockpile (NDS) to supply Materion's and the market demand. While beryllium-copper alloys and beryllia ceramics continued to be produced using hydroxide supplied by the Utah operations, all of Materion's metallic beryllium requirements had been supplied from beryllium acquired under long-term contract from the NDS and other small quantities of beryllium feedstock material purchased internationally.

Figure 9 – New Materion "Pebbles Plant" located in Elmore, Ohio, USA.

In 2003, Congress directed the DOD to study the strategic need for beryllium and the status of the domestic beryllium industrial base. In 2004, based on the DOD report, Congress appropriated money for the Defense Production Act, Title III Program for the preliminary design of a new primary beryllium facility. The new "pebbles plant" was designed and built next to the existing factory in Elmore Ohio, starting operations in 2011. The plant is 22 m tall, has a 4,742 m^2 footprint and contains 11,553 m^2 of total floor space over three levels. The plant has a demonstrated production capability of 72,500 kg per year.

Figure 10 – Inside one of the solution treatment rooms in the "Pebbles Plant."

Because beryllium metal is critical and strategic to the national security of the United States, details of the processing and control equipment cannot be made readily available.

Figure 11 – The Materion "Pebbles Plant" that produces primary beryllium is controlled by several systems as the one shown.

The new Materion pebbles plant is a state-of-the-art facility that has the latest technologies in terms of process controls and monitoring. It also includes the most advanced systems to ensure the safe handling of beryllium that minimizes the risk of exposure to workers of beryllium containing compounds. Remote operations are the core of the new technologies used in this plant.

Conclusions

Beryllium has and continues to be a key element in the advancement of technologies that improve the way we live. Many years ago William Kroll developed a process that still constitutes the pillar for beryllium extraction. With advances in controls, simulations and modeling, the modified Kroll Process takes a new life in the production of "complicated" elements. Many advances have been made in the safe handling of beryllium and its compounds, and have been implemented in the new primary beryllium facility operated by Materion.

References

1. Walsh, K.A., (2009) "Beryllium Chemistry and Processing," ASM International. Materials Park, OH, USA.
2. Hausner, H.H., (1965) "Beryllium: Its Metallurgy and Properties," University of California Press. Los Angeles, CA, USA.
3. White, D.W. and Burke, J.E., Eds., (1955) "The Metal Beryllium," The American Society for Metals. Cleveland, OH, USA.

Rare Metal Technology 2015
Edited by: Neale R. Neelameggham, Shafiq Alam, Harald Oosterhof, Shijie Wang, and David Dreisinger
TMS (The Minerals, Metals & Materials Society), 2015

How to Recover Minor Rare Metals from E-Scrap

Takashi Nakamura
Institute of Multidisciplinary Research for Advanced Materials, Tohoku University
980-8577 2-1-1 Katahira Aoba-ku Sendai, Miyagi, Japan

Keywords: Urban Mining, Resource Efficiency, Critical Metals, Recycling

ABSTRACT

Dismantling and detachment of parts from e-scrap are essential and involve higher cost techniques in pretreatment processes for physical separation. New techniques are desired for effective recycling of minor rare metals, therefore, we are trying to develop new detachment processes such as a new break down process. The principal of this new process is an electrical disintegration by electric pulse. E-scrap is broken down by high voltage electric pulses in water. IC chips and LSI can be detached from printed circuit boards under proper conditions. Ta- capacitors are also broken down to plastic parts and Ta sinter with lead wire. This is very useful in terms of pretreatment for e-scrap recycling.

Also in the case of metallurgical production, with its intrinsic potential of smelting, extraction, enrichment and separation methods, play an important role in the context of minor rare metals.

1 Introduction

Let us consider the economics of cyclical usage. We do not have a sufficient recycling ratio for minor metals, even those with high prices. The recycling ratios for minor rare metals and some non-ferrous metals, all of which are relatively expensive, have not always been high. The reason is partly that it is difficult to collect scrap from in-use markets, and that scrap containing unstable impurities is hardly used in mass-production processes.

The current state of natural resources supports our modern techniques and mass production system. It is difficult to give a general definition of natural resources, but mineral resources can be defined as follows. Minerals rich in target components used as resources and deposits that contain even high- level impurities can be good resources, if these impurities are constant and sufficient amounts of target minerals are to be found in one place. There are two characteristics required; one is a rich composition of target elements and the other is sufficient amounts for development as a good resource. Both are equally important. However, it is the latter, a sufficient amount at only a single location, that makes mass production possible. The average purity of copper ore is now less than 1% and sometimes near 0.5%. This is by no means a high purity but it is presently adequate, since we have effective methods to upgrade it. This is done via a flotation process that readily upgrades sulfide ore. However, this technique requires impurities to be at a constant level throughout the mineral. Having minerals with consistent impurities allows the ore to be put through a mass production process. If impurities are not consistent in the raw materials, then we must change the process for each resource. This is impossible within the metal production industry. Thus, even if a high-concentration ore exists, it is worthless as a resource if impurity of the ore is inconsistent. These points are summarized for both natural and artificial resources in Table 1. The average grade of artificial resources is superior, but these are inferior to natural resources in that they are not concentrated in particular locations (and thus are difficult to collect), and their grades vary (impurities are not consistent). These points cannot be applied to mass production. Recycling is often considered to make economic sense, but our current recycling does not only target items that are economically feasible. One way of solving the above problems is to first reserve, then stockpile and create future stocks of byproducts and waste products. This is because of the possibility that these items will one day be viewed as natural resources, even though they may not be considered so presently. We call these reserving and managing activities an artificial deposit [1]. Artificial deposits are storage sites for items mid-way through the process of becoming natural resources.

Table 1 Characteristics of Natural and Man-made Resources

	Natural Resources	Man-made Resources
Amount	Enough, though we must consider depletion of some	Insufficient, but increasing recently
Existing Forms	Natural ores, most of which can be extracted effectively by large scale processes	Scrap or waste, high collection cost
Content	Low	High
Impurity	Relatively high but constant	**Relatively low but has low stability**
Cost of Extraction	Low	**High**

This article explains the effect of E-waste treatments on environmental issues, from the viewpoint of metallic resources.

An important means for elevating sustainability to higher levels is the 3 Rs, and these should receive highest priority. In an economic sense, recycling is regulated by either law or self-action. If this is economically sensible then there are no problems. However, an effective social system that includes laws, is necessary if differences between resources and waste products are unclear. End-of-life products sometimes become waste and sometimes useful resources, even with the same compositions.

2. Urban Mining

Professor Hideo Nanjo of the Research Institute of Mineral Dressing and Metallurgy at Tohoku University (a forerunner of the current Institute of Multidisciplinary Research for Advanced Materials) made the following comment in 1988 in the journal *Bulletin of the Research Institute for Mineral Dressing and Metallurgy at Tohoku University:* "A key to the steady supply of minor rare metal, a scarce resource, is establishment of a system of recycling scraps gathered both domestically and overseas in international cooperation with different countries that produce primary resources" [2]. Professor Nanjo was predicting present conditions exactly. Further, he accurately stated that "Minor rare metals are extremely important for maintaining the high-tech industry and if their supply dried up it would spell the downfall of Japanese industry." This could be stated about the present day, without modification.

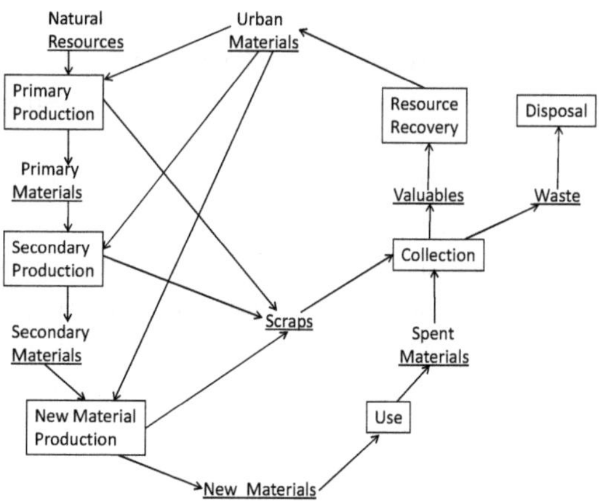

Figure 1 Resource Supply Chain in urban mine

Nanjo proposed a resource recycling chain as shown in Figure1. This shows a circular supply chain presently under discussion, with items flowing from arteries to veins in a cycle. When that paper was published, most people did not recognize the importance of the proposals therein, since the oil crisis had begun and prices had not risen to peak levels yet. However, those ideas have been passed to us, and we again proposed a new concept of "artificial deposit" [1]. Since rich deposits should be existed in good mines, we need rich deposits for mining operations. More detail of urban mining was published in recent [3].

3. Present status of minor rare metal recycling in non-ferrous smelters

Some minor rare metals are recovered using well established non-ferrous smelting system. In case of copper smelting, anode copper is purified in an electro-refining process and separated from more noble metals contained in the mineral including precious metals such as silver and gold. Emissions include the minor elements like arsenic, antimony, bismuths, selenium and tellurium. These accompanying metals are sometimes beneficial but sometimes represent cost. Cooperation among copper smelting, zinc smelting and lead smelting, therefore, becomes very important. Each smelting system has several processes to treat those minor elements. Schematic presentation for combination between each smelter is shown in Figure 2. Copper concentrate contains a certain amount of lead and zinc with minor elements and other zinc and lead concentrates also containing copper. Thus, a copper smelter produces by-products which contain lead and zinc, and a zinc smelter also produces by-products containing copper and lead, and same situation is found in a lead smelter. Therefore, minor elements are efficiently recovered with a combination of non-ferrous smelters. This is also important in the non-ferrous industry. To achieve this efficiency, we must make effective use of our existing recycling system and methods. It is difficult to recycle sufficient minor metals to justify creating new infrastructure only for recycling. We must consider new systems after thoroughly checking which elements are being recovered in the existing non-ferrous metal smelting industry. Substantial minor rare metal recovery is done during the processing of waste products that contain these and related elements. In particular, gold, silver, PGMs, selenium and tellurium are contained in minerals as non-ferrous metal smelting impurities, and these have been recovered for a long time. In addition, recovery of indium, gallium, antimony and bismuths is ongoing at present.

Precious metals have high value, and are being recovered outside of non-ferrous metal smelting. In addition, there has been discussion regarding recycling elements used in battery devices (such as minor metals) and in lithium batteries, which are anticipated to grow in volume. Considering the above, it is preferable to recycle used batteries and battery devices by selecting individual items, and separating components that have an existing recycling process for most of their elements from those that do not. Such a system would require the use of existing non-ferrous smelting sites. In summary, to turn by-products and waste products that appear unlikely to become natural resources into resources, and in order that we may recycle elements that do not have immediate economic value, we must initially reserve such items and then stockpile them as future stock.

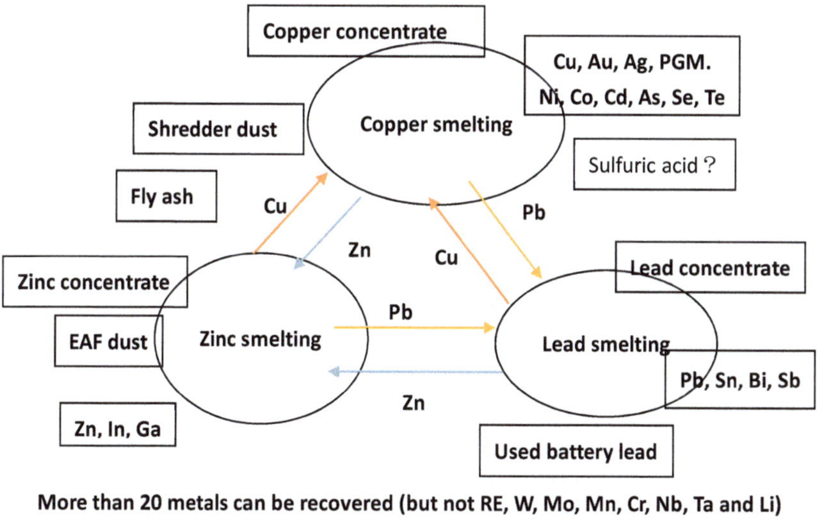

Figure 2 Base and minor metals recovered from primary and secondary resources in non-ferrous industry

Precious metals and PGM are recycled efficiently from non-ferrous smelters. Most recycling resources are put into smelting furnaces like copper converters, and they are ultimately recovered from copper slimes after copper electro-refining. Some pre-treatment processes are required for the furnaces.

Reactive rare metals can't be recovered in non-ferrous smelting processes because most of them are oxidized into slag and it is impossible to recover them from slag economically. Therefore, we have to separate reactive rare metals from e-scarp before treating in non-ferrous smelters like Figure 3. Key point of this flow is removal of parts which include high content reactive rare metals like Nd-Fe-B magnet in pre-treatment stage.

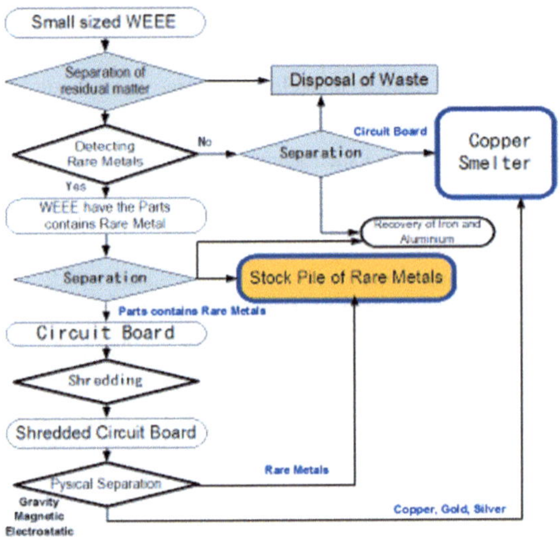

Figure 3 Flow of Recovering rare metals from small sized WEEE

Dismantling and detachment of parts from e-scrap are essential and involve higher cost techniques in pre-treatment processes for physical separation.

New techniques are desired for effective recycling of minor rare metals, therefore, we are trying to develop new detachment processes. Our attempt at a new break down process is shown on the right hand side of Figure 4 with the normal crushing process on the left hand side [4]. The principal of this new process is an electrical disintegration by electric pulse. E-scrap is broken down by high voltage electric pulses in water. IC chips and LSI can be detached from printed circuit boards under proper conditions. Ta- capacitors are also broken down to plastic parts and Ta sinter with lead wire. This is very useful in terms of pretreatment for e-scrap recycling.

The next step in the pre-treatment is a separation of each part and/or material. In this area various techniques have been available, including magnetic separation as a unit operation. The selection of a good combination of these techniques is important for high performance separation systems. The standard line up of these kinds of techniques is magnetic separation, gravity separation and eddy current separation used to separate iron scrap, non-ferrous scrap and plastics. Of course, each technique has been developed over time to enhance the separation quality and speed.

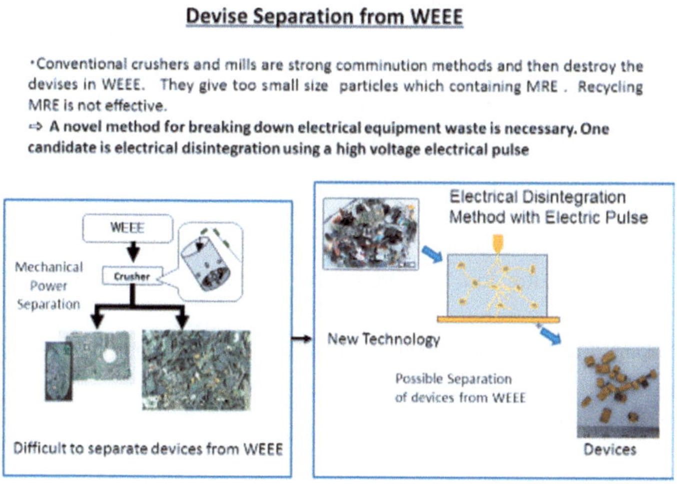

Figure 4 New developments in the technology of component break-down

Sorting techniques have been rapidly developed in recent years. The first development of this technique is as old as other physical separation processes. It had a slow separation speed at early stages of development. Now this technique has become one of the major processes in physical separation. The key point of this process concerns what characters are analyzed by using various sensors like color, shape, weight and chemical composition. Almost all sensing techniques such as IR, XRD and XRF are available for the characterization of target materials. High quality separation can be achieved by this technique especially for waste plastics. Even plastics containing brominated flame retardants can be separated from waste plastics.

4. Summary

E-scrap recycling is vital to maintaining a supply chain of minor rare metals. Non-ferrous smelters are important to advance recycling of basic and minor rare metals. A new system is necessary for this and the greater development of recycling technologies for minor rare metals. The above has considerable impact on environment from a resource perspective. Environmental problems will continue to be important but natural resource security will become equally important. Creating a zero-waste society aims to combat both of these issues and promotes sustainable development. Furthermore, natural resources depend on the land in which they are found and so can easily become involved in politics. Environmental sustainability is next to impossible without bold political leadership. Thus we need to consider both of these factors when creating our national strategy. Particularly in countries like Japan where advanced technology is everywhere but there are next to no resources, there is a need to thoroughly debate how to use technology to create a zero-waste society and create it.

References

[1] T. Shiratori and T. Nakamura. *J. MMIJ* **122**, 325–329 (2006) (in Japanese)

[2] M. Nanjyo. Bull. Res. Inst. Min. Dress. Met., Tohoku University, **43**, (1988), 239–243 (in Japanese)

[3] T.Nakamura and K.Halada, Springer e-Book , http://dx.doi.org/10.1007/978-4-431-55075-4

[4] http://tohoku-timt.net/rare-elements/

Rare Metal Technology 2015
Edited by: Neale R. Neelameggham, Shafiq Alam, Harald Oosterhof, Animesh Jha, David Dreisinger, and Shijie Wang
TMS (The Minerals, Metals & Materials Society), 2015

SOLVENT EXTRACTION OF Cu^{2+} WITH LAMINAR FLOW OF MICROREACTOR FROM LEACHANT CONTAINING Cu AND Fe

Jiang Feng[1,2,3,4], Li Chuanhua [1,2,3,4], Peng Jinhui [1,2,3,4], Zhang Libo[1,2,3,4] ,Ju Shaohua[1,2,3,4]*

[1]Yunnan Provincial Key Laboratory of Intensification Metallurgy, Kunming, Yunnan, 650093, China;
[2]National Local Joint Laboratory of Engineering Application of Microwave Energy and Equipment Technology, Kunming, Yunnan, 650093, China;
[3]Key Laboratory of Unconventional Metallurgy, Ministry of Education, Kunming University of Science and Technology, Kunming, Yunnan 650093, China;
[4]Faculty of Metallurgical and Energy Engineering, Kunming University of Science and Technology, Kunming, 650093, China;

Keywords: microfluid; solvent extraction; copper; microreactor

Abstract

Solvent extraction of Cu^{2+} from the leaching liquor contain Cu and Fe use a microfluidic technology. Currently, the conventional copper extraction process has many problems such as long distance and small driving force of mass transfer. In order to solve the problems, studies are carried out on the extraction process of low grade copper leaching liquid using Lix984N as an extractant with a new microfluidic solvent extraction technology. The effects of flow rate on the separation and extraction efficiency between Cu and Fe are studied, and the results show that a steady liquid-liquid interface of laminar flow is form in the micro-extractor, and extraction efficiency is 78 % and 2.5 % for Cu and Fe at the contact times of 0.84s respectively. Besides, the separation coefficient between Cu and Fe is as high as 140.04. The studied microfluidic technology has the advantages like improving extraction efficiency and reducing reaction time and providing the potential application in the hydrometallurgy extraction of Cu.

Introduction

Microreactor is a 3-dimensional component with solid substrate manufacturing, which with aid of a kind of special micro processing technology or precision machining technology. It can be used for chemical reaction. The microchannel size of the fluid in micro reactor is sub-micron to sub-millimeter scale. Required chemical reaction needs to be done in those microchannels[1-2]. Because of the micro reactor effective channel or chamber reduce the physical size of micron or nanometer level, making the fluid physical quantities such as temperature, pressure,

*Author: Feng Jiang, Tel:+86-087165138997;E-mail:jiangfeng-0311@126.com
Foundation item: Project (U1302271)supported by the National Natural Science Foundation of China.and Project (KKSY201452088)supported by Kunming University of Science and Technology
Corresponding author: Ju Shaohua,Tel:+86-087165138997; Fax: +86-65138997; E-mail: shj_200801@126.com

concentration and density gradient increased dramatically, leading the driving force of the heat and mass transfer increase greatly, it can improve an order of magnitude of the heat transfer coefficient and down an order of magnitude of mass transfer reaction time. Thus, microreactor application usually embodies a much smaller chemical device, shorter reaction time, higher efficiency [3], lower consumption of materials and energy [2] and safer operation conditions. Another advantage of a microreactor is that it can easily scale up by simply " numbering up ". Different microreactor procedures have been applied to organic synthesis [1] and material preparation [4-6]. Microreactors can now generate tons of product per year [7].

The extraction process is the key link of metallurgy separation purification solution system, it can separate and purify a variety of metal ions from liquor directly, and avoid the separation process of precipitation, crystallization .It's the key method of system of metallurgical solution separation and purification.

Microfluidic technology applied in the extraction process can avoid the traditional problems. In the field of microfluidic, solvent extraction is very efficient, because it is able to provide very high specific surface area and the characteristics of the short diffusion distance. It's very helpful to reduce the spread of the path length and increase the mass transfer rate of two phase interface, thus improve the chemical reaction rate. In addition, the mass transfer that under the control of laminar flow can avoid the emulsion. At present, the main application fields of microfluidic technology is chemistry and chemical engineering fields, such as gas processing, chemical synthesis and particle synthesis, etc. And it can achieve annual tons of production capacity[8].

Very few researchers have explored the potential of using microreactors in the field of solvent extraction of metal ions. In Australia, the possibility of extraction Cu^{2+} from a particle-laden solution was investigated, with a positive result [9]. In Japan, it was shown that Cs^{+} extraction from a complex solution in stable slug flow, by inserting a piece of glass bead into a microchannel, resulted in faster separation compared with conventional batch experiments [10]. In this paper, the extraction and separation efficiency of Cu^{2+} from low concentration leachant, was studied in a laminar flow microreactor. The process of diffusion and mass transfer in the microreactor was also analyzed, and the extraction effects were compared with traditional extraction.

Experimental

2.1 Materials and Devices

The experimental material, an aqueous solution in a sulfuric system at a pH value of 1.58, was obtained from a hydrometallurgy copper plant in Yunnan Province, China.

Its chemical composition was analyzed (listed in Table 1) . The organic phase for solvent extraction contained 4% LIX984N (volume fraction, Henkel, Germany) and 70% kerosene (DeZhong Chemical factory, ZhengZhou, HeNan province, China).

Table 1 Chemical composition and the pH value of the leachant.

Elements	Cu^{2+}	Fe^{3+}	pH
Unit	$g \cdot L^{-1}$	$g \cdot L^{-1}$	Value
Date	0.48	8.09	1.58

The microchip used in this study is shown in Figures 1, Figures 2 and Figures 3. It was made of Polymethyl Methacrylate (PMMA,Tsinghua University, China).The extraction reaction takes place in a 140 mm long microchannel (600 μ m × 100 μ m), which is shown in Figure 2.

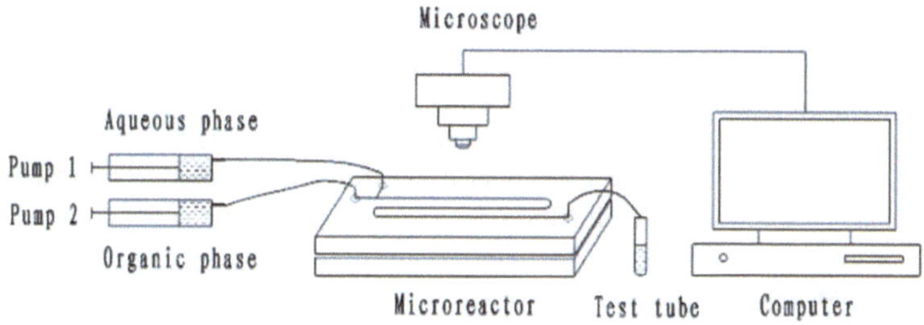

Fig 1 Schematic of microreactor system

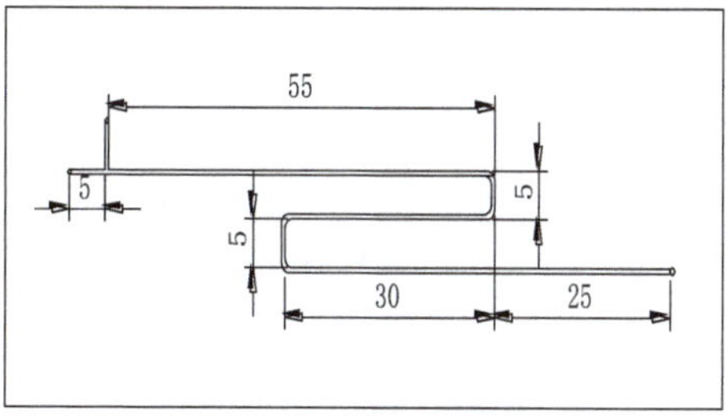

Fig 2 Schematic of Microchannel

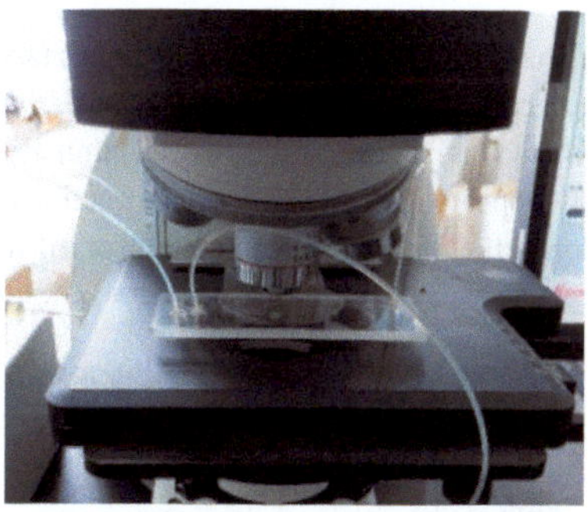

Fig 3Microreactor channel conditions monitored by optical microscopy.

2.2 Procedures

During the experiments of microreactor extraction, the aqueous and oil phases were, respectively, pumped into the PMMA microchip by two constant flow pumps (HLB-4015,

Yansan). The flow stability of the contacting interface of the aqueous and oil phases in the microchip was monitored with an optical microscope (Leica DM 4000 M). The flow rate of the aqueous phase was ranged from 0.05 to 2 ml · min^{-1} at a fixed organic/aqueous flow rate ratio of 1. Sample was collected and analyzed after System stability

The experiments of conventional extraction were carried out in a 125 ml separating funnel. The shaking time was 2 min,3 min,4 min,5 min and 6 min, respectively. Each time, 40 ml aqueous phase and organic phase (4% LIX984N in kerosene) were added to a 125 ml separating funnel. Then, the funnel was stirred on a shaking machine for different time at a speed of 200 rpm. After separation, the aqueous phase was diluted and analyzed by inductively coupled plasma atomic emission spectrometry (ICP-AES).

2.3 Analysis Method

The concentrations of Cu^{2+} , Fe^{2+} and Fe^{3+} in the aqueous phase before and after the extraction were determined by an ICP-AES (Leeman ICP-AES PS1000). The acidity of the aqueous solution was determined by an acidity meter. Equation (1) is an expression of extraction efficiency, where C_{ao} and C_{al} signify the elements contained before and after extraction, respectively:

$$E = \frac{C_{ao} - C_{al}}{C_{aO}} \times 100 \tag{1}$$

Results and Discussion

3.1 Stability of Liquid-Liquid Interface

Figure 4 is the photo of the phase interface in the microchannel s taken by an optical microscope. A stable liquid-liquid interface was observed during this experiment under the condition of optimal flow rate ratio (oil/aqueous), R = 1, and flow was laminar in all cases. Compared with the emulsions which easily appeared in conventional extractions, this type of micro-reactor extraction radically abandoned the formation of emulsions of oil-in-water.

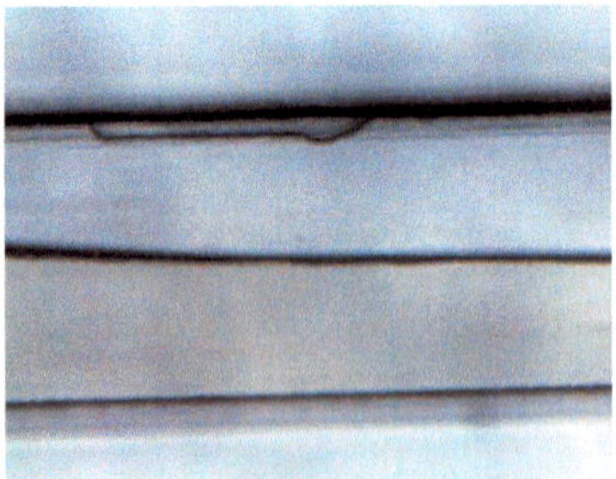

Fig 4 Photo of the phase interface in the microchannel

3.2 Effect of Extraction Experiment

3.2.1 Effect of Flow Rate As we can see from the Figure 5, extraction efficiency of copper reduce and extraction efficiency of iron is stable about 2% with the increase of flow rate. When flow rate is 0.03mL/min, the extraction efficiency of copper and iron is 7.8% and 2.5%, respectively.

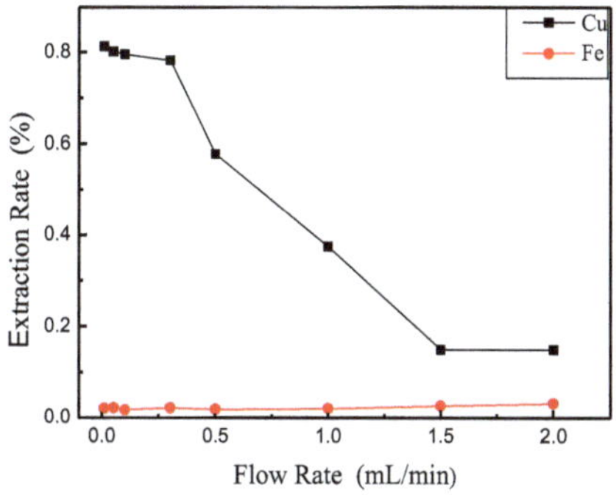

Fig 5 The effect of flow rate on extraction

3.2.2 Calculation of Contact Time in the Micro-Channel The changes of geometry structure of a micro-reactor have a great influence on the flow pattern and the shape of the interface between L-L phases [11]. According to the micro-channel geometry in Figure 2 , the volume of aqueous channel, V_c , was calculated by Eq. (2):

$$V_c = d \times l \times w \tag{2}$$

The V_c is $8.4\times10^{-9}m^3$ by calculating. When the flow rate of the aqueous phase and organic phase was 0.3 mL $\cdot min^{-1}$. The residence time, t , of the aqueous phase in the microchannel can be expressed as Eq. (3):

$$t = \frac{V_c}{v} \tag{3}$$

The t is 0.84 s by calculating, which indicated that the extract ability could be measured for a short contact time of both phases in the micro-reactor. In traditional solvent extraction plants with a mixing settler system, the extraction time is usually about 5 min (4 min of mixing and 1 min of phase separating).

Fig 6 shows that the effect of residence time on extraction in micro-channel is significant. The extraction efficiency of copper has a linear correlation with the increase of residence time before 0.84s. The characteristics of extraction efficiency increase with residence time are also reported in extraction of indium[12].Extraction efficiency of copper reach 78% in only 0.84s and extraction efficiency of iron is only 2.5% at the moment.

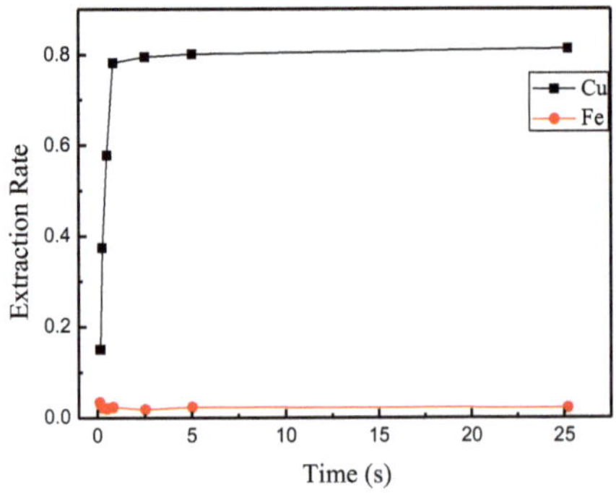

Fig 6 The effect of residence time on extraction in micro-channel

3.2.3 Separation Coefficient and Mass Transfer Character The aqueous-organic interfacial area A (m^2) can be calculated with follow equation:

$$A = w \times L \tag{4}$$

where w is the width of the micro-channel and L is length of the micro-channel. Thus, A is calculated to be 8.4×10^{-5} m^2 .when the value of flow rate is 0.3mL/min ,the contact time of 0.84 s, 4.2×10^{-9} m^3 of solution involved in the reaction and copper ion concentration in aqueous phase change from 0.46 g/L to 0.10 g/L. The mass transfer coefficient, J_M, can be calculated with follow equation:

$$J_M = \frac{V_{aq}\left(C_{M_0} - C_M\right)}{At} \tag{5}$$

The mass transfer coefficient is 0.0214 $g \cdot m^{-2} \cdot s^{-1}$. When v=0.3 mL/min , the raffinate was collected and analyzed by ICP-AES. The results are shown in Table 2 .

Table 2 The chemical composition of the original solution and the raffinate after microreactor extraction.

Composition	Cu^{2+}	Fe^{2+} and Fe^{3+}
Original solution (g/L)	0.46	7.98
Raffinate (g/L)	0.10	7.78
Extraction rate (%)	78.26	2.50

The value of distribution ratio D_{Cu} and D_{Fe} was calculated to be 3.6 and 0.0257, respectively.

The separation coefficient $\beta_{Cu/Fe}$ is 140.04 by the calculation.

It is shown in Table 2 that the extraction percentage of copper is much higher than that of iron. The micro-reactor provides a good separation effect of Cu^{2+} and impurities. The main reason lies in the fact that the mass transfer speed of copper is much higher than iron. During the microchannel extraction, the flow pattern is much different with the mixing settler system. The mass transfer of the former is by diffusion through a stable interface, which may lead to better selection of metal ions.

For checking the co-extraction effects of iron during the traditional mixing settler system, the extraction experiment was carried out in separating funnels. The relationship between the extraction efficiency and time is shown in Table 3 .

Time(min)	2	3	4	5	6
Extraction efficiency of copper(%)	47.86	70.84	80.32	83.20	81.41
Extraction efficiency of iron(%)	1.21	1.89	2.05	2.17	2.81

Table 3 Extraction in different time during separating funnel experiments.

Comparison with conventional extraction, we can see that the microreactor extraction in 0.84s can achieve the result of conventional extraction in 3 min. Thus it can be seen that microreactor reduce reaction time greatly. The reason for this is that extraction reaction of copper and mass transfer type is controlled by diffusion in microchannel and that extraction efficiency increased with the extension of time in stable laminar flow microchannel. The channel size is far less than mixer settler of conventional extraction, that is conducive to diffusion and mass transfer.

Conclusions

The extraction of Cu^{2+} from a low concentration leachant with LIX984N was conducted in laminar flow microreactor. According to the results, the following conclusions are drawn:

1.In the microfluid extraction experiment, organic phase and aqueous phase has maintained steady laminar phase interface which avoid the phenomenon of the emulsion.

2.In the traditional mixing settler process, the operation time of a single stage is over 3 min, while in the microreactor, the time can be as short as 0.84 s.

3. In the microreactor process, the average mass transfer speed is 0.0214 $g \cdot m^{-2} \cdot s^{-1}$. The separation coefficient $\beta_{Cu/Fe}$ is 140.04

Acknowledgements

The authors are grateful for supports by the National Natural Science Foundation of China(U1302271), and Kunming University of Science and Technology Personnel Training Fund (KKSY201452088).

References

[1] Fletcher P D I, Haswell S J, Pombo-Villar E, et al. Micro reactors: principles and applications in organic synthesis[J]. Tetrahedron, 2002, 58(24): 4735-4757.

[2] Hessel V, Löwe H, Schönfeld F. Micromixers—a review on passive and active mixing principles[J]. Chemical Engineering Science, 2005, 60(8): 2479-2501.

[3] Huh, Yun Suk; Jeon, Sang Jun; Lee, Eun Zoo et al. Microfluidic extraction using two phase laminar flow for chemical and biological applications[J].2011,28(3),633-642

[4] Peela, Nageswara Rao; Kunzru, Deepak. Oxidative steam reforming of ethanol over Rh based catalysts in a micro-channel reactor[J]. 2011, 36(5), 3384– 3396.

[5] Kralj, Jason G.; Sahoo, Hemantkumar R.; Jensen, Klavs F. Integrated continuous microfluidic liquid-liquid extraction[J].2007,7(2),256-263

[6] Yang Kuang; Chu Guangwen; Shao Lei et al. Micromixing Efficiency of Viscous Media in Micro-channel Reactor[J]. CHINESE JOURNAL OF CHEMICAL ENGINEERING.2009,17(4),546-551

[7] Pennemann H, Hessel V, Löwe H. Chemical microprocess technology - from laboratory-scale to production [J]. CHEMICAL ENGINEERING SCIENCE. 2004, 59(22-23),4789 – 4794.

[8] Nisisako, T., Torii, T. Microfluidic large-scale integration on a chip for mass production of monodisperse droplets and particles [J]. Lab Chip, 2008,(8):287.

[9] Priest, Craig; Zhou, Jingfang et al. Microfluidic extraction of copper from particle-laden solutions[J]. INTERNATIONAL JOURNAL OF MINERAL PROCESSING. 2011,98(3-4),168-173

[10] Tamagawa O, Muto A. Development of cesium ion extraction process using a slug flow microreactor[J] CHEMICAL ENGINEERING JOURNAL. 2011, 167(2-3), 700 – 704.

[11] Aota A, Mawatari K, Takahashi et al. Phase separation of gas-liquid and liquid-liquid microflows in microchips [J].Microchim. Acta 2009, 164(3.4),249 – 255.

[12] Shaohua Ju , Peng Peng , Yaqian Wei,et al. Solvent extraction of In^{3+} with microreactor from leachant containing Fe^{2+} and Zn^{2+}[J]. Green Process Synth 2014

Rare Metal Technology 2015

Precious Metals

Rare Metal Technology 2015
Edited by: Neale R. Neelameggham, Shafiq Alam, Harald Oosterhof, Animesh Jha, David Dreisinger, and Shijie Wang
TMS (The Minerals, Metals & Materials Society), 2015

EXTRACTION OF GOLD FROM A LOW-GRADE DOUBLE REFRACTORY GOLD ORE USING FLOTATION-PREOXIDATION-LEACHING PROCESS

Yongbin Yang, Shiqian Liu, Bin Xu, Qian Li, Tao Jiang, Peng Lv
School of Minerals Processing & Bioengineering, Central South University, Changsha, Hunan 410083, China

Keywords: Double refractory gold ore, Preoxidation, Flotation, Alkaline pressure oxidation.

Abstract

The treatment process of a low-grade refractory gold ore containing sulfide minerals and carbonaceous matters was studied in this work. The gold extraction of this ore by all-sliming cyanidation was only 11.75%. Hence, a pretreatment was necessary. In this paper, a flotation-preoxidation-cyanide leaching process was developed. The concentrate from the flotation procedure was processed using different preoxidation methods. A concentrate containing 34.08g/t of gold with the gold recovery ratio of 81.20% was obtained by flotation. After roasting, microwave heating or alkaline pressure oxidation, the gold cyanidation ratio reached 82.37%, 81.26% and 71.09% respectively. Roasting was an effective, reliable and mature technique to oxidize sulfide minerals and carbonaceous matters. Roasting and microwave heating both produced dangerous environmental pollutants while alkaline pressure oxidation was eco-friendly.

Introduction

In sulfidic refractory gold ores, fine gold particles may be highly disseminated and locked up in sulfide minerals such as pyrite and arsenopyrite. These largely unoxidized ores exhibit very low gold recoveries (typically<20%) by direct cyanidation [1]. This suggests that the primary ore is highly refractory in nature and will therefore have to be processed using an appropriate preoxidation method.

Pre-concentration methods play an important role in the process allowing economic gold production from ores with low grades [2]. Flotation is an economic and widely applied pre-concentration method [3, 4].

Roasting has been used for more than a century and has been applied successfully to a wide range of materials [5]. However, the process produces pollutants such as sulfur dioxide and arsenic trioxide that need to be removed before discharge. Stringent discharge standards make the roasting process very expensive. In addition, localized high temperatures during roasting may produce impervious and glassy fluxes that encapsulate the gold and reduce gold extraction [6, 7].

Microwaves could potentially be utilized as an alternative energy source for processing minerals. Some advantages of microwave heating include: rapid; heat is transferred efficiently; internal heating is possible, the energy source is clean; and microwave processing is compatible with continuous processing [8, 9]. Carbon and metal sulfides are known to be very good microwave absorbers and they can be rapidly and selectively heated [10].

Alkaline pressure oxidation, as a clean and efficient method for strengthening hydrometallurgical leaching, has been used in the pretreatment of complex and refractory materials, such as lead-containing copper matte, refractory gold ore, and nickel smelter slag [11, 12]. Alkaline pressure oxidation can be used to process refractory gold ores containing high alkaline gangues and carbonate. Under the alkaline conditions sulfur and arsenic in the refractory ores are completely dissolved as sulfates and arsenates which are harmless to the environment [13, 14]. However, high content of carbonaceous matters, especially elemental carbon could not be oxidized readily by using pressure oxidation.

In this paper, the treatment process of a low-grade refractory gold ore containing sulfide and carbonaceous matters was studied. The concentrate from the flotation procedure was processed using preoxidation methods (i.g. roasting, pressure oxidation, and microwave heating). Comparisons between the different flotation and preoxidation methods were made, and a flotation-preoxidation-cyanidation process was established.

Experimental work

Materials and reagents

A refractory gold ore containing 2.43g/t of gold was used for this study. The gold extraction of all-sliming cyanidation was only 11.75%. Chemical analysis, X-ray diffraction (XRD) and optical mineralogy by using the polished and thin sections were applied for ore characterization. Results of chemical compositions, gold phases and XRD are shown in Table I, Table II and Figure 1. Mineralogical characterization studies on this ore indicated that the ore was dominated by silicate minerals and clays including quartz, dolomite, magnesite, lizardite, phlogopite, and muscovite. Pyrite and arsenopyrite accounted for most of the sulfide content. Most of the gold occurred as sub-microscopic grains associated with fine-grained sulfide minerals. Butyl xanthate and butylamine dithiophosphate used in the flotation procedure were commercial pure. Other chemical reagents, such as sodium hydroxide, sodium carbonate, sodium cyanide and sulfuric acid used in this study were all analytically pure. The purity of oxygen gas used in the leaching tests was 99.9%.

Table I. Chemical compositions of the gold ore

Elements	Au *	Ag*	As	Sb	S	C	Cu
Compositions(%)	2.43	1.16	0.093	0.047	0.35	4.08	0.0052
Elements	Ni	Fe	SiO_2	CaO	MgO	Al_2O_3	K_2O
Compositions(%)	0.083	4.06	43.77	8.79	11.66	7.8	2.72

* Unit g/t.

Table II. Chemical phase analysis of the gold

Gold phase	Compositions (g/t)	Distributions (%)
Monomer and exposed	0.05	2.06
Wrapped by sulfide minerals	1.47	69.96
Wrapped by oxide minerals	0.54	22.22
Wrapped by silicate minerals	0.14	5.76
Total	2.43	100.00

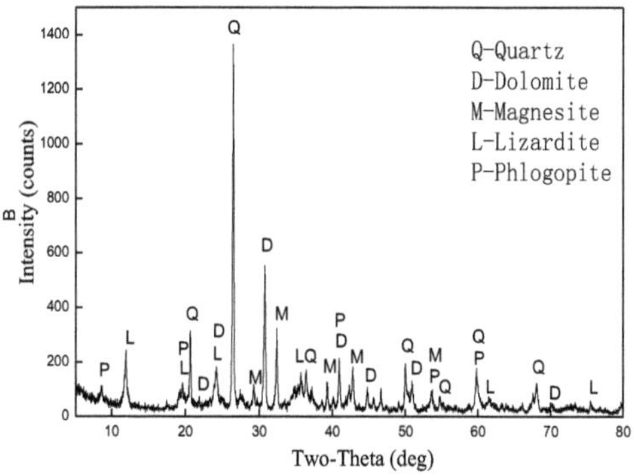

Figure 1. XRD analysis of the gold ore

Flotation tests

500g of the raw ore was used for the floatation tests. 85% of the flotation feed were finer than 74 micron. A lab-used flotation machine of XFDIII, 1.5L and 0.5L floatation cells and tap water were used for the flotation tests. The flotation tests were carried out at ~ 33% solids. The volume of the pulp in the floatation cell was kept constant by adding tap water. The impeller speed was 2400r/min. Methyl isobutyl carbinol (MIBC) was applied as the frother during the tests. Conditions of the flotation test as a pre-concentration method are shown in Table III. Sodium carbonate was used as ph regulator, water glass was applied as a depressor to gangue, copper sulfate was used as activator, and butyl xanthate and butylamine dithiophosphate were used together as a combined collector.

Table III. Conditions of the flotation test

Stage	Reagents added (g/t)						Time (min)
	Sodium carbonate	Water glass	Copper sulfate	Butyl xanthate	Butylamine dithiophosphate	MIBC	
Rougher 1	1000	1000	500	200	100	20	6
Rougher 2				100	50		4
Scavenger				50	25		4
Cleaner		200		50	25		4

Pre-oxidation tests

The concentrate from the flotation tests needed oxidative pre-treatments in order to improve recoveries and achieve the economic feasibility. Screen analysis showed that 90% of the concentrate was finer than 74μm.

Roasting In this study, roasting of the concentrate was performed in a horizontal tube furnace. The concentrate was put in a porcelain boat at a constant temperature. Air flow was sent in to the furnace. At the end of the test, the air flow was interrupted and the solid material was cooled. Roasted samples were weighed to evaluate mass change and volatilization of oxides. The gas effluent, mainly consisting of excess air, SO_2 and other volatilized oxides, was gurgled in a saturated solution of NaOH to reduce the emission of pollutants. Experimental conditions are reported in Table Ⅳ. Different tests were carried out to check the effects of temperature and roasting time.

Table Ⅳ. Conditions of the roasting tests

Temperature (℃)	Time (h)	Au leaching extraction (%)	Temperature (℃)	Time (h)	Au leaching extraction (%)
0	0	14.04	750	2	82.37
500	2	59.46	800	2	80.21
550	2	63.02	750	1	75.57
600	2	65.28	750	3	81.78
650	2	69.39	750	4	79.49
700	2	76.33			

Microwave heating The microwave heating tests were performed with 20g samples in a quartz crucible. The crucible was placed at the centre of the oven. During the tests O_2 was fed to maintain the oxidative atmosphere. The variables studied were: heating temperature, heating time and O_2 flow rate. Experimental conditions are reported in Table Ⅴ.

Alkaline pressure oxidation The flotation concentrate (20g) was pulped to ~20% solids with tap water. Afterwards, a certain amount of NaOH was added into the pulp.The 0.5-liter autoclave vessel was sealed and then heated. Once the temperature arrived at desired value, desired O_2 partial pressure was applied. The stirring speed was 700r/min. At the end of the test the autoclave was closed and the pressure was relieved. The conditions of alkaline pressure oxidation tests are reported in Table Ⅵ. Different tests were carried out to check the effects of the process parameters including temperature, oxygen partial pressure, NaOH dosage and oxidation time.

Cyanidation tests Cyanidation tests were carried out with the solid coming from the preoxidation treatments. The oxidation product (20g) was pulped with water to ~20% solids in a 250ml beaker. The pH was adjusted to 10.5~11 with sodium hydroxide and maintained stable during the whole leaching time (i.e. 48h). NaCN was then added at a concentration of 0.2%. The agitator rotation rate was 400r/min. The leaching tests were conducted at room temperature (~25℃). At the end of the test, pulp was filtered. The filtercake was washed several times with water. The leaching residue was dried for assay. The results are given in Tables Ⅳ ~ Ⅵ.

Table Ⅴ. Conditions of the microwave heating tests

Heating stage	Temperature (℃)	Time (min)	O_2 flow (L/h)	Au leaching extraction (%)
First Second	500 700	20 40	180 180	68.19
First Second	500 700	60 20	96 160	72.24
First Second	500 700	90 60	96 200	78.58
First Second	500 700	90 90	96 200	81.26

Table Ⅵ. Conditions of the alkaline pressure oxidation tests

Temperature (℃)	P_{O_2} (MPa)	NaOH (kg/t)	Time (h)	Au leaching extraction (%)
180	0.80	50.00	3.00	53.64
200	0.80	50.00	3.00	67.23
210	0.80	50.00	3.00	69.59
220	0.80	50.00	3.00	71.09
230	0.80	50.00	3.00	71.14
220	0.60	50.00	3.00	67.56
220	0.80	50.00	3.00	71.09
220	1.00	50.00	3.00	71.15
220	0.80	30.00	3.00	54.37
220	0.80	40.00	3.00	65.36
220	0.80	50.00	3.00	71.09
220	0.80	60.00	3.00	69.08
220	0.80	50.00	2.00	63.25
220	0.80	50.00	3.00	71.09
220	0.80	50.00	4.00	71.12
220	0.80	50.00	5.00	71.04

Results and discussions

Flotation as a pre-concentration method

According to the results, a gold concentrate containing 34.08g/t of gold with the gold recovery ratio of

81.20% was obtained. The concentrate yield was 5.79%. Chemical compositions of the flotation concentrate are shown in Table Ⅶ. The contents of S, C and As were high. Meanwhile, alkaline gangues in the flotation concentrate were also relatively high. Gold extraction of the direct cyanidation was only 14.04% (see Table Ⅳ) which indicated that the concentrate required pre-oxidation before leaching.

Table Ⅶ. Chemical compositions of the concentrate

Elements	Au *	Ag*	As	Sb	S	C	Cu
Compositions(%)	34.08	6.03	1.70	0.94	6.15	3.61	0.073
Elements	Ni	Fe	SiO_2	CaO	MgO	Al_2O_3	K_2O
Compositions(%)	0.177	10.85	21.37	10.30	20.38	8.27	1.33

* Unit g/t.

Comparisons between different preoxidation methods on the gold concentrate

The refractory concentrates needed oxidative pre-treatment in order to improve recoveries and achieve the economic feasibility. In this study, roasting, microwave heating and alkaline pressure oxidation tests were performed with the flotation concentrate. The control factors and their levels used in the different oxidation experiments are given in Table Ⅳ ~ Ⅵ.

Roasting Roasting is a reliable and mature technique to oxidize both sulfide minerals and carbonaceous matters at high temperatures using air or oxygen as oxidant gas. It could improve the leaching efficiency of gold and has strong adaptability. The main defects are: 1) the energy consumption is high because it is necessary to maintain the desired temperature for hours; 2) some process wastes such as sulfur dioxide and arsenic trioxide are dangerous environmental pollutants; 3) conditions demand to be properly and carefully controlled. The temperature for the oxidization of graphite carbon is higher than that for the sulfide minerals, while excessive roasting may form physical encapsulation. As can be seen from Table Ⅳ, exceedingly high roasting temperature and long time led to lower gold extraction.

According to Table Ⅳ, both roasting temperature and time had significant effects on gold leaching ratio. The highest gold extraction by roasting was found to be 82.37% at 750℃ and roasting time of 2h. Roasted overtime or under exceedingly high temperature would lead to an decrease in gold extraction. The overall recovery ratio of gold reached 66.88% using flotation-roasting-cyanidation process.

Microwave heating The gold extraction of the concentrate by cyanidation reached 81.26% after microwave heating. Microwave heating usually cost shorter time compared to conventional roasting. However, the heating time and temperature required in this case were similar with conventional roasting. In addition its processing capacity was small, and the pollutants produced in this process also needed to be removed. The overall recovery ratio of gold was 65.98% using flotation-microwave heating-cyanidation process.

Alkaline pressure oxidation As alkaline gangues in the gold concentrate were relatively high, alkaline pressure oxidation might be a proper method for pre-oxidation. Under the alkaline conditions sulfur and arsenic in the refractory ores are completely dissolved as sulfates and arsenates which are harmless to the environment. Compared to roasting and microwave heating, the methods used in flotation-pressure oxidation-cyanidation process are all hydrometallurgical.

The most significant parameter relating to gold leaching ratio was found to be temperature as shown in Table Ⅵ. Under the conditions of oxygen partial pressure 0.8Mpa, oxidation time 3h and NaOH dosage 50kg/t, gold leaching ratio increased dramatically from 53.64% to 71.09% when temperature raised from 180℃ to 220℃, and then the leaching ratio remained constant at a higher temperature of 230℃.

The optimal gold leaching ratio of 71.09%was obtained under the condition of temperature 220℃, oxygen partial pressure 0.8MPa, oxidation time 3h and NaOH dosage 50kg/t. According to the results, the gold extractions of pressure oxidation tests did not yield gold extraction as high as the extractions of the other two methods did. The overall recovery ratio of gold reached 57.72% using flotation-alkaline pressure oxidation-cyanidation process. It is not easy to oxidize graphite carbon in alkaline pressure oxidation process. These carbonaceous matters can consume large amount of reagents and adsorb leached gold from the solution during cyanide leaching, which may be the reason for the low leaching ratio of gold. On the other hand, the iron oxides and hydroxides formed during alkaline pressure oxidation can coat the surfaces of gold and sulfide minerals and prevent the ore from further oxidation [15, 16]. These may be resolved by using thiosulfate leaching which is less sensitive to carbonaceous matters or adding some surfactants. However, it needs further study to make alkaline pressure oxidation an effective preoxidation method.

Conclusions

A concentrate containing 34.08g/t of gold with the gold recovery ratio of 81.20% was obtained by flotation.

The maximum gold leaching ratio of 82.37%% was obtained by roasting-cyanidation. Roasting was an effective, reliable and mature technique to oxidize both sulfide minerals and carbonaceous matters. The main defects of roasting were: high energy consumption and environmental pollution.

Gold extraction of the concentrate by cyanidation reached 81.26% after microwave heating. The processing capacity of microwave heating was small and it also produced pollutants which needed to be removed before discharge.

The gold leaching ratio of 71.09% was obtained using alkaline pressure under the conditions of temperature 220℃, oxygen partial pressure 0.8Mpa and oxidation time 3h. The low extraction ratio of gold may be caused by the carbonaceous matters in the concentrate. It may be resolved by using thiosulfate leaching instead of cyanide leaching. Alkaline pressure oxidation is eco-friendly, and might be more effective after further study.

Acknowledgments

Financial supports from theNational Natural Science Foundation of China (Grant No.51074182) and China Postdoctoral Science Foundation (Grant No.2014M550422) are gratefully acknowledged.

References

1. R. Badri and P. Zamankhan, "Sulphidic refractory gold ore pre-treatment by selective and bulk flotation methods," *Advanced Powder Technology*, 24 (2) (2013), 512–519.
2. D. S. R. Murthy, V. Kumar, and K. V. Rao, "Extraction of gold from an Indian low-grade refractory gold ore through physical beneficiation and thiourea leaching," *Hydrometallurgy*, 68 (1–3) (2003), 125–130.
3. C. T. O'Connor and R. C. Dunne, "The flotation of gold bearing ores — A review," *Minerals Engineering*, 7 (7) (1994), 839–849.
4. M. B. M. Monte, A. J. B. Dutra, C. R. F. Albuquerque Jr., L. A. Tondo, and F. F. Lins, "The influence of the oxidation state of pyrite and arsenopyrite on the flotation of an auriferous sulphide ore," *Minerals Engineering*, 15 (12) (2002), 1113–1120.
5. J. O. Marsden, C. I. House, and I. House, *The Chemistry of Gold Extraction*. (Society for Mining Metallurgy, 2006).
6. C. Komnitsas and F. D. Pooley, "Mineralogical characteristics and treatment of refractory gold ores," *Minerals Engineering*, 2 (4) (1989), 449–457.
7. K. S. Fraser, R. H. Walton, and J. A. Wells, "Processing of refractory gold ores," *Minerals Engineering*, 4 (7–11) (1991), 1029–1041.
8. B. Nanthakumar, C. A. Pickles, and S. Kelebek, "Microwave pretreatment of a double refractory gold ore," *Minerals Engineering*, 20 (11) (2007, 1109–1119.
9. S. J. Ma, W. J. Luo, W. Mo, X. J. Su, P. Liu, and J. L. Yang, "Removal of arsenic and sulfur from a refractory gold concentrate by microwave heating," *Minerals Engineering*, 23 (1) (2010), 61–63.
10. R. K. Amankwah and C. A. Pickles, "Microwave roasting of a carbonaceous sulphidic gold concentrate," *Minerals Engineering*, 22 (13) (2009), 1095–1101.
11. H. Long and D. G. Dixon, "Pressure oxidation kinetics of orpiment (As2S3) in sulfuric acid," *Hydrometallurgy*, 85 (2–4) (2007), pp. 95–102.
12. W. Liu, T. Yang, D. Zhang, L. Chen, and Y. Liu, "Pretreatment of copper anode slime with alkaline pressure oxidative leaching," *International Journal of Mineral Processing*, 128 (2014), 48–54.
13. H. Huang, C. Jiang, and P. Sun, "A mechanism study on technology of pretreatment of refractory gold ores by alkaline hot-pressure oxidation," *Rare Met*, 02 (2003), 249–253.
14. S. Gredelj, M. Zanin, and S. R. Grano, "Selective flotation of carbon in the Pb–Zn carbonaceous sulphide ores of Century Mine, Zinifex," *Minerals Engineering*, 22 (3) (2009), 279–288.
15. K. G. Thomas, "Alkaline and acidic autoclaving of refractory gold ores," *JOM*, 43 (2) (1991), 16–19.
16. S. R. La Brooy, H. G. Linge, and G. S. Walker, "Review of gold extraction from ores," *Minerals Engineering*, 7 (10) (1994), 1213–1241.

Rare Metal Technology 2015
Edited by: Neale R. Neelameggham, Shafiq Alam, Harald Oosterhof, Animesh Jha, David Dreisinger, and Shijie Wang
TMS (The Minerals, Metals & Materials Society), 2015

GOLD EXTRACTION FROM A HIGH CARBON LOW-GRADE REFRACTORY GOLD ORE BY FLOTATION-ROASTING-LEACHING PROCESS

Yongbin Yang, Zhaohui Xie, Bin Xu, Qian Li, Tao Jiang
School of Minerals Processing & Bioengineering, Central South University,
Changsha, Hunan 410083, China

Keywords: Carbonaceous refractory gold ores, Flotation, Roasting, Cyanide leaching

Abstract

A high carbon low-grade gold ore, in which gold mainly occurs in carbonaceous matters, is a typical refractory gold ore whose conventional cyanide leaching extraction is only 12.9 %. In this paper, by the process of flotation-roasting-leaching, gold was effectively enriched and extracted. The flotation results showed that the enrichment ratio of gold was relevant to that of carbon, indicating that gold occurred in close association with carbon. A concentrate containing gold 7.26 g/t and carbon 12.37 % was obtained, with gold recovery 83.10 % and carbon recovery 90.24 %. Subsequently, the roasting of flotation concentrate favorably made the removal rates of carbon and sulfur reach 99.81% and 98.79%, respectively. After roasting, the gold in the roasted product was further enriched to 29.70 g/t and the optimal gold cyanide leaching extraction reached 93.66%.

Introduction

With the rapid depletion of non-refractory gold ores, refractory gold ores have become the main materials of gold industry, and nearly one third of gold comes from refractory gold ores all around the world. In particular, carbonaceous gold ores, accounting for over 20% of the refractory gold ore resources, obtain increasing concerns about how to utilize them economically and effectively [1].

Generally, the poor gold extraction of carbonaceous gold ores is attributed to carbonaceous matters which mainly have two aspects of harmful effects on gold extraction. On one hand, the carbonaceous matters can inhibit gold leaching because of locking gold up in themselves. On the other hand, the carbonaceous matters can absorb dissolved gold from solution and this phenomenon is termed as preg-robbing [2]. The ordinary carbonaceous gold ores are refractory due to preg-robbing. Whereas, there is a very unusual kind of carbonaceous gold ores in which both of the two adverse factors on gold extraction exist.

In order to extract the gold from the carbonaceous gold ores, basic approach either whole ores treatment process or beneficiation-metallurgy process is available [3]. For a low grade refractory gold ore, the costs involved in whole ores treatment process are certainly not economically. However, beneficiation-metallurgy process has an advantage that a concentrate with higher grade value of gold can be produced by beneficiation techniques and thus only a lesser quantity of the material will be treated in the subsequent process [4].

At present, flotation is commonly used as a preconcentration step to recover gold from the low grade carbonaceous sulfidic refractory gold ores when the gold mainly occurs in sulfides, followed by pretreatment methods including roasting oxidation, chemical oxidation and bio-oxidation [5,6]. Among them, due to its mature process and strong adaptability, roasting becomes a reliable technique to decompose carbonaceous matter and improve the leaching efficiency of gold. Therefore, this method has been most widely used.

Extensive studies were conducted on carbonaceous sulfidic refractory gold ores. Numerous flotation preconcentration experiments on recovering sulphide minerals and gold indicated that high reagents consumption was caused because of the strong adsorbability of organic carbon. Thereby, some researchers proposed priority flotation for carbon to reduce the adsorption of reagents, followed by flotation for gold from its tailings [7-9]. Thereafter, both of the two concentrates were processed by roasting-cyanidation method. In this way, gold was economically and effectively extracted.

However, there are fewer researches on processing the carbonaceous refractory gold ores when gold is hosted in carbonaceous matters. Liu [11,12] put forward the process flow of branch extracting gold when the gold partially occurred in carbonaceous matters, i.e. carbon flotation-oxidizing roast of carbonaceous gold concentrate-branch cyanidation of roasted ore and tailing. As a result, the gold recovery of flotation concentrate was 69.47% and the total gold recovery was 88.65%. Although the costs are reduced by roasting lesser concentrate, the branch process has the disadvantages of low gold recovery of concentrate, long process and massed operations.

In the present work, the gold extraction from a high carbon low-grade refractory gold ore whose gold largely occurred in carbonaceous matters was studied. The aim of this study was to recover the gold effectively by the flotation-roasting-leaching process.

Experimental

Raw materials

The high carbon low-grade refractory gold ore used in this study was taken from Jiangxi province in China. It is a typical refractory gold ore whose conventional cyanidation extraction is only 12.9 %. The chemical compositions of the ore shown in Table I indicated that the total carbon content reached 4.34% resulting in adverse effects on gold extraction. The XRD pattern of the ore was given in Figure 1. The major mineralogical constitutes of the ore were carbon and hematite, with quartz, muscovite and kaolinite as the major gangue minerals. Seen from Table II, the distribution of gold had a close association with carbon. It could be concluded from these results that gold could be concentrated by carbon flotation. In order to maximize the gold recovery of this ore, it was necessary to improve the recovery of carbon.

Methods

Preparation of sample The ore sample was crushed and passed through a 1 mm size screen thoroughly.

The crushed sample was mixed completely and then split into 500g samples by cone and quartering technique. Subsequently, these samples were stored in bags respectively.

Table I. Chemical compositions of raw materials

Element	Au*	C	S	Cu	As	Sb	TFe	SiO_2	Al_2O_3	CaO	MgO
Content (%)	3.05	4.34	0.30	0.006	0.005	0.05	4.94	66.47	13.94	0.34	0.22

*The unit of Au is g/t.

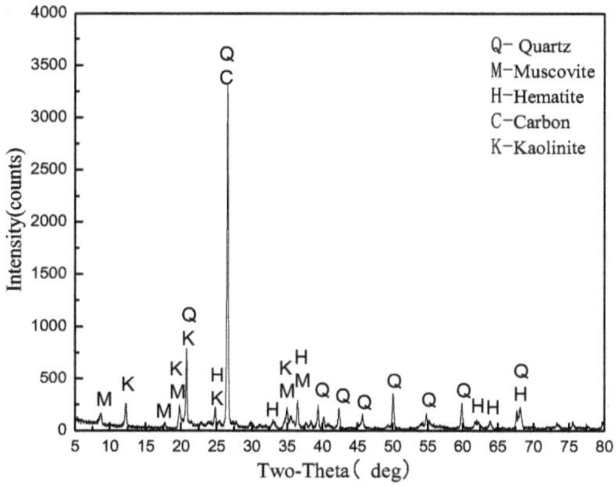

Figure 1. XRD pattern of raw materials

Table II. Particle size distribution of raw materials

Size (mm)	Weight (%)	Gold grade (g/t)	Carbon content (%)	Distributive rate (%)	
				gold	carbon
+0.15	30.16	0.99	0.87	9.77	6.04
-0.15+0.074	6.35	1.12	1.17	2.33	1.71
-0.074+0.045	3.85	1.36	1.48	1.71	1.31
-0.045+0.038	2.44	4.29	6.38	3.43	3.59
-0.038	57.20	4.42	6.63	82.76	87.35
Total	100.00	3.05	4.34	100.00	100.00

Flotation experiments The flotation reagents included collector diesel oil, inhibitor sodium hexametaphosphate and frother secondary octanol [12].

For each test, 500g sample was ground in a ball mill, at slurry density of 50% solids by weight, to obtain a certain grind size. To perform flotation experiment, the ground sample was transferred into a 1.5L flotation cell and diluted to about 30% solids using tap water. It was agitated for 2 min before any reagents were added. Then, the inhibitor and collector were added and conditioned for 2 and 3 min respectively, followed by the addition of frother with 2 min of conditioning. Thereafter, the pulp was aerated and the concentrate was collected for 15 min. Throughout the experiment, the froth height was maintained by adding an appropriate amount of water. At the end of the each test, concentrates and tailings were filtered, dried and analyzed for gold grade and carbon content. The gold was analyzed using AAS. A series of experiments were carried out to obtain the optimum optimal values of grind size, dosages of collector and inhibitor required for flotation. In addition, the frother dosage was decided by experiment phenomena.

Roasting-leaching experiments The flotation concentrate was utilized in roasting experiments. All the roasting experiments were performed in the experimental apparatus shown in Figure 2. In each experiment, the sample was loaded in two small fireclay crucibles. Each crucible contained 10g ore. Both of the crucibles were placed in a quartz tube which was put in an electrically heated horizontal tube furnace. The gas flow was controlled at 120 L/h by adjusting the flowmeter. Simultaneously, the temperature was measured using a Type K thermocouple. The sample was roasted for a certain time under these conditions, and then cooled in air. All the roasted samples were weighed and analyzed for the content of carbon and sulfur. The studied roasting variables were temperature and time.

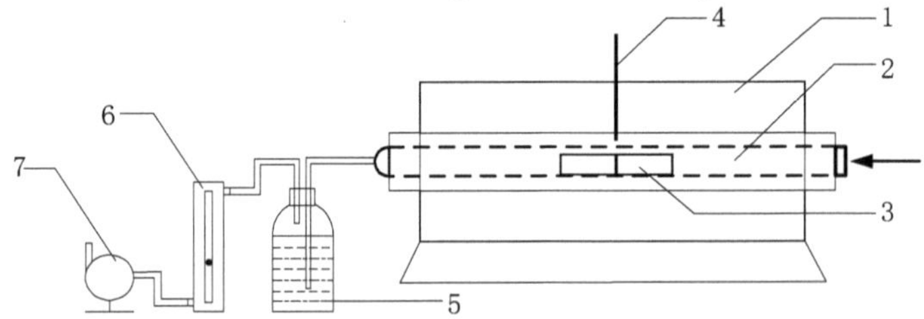

1- horizontal tube furnace; 2-quartz tube; 3-fireclay crucible; 4-thermocouple;
5-exhaust gas absorption equipment; 6-flowmeter; 7- pump
Figure 2. Schematic diagram of the equipment for roasting

After roasting, the roasted product was leached by sodium cyanide. In each leaching test, 20 g roasted product was put into a 300 ml beaker and diluted using certain water (solid-liquid ratio 1:3). The pH value of the leaching solution was kept at 10.5-11 by adding a moderate amount of sodium hydroxide. Next, appropriate sodium cyanide was added to obtain the cyanide concentration of 0.2%. The pulp was stirred on an agitator at 600rpm for 48h. After the test, the residual was filtered, dried and analyzed for gold grade. In roasting-leaching experiments, the removal rates of carbon and sulfur and the gold leaching extraction were adopted to test the effect of roasting.

Results and Discussion

Flotation studies

Effect of grind size To test the effect of grind size on flotation, a series of experiments were carried out by increasing the grind size with -0.074 mm from 63% to 85%. In each test, the additions of inhibitor, collector and frother were 500 g/t, 600 g/t and 40 g/t, respectively. The effect of grind size on flotation shown in Fig.3 indicates that the gold grade and carbon content decreases gradually (gold from 9.52 g/t to 8.27g/t and carbon from 16.70% to 13.90%) with the increase of grind size, and the enrichment ratios of gold and carbon are nearly equal under the same condition. These results prove that gold occurs in the carbonaceous matters. It can also be seen that both the recoveries of gold and carbon tend to peak at the grind size with -0.074 mm of 71%, with the values of 58.96% and 65.02%. This phenomenon could be explained that appropriate grind size can make target minerals fully liberated. However, the ores are extremely apt to be slimed when ground for too long, and the flotation recovery rate decreases. Hence, the grind size with -0.074 mm of 71% is recommended to the subsequent experiments.

Effect of collector Diesel oil is widely used as a collector in the flotation of coal and it was therefore speculated to recover the carbonaceous matter from the high carbon low-grade refractory gold ore as well. The effect of collector on flotation was tested by increasing the collector dosage from 600 g/t to 3400 g/t. The experiments were performed under the conditions of grind size with -0.074 mm of 71%, inhibitor 500 g/t and frother 40 g/t. Figure 4 describes the effect of collector on flotation that the grade of gold grade and carbon content increases with collector addition increasing collector addition to 2600 g/t and then declines to a lesser extent, and the enrichment ratios of gold and carbon are almost the same in each test. Furthermore, as the collector increases to 3000 g/t, the recoveries of gold and carbon rise sharply at first, after which the carbon recovery increases slightly while the gold recovery remains constant. Thus, a collector of 3000 g/t is considered to be the most favorable value.

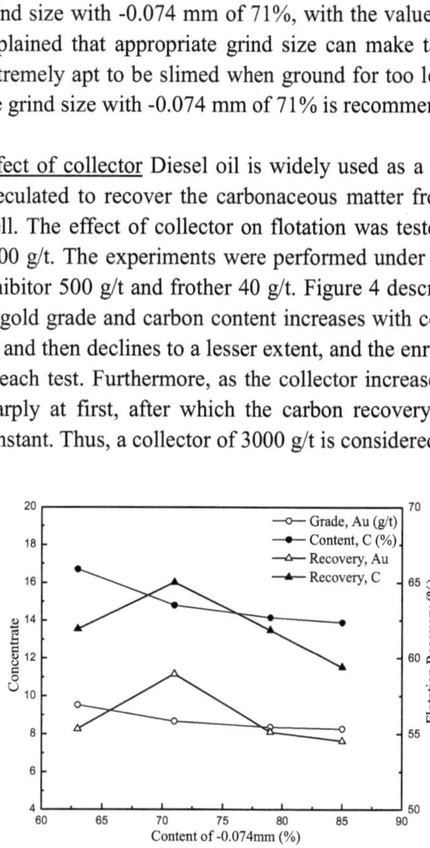

Figure 3. Effect of grind size on flotation

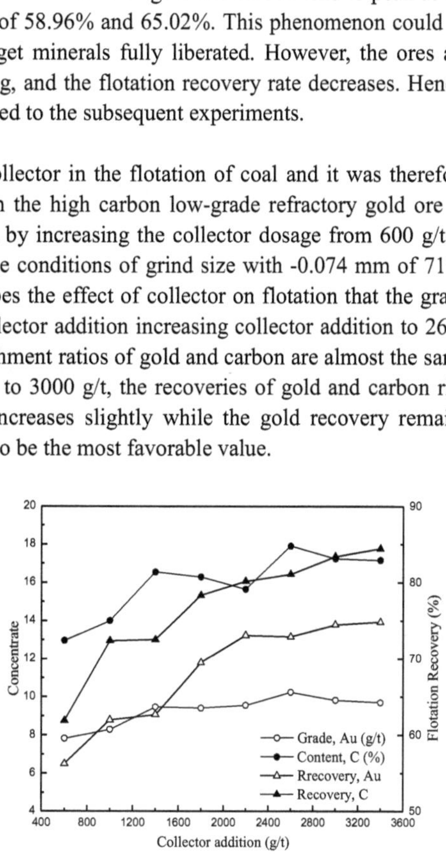

Figure 4. Effect of collector on flotation

Effect of inhibitor Sodium hexametaphosphate, which is deemed as an effective inhibitor, plays an important role in inhibiting gangue minerals and dispersing the slime. A series of experiments were conducted under the conditions of grind size with -0.074 mm of 71%, collector 3000 g/t and frother 40 g/t. The effects of increasing inhibitor dosage from 0 to 700 g/t are given in Figure 5. It clearly displays that the grade of gold grade and carbon content almost linearly increases with inhibitor addition increasing inhibitor addition to 500 g/t and then tends to be stable. Likewise, the variation regularity of the enrichment ratios of gold and carbon is the same as shown in Figure 3 and Figure 4. Moreover, the optimal recoveries of gold and carbon are achieved at an inhibitor dosage of 50 g/t. Under this condition, a concentrate containing gold 7.26 g/t and carbon 12.37 % is obtained, with gold recovery 83.10% and carbon recovery 90.24%. Consequently, an inhibitor of 50 g/t is chosen for the flotation process.

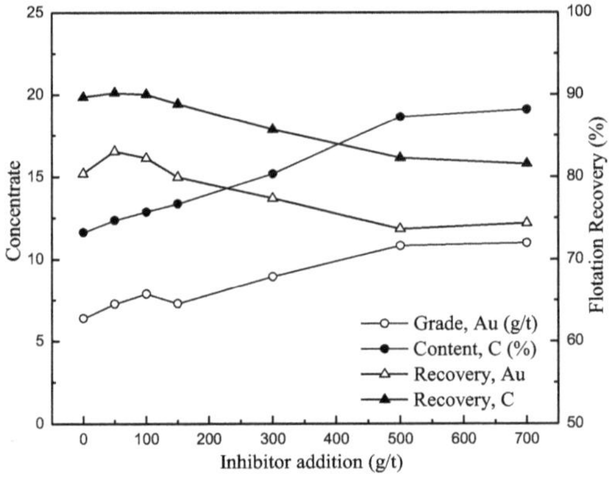

Figure 5. Effect of inhibitor on flotation

Roasting-Leaching studies

Production of the concentrate The flotation concentrate, which containing gold 17.82 g/t, carbon32% and sulfur 1.65%, is obtained by adopting one roughing stage under the optimum optimal conditions established earlier and one cleaning stage without using any reagents. The flow sheet of producing the concentrate is showed in Figure 6. The carbon and sulfur of concentrate should be removed to release the adverse effects on gold leaching.

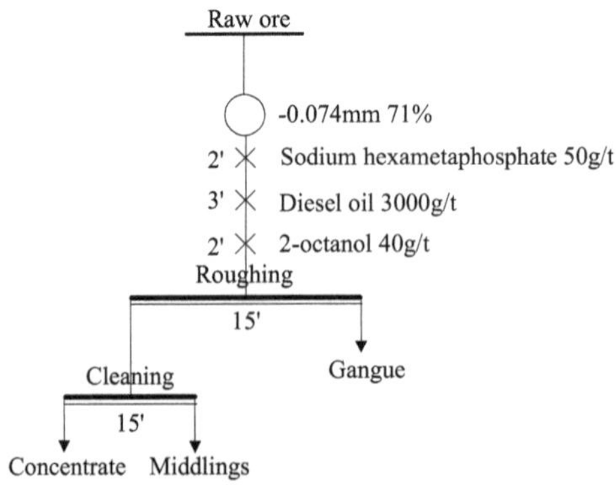

Figure 6. The flow sheet of producing the concentrate

Effect of roasting temperature To optimize the roasting temperature required for the process, the experiments were carried out by varying roasting temperature from 500°C to 800°C for 2h. Figure 7 illustrates the removal rates of carbon and sulfur and gold leaching extraction as a function of roasting temperature. As observed from Figure 7, the removal rates of carbon and sulfur increase from 95.76% and 95.89% to 99.83% and 97.82%, respectively, within the temperature ranging from 500°C to 600°C, and both of them are totally removed above 600°C. Simultaneously, the mass loss almost remains constant about 40%, indicating that carbon and sulfur are mostly removed and gold is further enriched. It can also be seen from the curve of gold leaching extraction that the optimal value of 91.22% is achieved at 600°C. Beyond 600°C the gold leaching extraction decreases markedly and this is because that some local melting occurs and there is some sintering and formation of some glassy material. Therefore, it is concluded that 600 °C is the optimal roasting temperature.

Effect of roasting time Figure 8 presents the effect of roasting time on roasting-leaching process at 600°C. Seen from it, the curve of mass loss is similar to that of Figure 7. The removal rates of carbon and sulfur increase from 91.50% and 94.25% to 99.81% and 98.79%, respectively, with the increase of roasting time from 1.5 h to 2h, and then keep constant. Meanwhile, the gold leaching extraction sharply rises from 1.65% to 92.27% as the roasting time ascends from 1.5 h to 2h, since the carbon and sulfur were largely removed. Thereafter, the best gold leaching extraction of 93.66% is achieved at 2.5h. As a result, the suitable roasting time is 2.5h.

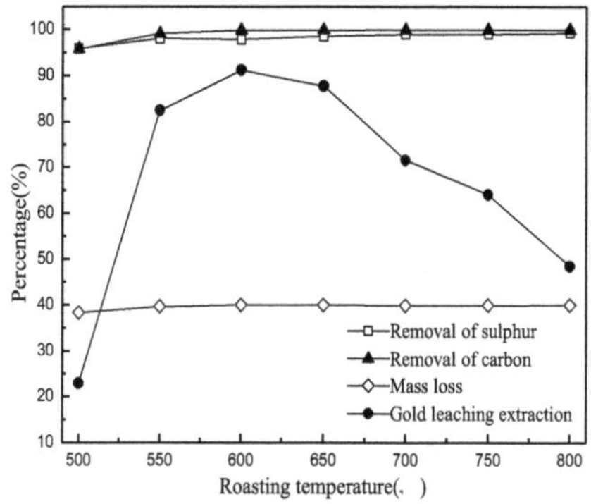

Figure 7. Effect of roasting temperature on roasting-leaching process

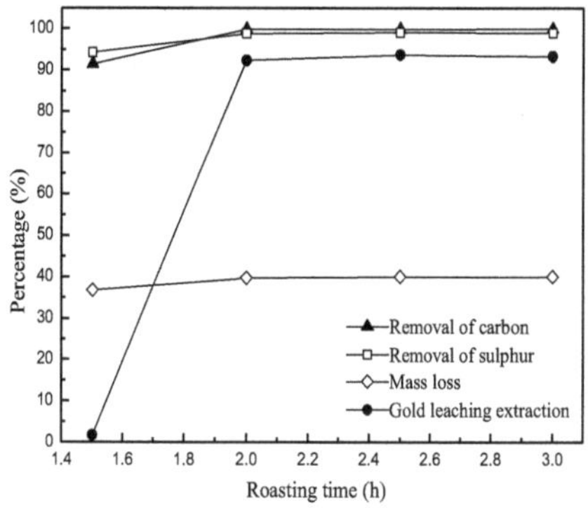

Figure 8. Effect of roasting time on roasting-leaching process

Conclusions

The combined flotation-roasting-leaching process successfully treated the high carbon low-grade refractory gold ore. It was found that beneficiating gold by the flotation of carbon was possible. The enrichment ratio of gold was relevant to that of carbon, indicating that gold occurred in a close association with carbon. Under the conditions of grind size with -0.074 mm of 71%, diesel oil of 3000 g/t and sodium hexametaphosphate of 50 g/t, the optimal recoveries of gold and carbon reached 83.10% and 90.24%, with gold grade 7.26 g/t and carbon content 12.37%, respectively. Thereafter, the removal rates of carbon and sulfur from the flotation concentrate reached 99.81% and 98.79%,respectively, after roasting at 600 ℃ for 2.5h. Finally, the gold in the roasted product was further enriched to 29.70 g/t and the optimal gold cyanide leaching extraction reached 93.66%.

Acknowledgments

Financial supports from the National Natural Science Foundation of China (Grant No.51074182) and China Postdoctoral Science Foundation (Grant No.2014M550422) are gratefully acknowledged.

References

1. Xiaoyang Xu, "Review of research on leachin process of carbonaceous refractory gold ore," *Gold Science and Technology*, 21(1) (2013), 82-88.
2. Hongying Yang et al., "Research status of carbonaceous matter in carbonaceous gold ores and bio-oxidation pretreatment," *Trans. Nonferrous Met. Soc. China*, 23(2013), 3405-3411.
3. Faraz Soltani et al., "Improved recovery of a low-grade refractory gold ore using flotation-preoxidation-cyanidation methods," *International Journal of Mining Science and Technology*, 24 (2014), 537-542.
4. D.S.R. Murthy, Vinod Kumar, and K.V. Rao, "Extraction of gold from an Indian low-grade refractory gold ore through physical beneficiation and thiourea leaching," *Hydrometallurgy*, 68 (2003), 125-130.
5. Reyhaneh H et al., "The effect of non-sulphide gangue minerals on the flotation of sulphide minerals from Carlin-type gold ores," *Minerals Engineering*, 60 (2014), 26-32.
6. Rezgar Badri, Piroz Zamankhan, "Sulphidic refractory gold ore pre-treatment by selective and bulk flotation methods," *Advanced Powder Technology*, 24 (2013), 512-519.
7. Renge Yang, Yongdong Zhang, "Research on flotation process of carbonous gold ore," *Gold*, 33(6) (2012), 47-49.
8. Shuiwang Zhang, Qinglin Li, and Yong Feng, "Process test study of contained carbon and micro-fine-grain complex gold ore," *Nonferrous Metals (Minerals Processing Section)*, 4(2010), 20-24.
9. Xue Yu, "Study on floatation process of a fine-particle gold ore containing carbon," *Gold*, 32(10) (2011), 52-55.
10. Shengming Liu, "Experimental study on process flow of branch extraction of gold from carbonaceous gold ore in a gold mine," *Gold*, 19(10) (1998), 37-39.
11. Shengming Liu, "The comparison of gold leaching processes on a carbonaceous gold ore," *Multipurpose Utilization of Mineral Resources*, 2(1999), 9-11.
12. S. Gredelj, M. Zanin, and S.R. Grano., "Selective flotation of carbon in the Pb-Zn carbonaceous sulphide ores of Century Mine, Zinifex," *Minerals Engineering*, 22 (2009), 279-288.

Rare Metal Technology 2015
Edited by: Neale R. Neelameggham, Shafiq Alam, Harald Oosterhof, Animesh Jha, David Dreisinger, and Shijie Wang
TMS (The Minerals, Metals & Materials Society), 2015

GOLD LEACHING FROM A REFRACTORY GOLD CONCENTRATE BY THE METHOD OF LIQUID CHLORINATION

Chao Li[1,2], Hongxu Li[*1,2], Xie Yang[1,2], Shuai Wang[1,2], Lifeng Zhang[2]

[1]Beijing Key Lab of Green Recycling and Extraction of Metals, University of Science and Technology, 30# Xueyuan Road, Beijing, 100083,China
[2]School of Metallurgical and Ecological Engineering, University of Science and Technology Bejing, 30# Xueyuan Road, Beijing, 100083, China

Keywords: Non-cyanide method, gold leaching, refractory gold concentrate

Abstract

The method of chloride-hypochlorite leaching has the advantages of short process, low pollution and higher in reaction rate compared with the traditional leaching method. In this research, NaCl - NaClO solution is used to leaching gold from a refractory gold concentrate, and the influence of NaCl and NaClO concentration, reaction time to the leaching rate were investigated in this experiment. The raw materials and the leaching residue were characterized by a series of analysis methods such as XRD, XRF, SEM and EDS. Results indicate that the leaching rate and the desulfurization rate can be achieved at 96.26% and 98% with most optimum conditions as follows: NaCl concentration 50g/L, NaClO concentration 200g/L, Liquid-solid ratio 10:1, the best reaction time was 150 mins. And these conditions should be guaranteed pH<8.0, ORP>900mV.

Introduction

Gold has been widely used in the high-tech industry due to its excellent physical and chemical properties [1]. Gold production in China has ranked the first in the world for several years. At present, leaching of gold by cyanide is still the most widely used technology for gold extraction, but its long-term application prospect will not be positive, by reason of some obvious shortcomings of cyanidation such as susceptible to impurity ions, the poor results for refractory ores, and the toxic cyanide have a damaging effect on the environment [2,3].
With the increasingly exhaustion of easily leachable gold ores and environmental concerns, numerous research projects have been conducted to search for a non-cyanide hydrometallurgical process for gold extraction [4, 5, and 6]. Noncyanide lixiviants include the halide [7,8], thiourea [9], thiosulfate [10,11] and thiocyanate systems [12], the chlorination method attracted people's attention by the advantages of much faster gold dissolving speed, low price and environment friendly.

Chloride-hypochlorite solution is a completely non-cyanide system, the presence of chloride ions can greatly reduce the dissolution potential of gold [13, 14]. Gold dissolved and formed $[AuCl_4]^-$ complex ions. The leaching reaction is as follows:

$$2Au + 3ClO^- + 6H^+ + 5Cl^- = 2AuCl_4^- + 3H_2O \quad (1)$$

By the Eh-pH diagram of $Au\text{-}Cl\text{-}H_2O$ system, $[AuCl_4]^-$ can be stably exist in the acid solution under the condition of redox potential is greater than 900mV [15].

Materials and experiment

Properties of materials

The experimental materials is flotation gold concentrate from Hebei province in China, which have a gold content of 107g/t. The results of XRD and petrographic analysis of the concentrate shows the major gold minerals are native gold, petzite and kustelite, mostly are wrapped in pyrite or as growth in mineral fissures in the shape of granular and long strips. Part of coarse native gold is included in albite and potassium feldspar. The major metallic minerals are hematite, pyrite, limonite, chalcopyrite, galena and trace of tellurium. The gangue minerals mainly are potassium feldspar, albite, microcline and quartz phase, as well as a small amount of calcite, mica, and limestone.

Experimental procedure

Before the experiments, gold concentrate was ground to more than 95% of -0.074mm by ball grinder and flour mill. The flow diagram of experiments as shown in Fig.1.

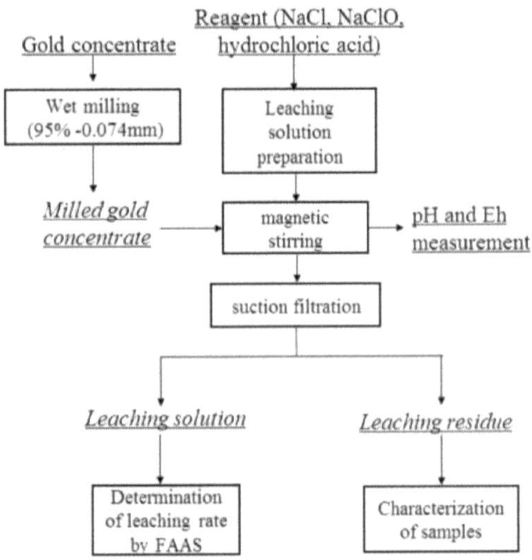

Fig.1 The flow diagram of gold leaching experiments

Experiments were carried out in a 100mL conical flask at room temperature. The experimental reagents mainly included sodium chloride, sodium hypochlorite solution (Active chlorine is 10%) and hydrochloric acid, the initial pH was 6.5, the liquid-solid ratio is 10:1, the solution was stirred by magnetic stirring apparatus and rotating Speed was 600r/min. The pHS-3B acidometer was used for pH measurement, platinum electrode was taken as indicator electrode and calomel electrode as reference electrode. The leaching residue and leaching solution were obtained by using filtration method after the completion of the reaction, flame atomic absorption spectrophotometry (faas) was used for determination gold concentration in solution, and then calculated to the leaching rate. The method of zinc dust replacement was taken in this experiment to recovering the gold in leaching solution.

Results and discussion

Experimental results

Fig.2 shows that the gold leaching rate with different concentration of NaCl and NaClO.

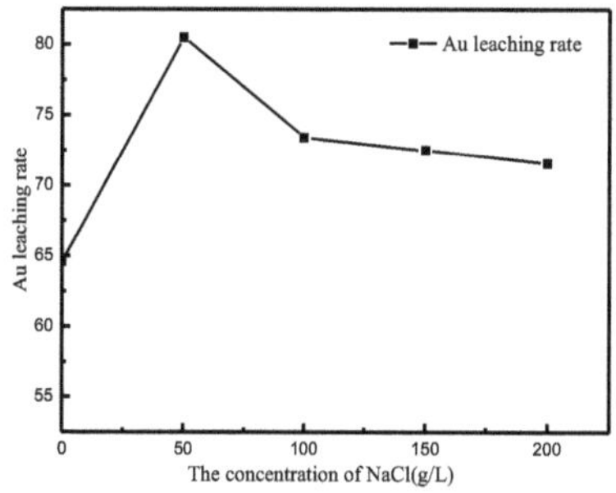

Fig.2 The change of gold leaching rate of different NaCl concentration

As can be seen from Fig.2, suitable concentration of sodium chloride can increase the leaching rate of gold, but too much NaCl may cause the solution potential to decrease; the optimum concentration is 50g/t. It is worth noting that the leaching rate can reach 60% without the addition of NaCl because of the presence of Cl^- in hydrochloric acid.

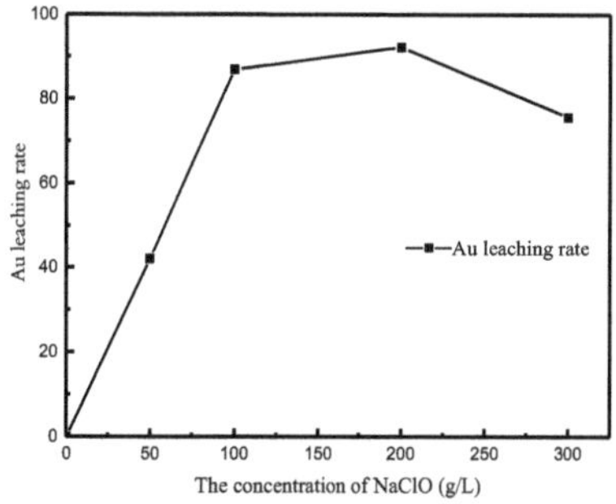

Fig.3 The change of gold leaching rate of different NaClO concentration

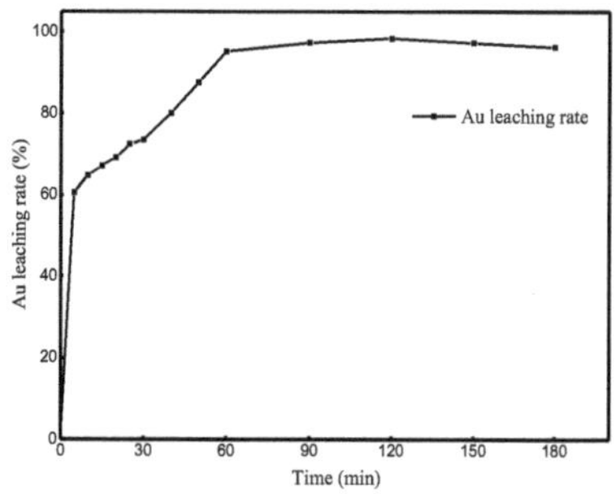

Fig.4 The gold leaching rate changes over time (25°C, Liquid-solid ratio 10:1, 50g/L NaCl, 200g/L NaClO)

NaClO is strong oxidant, high concentration of ClO^- can promote the gold leaching and the oxidative decomposition of sulfide ores such as pyrite and arsenopyrite. The leaching rate reached the peak when the concentration of NaClO is 200g/L (Fig.3), a greater amount of NaClO didn't have a positive effect on the leaching.

The Fig.4 shows the time-varying curve of gold leaching rate on the condition of 50g/L NaCl,200g/L NaClO, liquidated ratio 10:1, 25°C and the initial pH is 6.5. The vigorously reaction in the initial stage demonstrated good leaching kinetics in NaCl-NaClO solution system. After two hours the reaction was almost complete, while the leaching rate had stabilized, and the best leaching rate that can be achieved is 96.26%.

The change of Eh and pH.

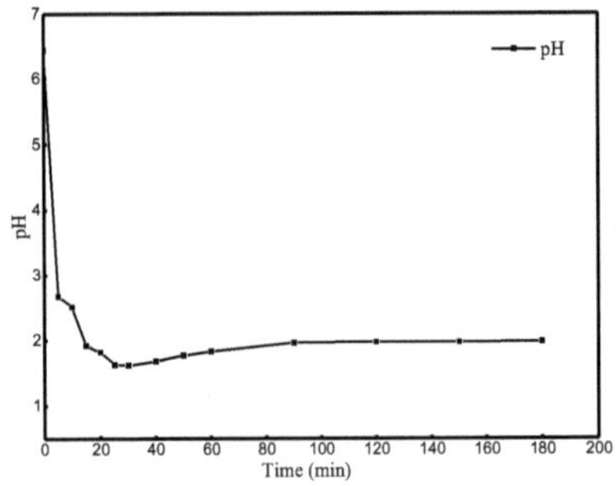

Fig.5 pH changes over time
(25°C, Liquid-solid ratio 10:1, 50g/L NaCl, 200g/L NaClO)

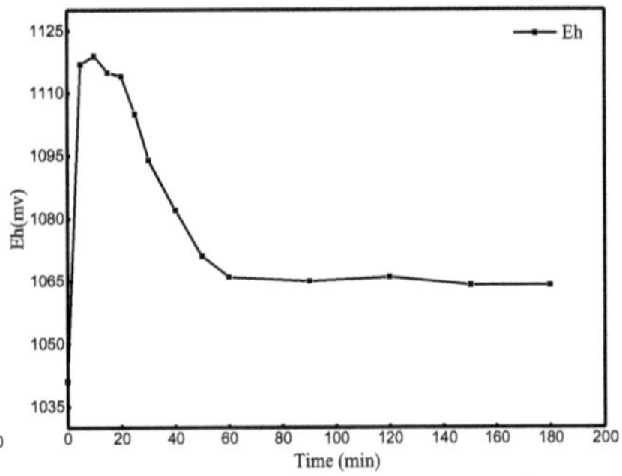

Fig.6 Eh changes over time
(25°C, Liquid-solid ratio 10:1, 50g/L NaCl, 200g/L NaClO)

The curve of Eh and pH change over time is shown in Fig.5 and fig.6. In the early stage of the reaction, the value of pH decreased sharply as minerals are gradually oxidized. In early stages of reaction process, HClO played a major role of oxidation, after 30 minutes the leaching rate tended to be stable. When the pH less than 3.5, chlorine had become the main oxidant, the potential gradually increased to more than 1000mv with the decreased of pH.

As shown in Fig.6, in the case of excess NaClO reagent was added, the potential can be kept at about 1065mv in the later stage of oxidation, and this would ensure the $[AuCl_4]^-$ complex ion stable exist in solution.

XRD analysis

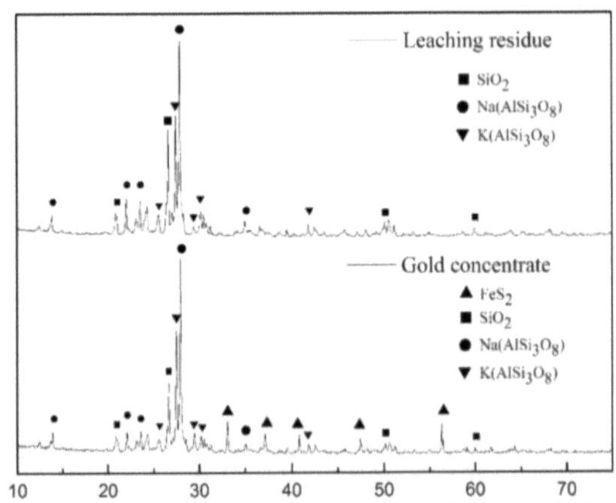

Fig.7 X-ray diffraction spectra of gold concentrate before and after leaching

Compare the X-ray diffraction patterns of gold concentrate before and after leaching (Fig.7), it can be seen obviously that the diffraction peaks of pyrite disappear completely, and the peaks of gangue components have no significant change. XRD pattern shows that pyrite is decomposed completely, the micro-fine gold wrapped in sulfide minerals can be exposed, then react with the leaching agent to generate complex ions. The oxidation and leaching reaction equation as follows:

$$2FeS_2+15HOCl+7H_2O \rightarrow 2Fe(OH)_3+4H_2SO_4+15HCl \quad (2)$$

$$2Au+3HOCl+3H^++5Cl^- \rightarrow 2[AuCl_4]^-+3H_2O \quad (3)$$

X-ray fluorescence analysis

The results of XRF quantitative analysis for gold concentrate before and after leaching are reported in table 1 and table 2.

Table.1 The main chemical composition of gold concentrate

Element	Au(g/t)	Fe	S	Si	Al	Ca	Mg	K
Content (%)	107	20.30	7.11	27.09	17.42	6.96	0.97	10.31
Element	Na	Cr	Mn	Zr	Sr	Pb	Cu	Ti
含量(%)	6.57	0.05	0.19	0.06	0.46	0.22	0.22	0.27

Table.2 The main chemical composition of leaching residue

Element	Au	Fe	S	Si	Al	Ca	Mg
Content (%)	3.97	3.79	0.08	60.56	14.98	0.89	0.07
Element	K	Na	Cr	Mn	Zr	Sr	Ti
Content (%)	11.46	6.44	0.04	0.15	0.08	0.21	0.29

The gold content is obtained by fire assaying method. It can be seen Fe and S content decreased obviously by comparison the data in the table, the content of S dropped from 7.11% to 0.08%, and the desulfurization rate reached 98.87%. Most of the Fe element in pyrite was dissolved, and existed in leaching solution in form of Fe ions, in addition, part of Fe formed $Fe(OH)_3$ precipitate into the leaching residue. Pb element in the residue was not detected, it proved a small amount of galena was decomposed completely.

SEM and EDS Characterization

The samples were characterized by scanning electron microscopy for observing its microstructure. Fig.8 and Fig.9 are SEM photograph of gold concentrate before and after leaching, the bright spots is micro-fine gold wrapped in pyrite by EDS spectrum detection. From the graph can clearly see the oxidation process of minerals, the minerals were eroded gradually from outside to the inside. The pyrite was oxidized to $Fe(OH)_3$ and showed porous structure.

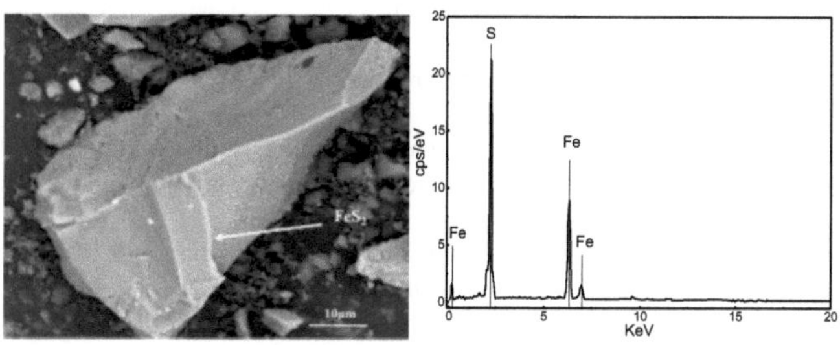

Fig.8 SEM photograph and EDS Energy spectrum diagram of gold concentrate

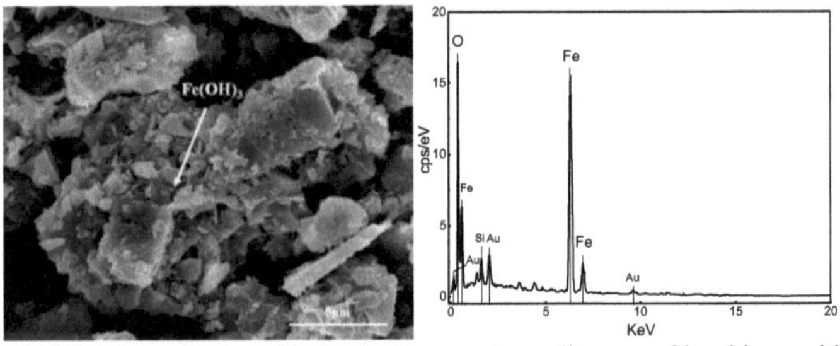

Fig.9 SEM photograph and EDS Energy spectrum diagram of leaching residue

Conclusion

Experimental results show that NaCl-NaClO leaching system is a stable gold leaching system, it have some major advantages of low pollution, fast reaction speed, short process and the reagents is easy to get. Under appropriate reaction conditions, the gold wrapped in sulfide ores completely dissolved into NaCl-NaClO leaching system to form $[AuCl_4]^-$ complex ions which can be stable exist in acid solution. The leached gold in solution can be recovered by means of ion exchange resin adsorption method or zinc dust replacement.

The concentrate reacted violently with leaching agent at the beginning of leaching, then gradually stabilized. The characterization results showed sulfide minerals were broken down completely after 150 minutes, the best gold leaching rate and desulfurization rate up to 96.26% and 98.87%.

The leaching agent and oxidizer of NaCl-NaClO leaching system are low-cost common chemical reagent, it has a great potential for application. But it still have some problems remain, the chlorine released during oxidation process is difficult to control, the corrosion problem is also an obstacle to the development of chlorination leaching method.

As described in the end, chlorination leaching system plays an important role in non-cyanide gold extraction methods, but there are still many problems to be solved for industrial applications, more theoretical and reactors studies is needed for the better development of gold industry.

ACKNOWLEDGMENT

The authors gratefully acknowledge the financial support of the National Science Foundation of PRC for the Research Project (2012-51234008), also the financial support of Beijing technical development project (00012132).

References

1. Shanmugam Manivannan, Ramasamy Ramaraj. "Synthesis of cyclodextrin-silicate sol–gel composite embedded gold nanoparticles and its electrocatalytic application," *Chemical Engineering Journal*, 2012, 210: 195-202

2. Wong Wai, Leong Eugene, Arun S. Mujumdar. "Gold Extraction and Recovery Processes," (Minerals, Metals and Materials Technology Centre (M3TC), 2009).

3. Jarnes L. Hendrix. "IS THERE A GREEN CHEMISTRY APPROACH FOR LEACHING GOLD?" *Papers in Sustainable Mining*. 2005: 2

4. S. Syed. "Recovery of gold from secondary sources—A review," *Hydrometallurgy*, 2012: 30-51
5. G. Senanayake. "Gold leaching in non-cyanide lixiviant systems: critical issues on fundamentals and applications," *Minerals Engineering*, 2004. 17（6）: 785-801
6. Rong Yu Wan, Marc Le Vier, Jan D. "Miller. Research and development activities for the recovery of gold from noncyanide solutions," *Hydrometallurgy Fundamentals, Technology and Innovation* , 1993: 415-436
7. Morteza Baghalha. "The leaching kinetics of an oxide gold ore with iodide/iodine solutions," *Hydrometallurgy*. 2012. 113–114: 42-50
8. R. K. Mensah-Biney, K. J. Reid, M. T. Hepworth. "Kinetics of gold loading from bromide solution onto an anion exchange resin," *Minerals Engineering*. 1994. 7(7): 865-887
9. N. Gönen, E. Körpe, M. E. Yıldırım, et al. "Leaching and CIL processes in gold recovery from refractory ore with thiourea solutions," *Minerals Engineering*. 2007. 20(6): 559-565
10. D. Feng,J. S. J. van Deventer. "The effect of iron contaminants on thiosulphate leaching of gold," Minerals Engineering, 2010, 23(5): 399-406
11. S. Syed. "A green technology for recovery of gold from non-metallic secondary sources," *Hydrometallurgy*. 2006. 82(1-2): 48-53
12. S. Vukcevic. "A comparison of alkali and acid methods for the extraction of gold from low grade ores," *Minerals Engineering*, 1996. 9(10): 1033-1047
13. Mehdi Ghobeiti Hasab, Shahram Raygan, Fereshteh Rashchi. "Chloride–hypochlorite leaching of gold from a mechanically activated refractory sulfide concentrate," *Hydrometallurgy*. 2013. 138: 59-64
14. Mehdi Ghobeiti Hasab, Fereshteh Rashchi, Shahram Raygan. "Simultaneous sulfide oxidation and gold leaching of a refractory gold concentrate by chloride–hypochlorite solution," *Minerals Engineering*, 2013. 50-51: 140-142
15. R. Y. Wan, J. D. Miller. "Research and Development Activities for the Recovery of Gold From Alkaline Cyanide Solutions," *Mineral Processing and Extractive Metallurgy Review*, 1990. 6(1-4): 143-190

Rare Metal Technology 2015
Edited by: Neale R. Neelameggham, Shafiq Alam, Harald Oosterhof, Animesh Jha, David Dreisinger, and Shijie Wang
TMS (The Minerals, Metals & Materials Society), 2015

THE EFFECTS OF COMMON ASSOCIATED SULFIDE MINERALS ON THIOSULFATE LEACHING OF GOLD

Yong-bin Yang, Xi Zhang, Bin Xu, Qian Li, Tao Jiang, Ya-xuan Wang

School of Minerals Processing and Bioengineering, Central South University, Changsha 410083, China

Keywords: Thiosulfate consumption, gold leaching, sulfide minerals, CMC

Abstract

The effects of common associated sulfide minerals including pyrite, arsenopyrite, galena and sphalerite on thiosulfate leaching of gold were studied in this work. Sulfide minerals had significant effects on gold leaching because that the dissolved products of sulfide minerals precipitated on gold surface to form coating layer. The degree of sulfide minerals impact on gold dissolution decreased in the order of pyrite ≈ galena > arsenopyrite > sphalerite, while the degree of effects on thiosulfate consumption decreased in the order of pyrite > arsenopyrite > sphalerite > galena. The additive CMC (Carboxymethyl cellulose) could not only reduce thiosulfate consumption but also increase gold dissolution. The mechanism of this might be that CMC could restrict the decomposition of thiosulfate by cupric ions and the precipitates on gold surface.

Introduction

Thiosulfate is one of the most promising alternative lixiviant in gold leaching. But, the industrial application of thiosulfate method is rare [1]. One of the limitations is the high consumption of thiosulfate by copper ammonia, as shown in Eq. (1).

$$2[Cu(NH_3)_4]^{2+}+8S_2O_3^{2-}=2Cu(S_2O_3)_3^{5-}+8NH_3+S_4O_6^{2-} \tag{1}$$

Tetrathionate generates in Eq. (1) will undergo further reactions to produce higher or lower polythionates, sulfide ions, sulfur, etc [2]. These products originated from thiosulfate decomposition could form a compact film on gold surface, and hence passivated gold leaching [3].

Since gold always associates with sulfide minerals, the dissolution behavior of common sulfide minerals could influence thiosulfate solution and hence influence thiosulfate leaching of gold [4].

D. Feng studied the dissolution rates of sulfide minerals in ammonia thiosulfate solution, which decreased in the order of chalcopyrite > pyrrhotite > arsenopyrite > pyrite [5]. Step further, Chen Xia studied the effects of various minerals on thiosulfate leaching of gold. The study showed that ferrous minerals, lead-bearing minerals and common copper minerals all had detrimental effects on gold leaching [6]. Aimed at the detrimental effects, some chemical additives were proposed such as CMC.

Additive CMC could not only reduce thiosulfate consumption but also enhance gold dissolution in thiosulfate leaching in the presence of sulfide minerals [1]. The mechanism for this might be that CMC could reduce the coordination of thiosulfate by copper and the dissolution of sulfide minerals.

Above all, the effects of sulfide minerals on thiosulfate leaching have not systematically studied. The purpose of this paper is to study the effects of four types of sulfide minerals on thiosulfate leaching of gold. The intensification effect of CMC on thiosulfate leaching of gold was also studied.

Experimental

Minerals and reagents

Gold foils (99.99%Au, thickness 0.6mm) were used in the experiments with a surface area of $0.38cm^2$. Pyrite, arsenopyrite, galena, sphalerite and quartz samples were obtained from Mineral Specimen Market, Wuhan, China. All minerals were dry milled in a planetary ball mill to the size of 45-75μm, and then stored in air-tight plastic bags to avoid further oxidation. XRF results showed that all these minerals were over 93% purity. Laboratory grade sodium thiosulfate pentahydrate, ammonia water (25%) were provided by Qingzhong Chem Supply Pty Ltd. Analytically pure cupric sulfate, hydrogen peroxide (30%) and nitric acid were used. De-ionized water was used during all experiments.

Detection techniques

Elemental concentrations in solutions were determined by atomic absorption spectrum (AAS), it is necessary to oxidize sulfur species to stable sulfates before elemental analysis. The thiosulfate concentration was determined by iodometric method. In order to eliminate the effects of the cupric tetra-amine complex and sulfite on iodine titration, a certain amount of EDTA and formaldehyde were added prior to the titration with the indicator Vitex.

Experimental method

Experiments were conducted in a 250 ml beaker with 200 ml leaching solution. Then a gold foil was suspended in the middle of the solution for leaching. The stirring speed was maintained at 200 rpm by electric blender (IKA EURO-STPCUS25). All experiments were performed at 25℃. Samples were taken continuously at certain intervals during a total 24 h leaching. The samples were centrifuged and filtered for the subsequent iodine titration and AAS analysis. All experiments were conducted with the reagent dosages of 0.1 M $Na_2S_2O_3 \cdot 5H_2O$, 0.012 M $CuSO_4 \cdot 5H_2O$ and 0.5 M NH_4OH at pH 10.3 with minerals at 10 g/L.

Results and discussion

A comparison of the effects of sulfide minerals on thiosulfate leaching

Based on the inert properties of quartz, it can be used as standard in thiosulfate leaching of gold [5]. As it exhibited in Figure 1a, compared with quartz system, all sulfide minerals had detrimental effects on gold dissolution in thiosulfate leaching of gold. Therein, pyrite and galena had the largest detrimental effects on gold dissolution, in comparison, the presence of arsenopyrite and sphalerite were less detrimental. In specific, sphalerite had minor detrimental effect and arsenopyrite appeared to be moderately detrimental.

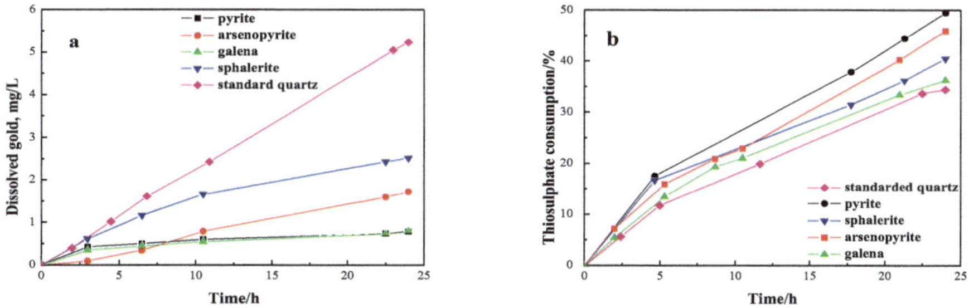

Figure 1 Effects of various sulfide minerals on (a) gold dissolution and (b) thiosulfate consumption.

As it depicted in Figure 1a, the degree of sulfide minerals impact on gold dissolution decreased in the order of pyrite ≈ galena > arsenopyrite > sphalerite. It was found that it might be the coated layer on gold surface which would act as a physical barrier that passivated gold leaching [8]. In addition, the heavy metal ions, such as, iron ions or lead ions would compete with gold to complex thiosulfate and thus reduced gold leaching [9].

The effects of sulfide minerals on thiosulfate consumption are shown in Figure 1b. The standard quartz system consumed about 33% of thiosulfate after 24 h leaching. In the presence of galena, thiosulfate consumption was almost the same as that in standard quartz system, i.e., galena had the minimal detrimental effect on thiosulfate consumption among these four minerals. In comparison, thiosulfate consumption was almost up to 50% in the presence of arsenopyrite or pyrite. The effect of sphalerite was moderate. As formerly suggested by Moses, sulfide minerals were semi-conductive [7]. It is the absorption of thiosulfate on the surface of sulfides that accelerate the transfer of electrons from thiosulfate to oxygen, and thus, contribute to the increased consumption of thiosulfate.

To sum up, in thiosulfate leaching of gold, pyrite is the most detrimental mineral both on gold dissolution and on thiosulfate consumption. On the contrary, sphalerite is the mildest among these four minerals.

<u>The dissolution behavior of sulfide minerals</u>

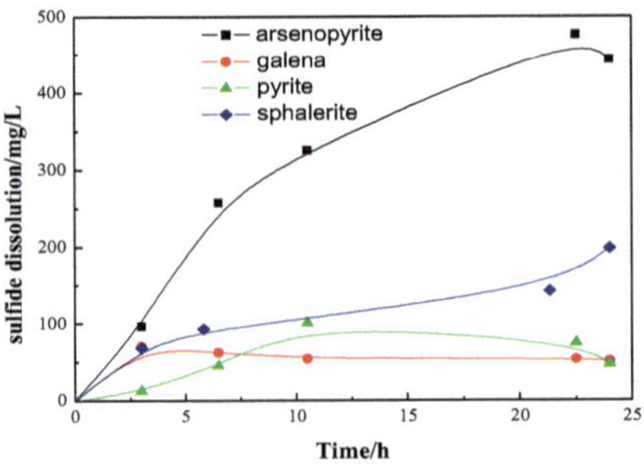

Figure 2: Dissolution behavior of sulfide minerals in thiosulfate leaching-a comparison.

Figure 2 presents the dissolution behavior of four types of sulfide minerals in thiosulfate leach solution. The concentration of Fe, Pb, Zn and As can be used to represent the dissolution behavior of pyrite, galena, sphalerite and arsenopyrite in the leaching solution, respectively. As it shown in Figure 2, the initial dissolution kinetics of sulfide minerals is fast but, subsequently slows down with time. The dissolution of sulfide minerals in thiosulfate solution decreased in the order: arsenopyrite > sphalerite > galena ≈ pyrite. The degree of sulfide minerals impacted on gold dissolution decreased in the order of galena ≈ pyrite > arsenopyrite > sphalerite.

Above all, the lowest dissolution of pyrite had the largest significant detrimental effect on thiosulfate gold leaching. It was found that the dissolution of pyrite would ultimately produce the precipitates of iron hydroxide [10]. The precipitates can not only coat on the surface of sulfides but also on the leached surface of gold, and thus, passivated gold leaching. Similar to the effect of pyrite, the dissolution of galena would form the precipitate of lead oxide, which would also coat on the leached gold and thus hindered gold leaching [9].

More specifically, the reason why a smaller dissolution of pyrite or galena plays a big role on gold leaching could be summed up in the following: firstly, the micro scale iron hydroxide or lead oxide precipitates have a strong affinity for gold surfaces [11]; secondly, the heavy metal ions originates from sulfides dissolution would compete with gold to complex thiosulfate.

The dissolution of sphalerite and arsenopyrite were larger in thiosulfate leach solution, compared with the dissolution of pyrite. As it formerly referred, zinc oxide is the product of thiosulfate leaching in the presence of sphalerite [12]. As it shown in Figure 1 and 2, the largest dissolution of sphalerite had the lowest detrimental effects on gold leaching. The reason for this might be that zinc oxide layer is porous and less detrimental. Thus, sphalerite is the mildest on thiosulfate leaching of gold.

The intensification effect of CMC

Effect of CMC on thiosulfate leaching in the presence of arsenopyrite Figure 3a shows that the kinetics of gold leaching increases in a large scale initially and basically unchanged afterward with additive CMC in thiosulfate leaching of gold in the presence of arsenopyrite. With additive CMC, the content of gold dissolution was even larger than that in standard quartz system. It was found that thiosulfate leaching in standard quartz system produces the precipitates of copper sulfides, sulfur etc on gold surface that inhibited gold leaching [12]. While in the presence of arsenopyrite, the precipitates of cuprous sulfide, sulfur, iron arsenate and iron hydroxide would deposit on the leached gold surface and then hindered gold leaching [6]. Thus, it is reasonable to suggest that CMC could effectively reduce the precipitate of sulfur, copper, arsenate and iron species on gold surface and thus largely improved thiosulfate leaching of gold.

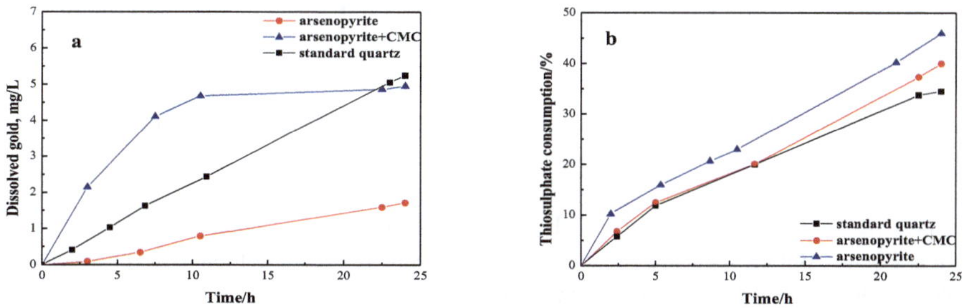

Figure 3 Effect of CMC on (a) gold dissolution and (b) thiosulfate consumption in thiosulfate leaching in the presence of arsenopyrite. Dosages of CMC: 100mg/L.

Figure 3b shows that the additive CMC preferably reduced thiosulfate consumption in the presence of arsenopyrite. The intensification effect of CMC on reducing thiosulfate consumption was obvious at the initial period but came down afterwards. As shown in Figure 3b, after 24 h leaching in the presence of arsenopyrite, thiosulfate consumption was about 40% with additive CMC, but, it was 47% without additive.

The mechanism of CMC on reducing the negative effects of sulfides in thiosulfate leaching of gold will be summed up together below.

Effect of CMC on thiosulfate leaching in the presence of pyrite The influence of CMC on thiosulfate leaching in the presence of pyrite is reflected in Figure 4. Additive CMC could improve the dissolution of gold in some extent, but the improving effect on gold dissolution is not as large as that in arsenopyrite system. CMC is a traditional additive that plays a role of mineral surface passivation agent. Deventer suggested that the adsorption of CMC on the surface of pyrite largely reduced the catalysis effect of it [1]. So, the reductive surface of pyrite would be passivated and then the detrimental effect of pyrite was hindered.

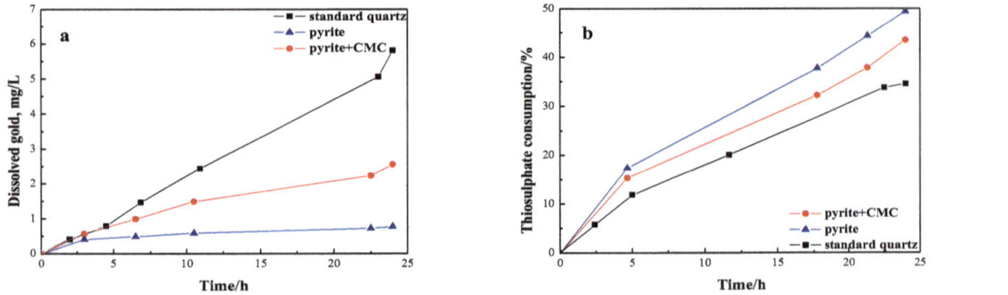

Figure 4: Effect of CMC on (a) gold dissolution and (b) thiosulfate consumption in the presence of pyrite. Dosages of CMC: 100mg/L.

In the presence of pyrite, the addition of CMC decreased thiosulfate consumption from 49.5% to 43.5%, shown in Fig. 4b. But the reducing effect of CMC on thiosulfate decomposition in the presence of pyrite was not as effective as that on arsenopyrite system.

Compared with pyrite, CMC was more effective in reducing the detrimental effect of arsenopyrite both on gold dissolution and on thiosulfate consumption.

Effect of CMC on thiosulfate leaching in the presence of galena Figure 5 shows the influence of CMC on thiosulfate leaching of gold in the presence of galena. Compared with Figure 4, the improving effect of CMC on gold dissolution was more prominent in the presence of galena. Figure 5b shows that the effect of CMC on reducing thiosulfate consumption in the presence of galena was significant, despite of the minimal detrimental effect of galena on thiosulfate consumption, formerly discussed in 3.1.

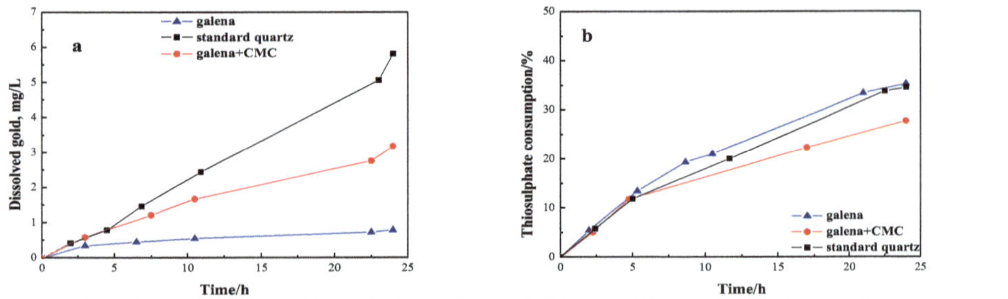

Figure 5: Effect of CMC on (a) gold dissolution and (b) thiosulfate consumption in the presence of galena. Dosages of CMC: 100mg/L.

Effect of CMC on thiosulfate leaching in the presence of sphalerite Figure 6a shows that the additive CMC could slightly promote the dissolution of gold in thiosulfate leaching of gold in the presence of sphalerite. As it shown in Figure 6b, thiosulfate consumption was reduced from 40.6% to 35.6% in the presence of sphalerite by adding additive CMC. Even though the reducing effect on gold dissolution was minimal, but considering the reducing effect on thiosulfate consumption, it is reasonable to conclude that CMC was also an effective additive in thiosulfate leaching of gold in the presence of sphalerite.

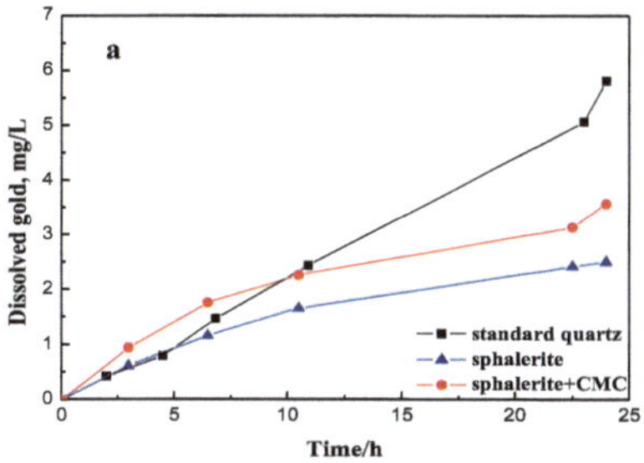

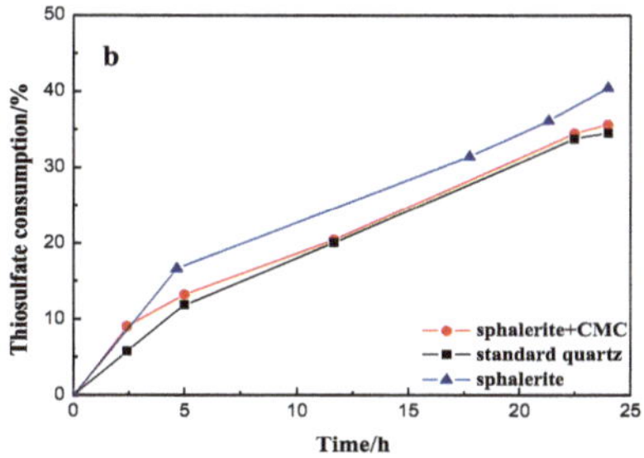

Figure 6a: Effect of CMC on gold dissolution in the presence of sphalerate. Figure 6b: Effect of CMC on thiosulfate consumption in the presence of sphalerite. CMC: 100mg/L.

Above all, the additive CMC could relive the detrimental effects of sulfide minerals in thiosulfate leaching. The effectiveness of CMC might be like this: CMC obviously depressed thiosulfate decomposition by weakening the interactions between cupric ions and thiosulfate and by reducing the catalysis effects of sulfide minerals; CMC could reduce the precipitates on gold surfaces by its negatively charged functional groups and its non-selective adsorption.

Conclusions

(1) Different sulfide minerals have different detrimental effect on gold dissolution as well as on thiosulfate consumption in thiosulfate leaching of gold. The effects of sulfide minerals on gold dissolution decreased in the order of arsenopyrite > sphalerite > galena ≈ pyrite, while the descending order on thiosulfate consumption follows pyrite > arsenopyrite > sphalerite > galena. Synthesizes these two effects, pyrite is the most detrimental one among the four studied minerals in thiosulfate leaching of gold.

(2) The relative leaching rates of sulfides in thiosulfate leaching were in the order of arsenopyrite > sphalerite > galena > pyrite. It is likely because of the products on gold surface originated from sulfide dissolution and thiosulfate decomposition that largely detriments thiosulfate leaching of gold.

(3) Additive CMC could reduce the detrimental effects caused by sulfide minerals at varying degrees.

Acknowledgment

Financial supports from the National Natural Science Foundation of China (Grant No.51074182) and Financial supports from China Postdoctoral Science Foundation (Grant No. 2014M550422) are gratefully acknowledged.

References

1. D. Feng and J.S.J. van Deventer, "Thiosulphate leaching of gold in the presence of carboxymethyl cellulose," *Mineral Engineering*, 24 (2011), 115-121.

2. G. Senanayake, "Analysis of reaction kinetics, speciation and mechanism of gold leaching and thiosulfate oxidation by ammoniacal copper (II) solutions," *Hydrometallurgy*, 75 (2004), 55-75.

3. M.I. Jeffrey, K. Watling, G.A. Hope and R. Woods, "Identification of surface species that inhibit and passivate thiosulfate leaching of gold," *Mineral Engineering*, 21 (2008), 443-452.

4. Y. Xu and M.A.A. Schoonen, "The stability of thiosulfate in the presence of pyrite in low-temperature aqueous solutions," *Geochimica et Cosmochimica Acta*, 59 (1995), 4605-4622.

5. D. Feng and J.S.J. van Deventer, "Ammoniacal thiosulphate leaching of gold in the presence of pyrite," *Hydrometallurgy*, 82 (2006), 126-132.

6. Chen Xia, "Associated sulfide minerals in thiosulfate leaching of gold: problems and solutions" (Ph.D. thesis, Queen's University, 2008), 58-61.

7. C.O. Moses, D.K. Nordstorm, J.S. Herman and A.L. Mills, "Aqueous pyrite oxidation by dissolved oxygen and by ferric iron," *Geochimicaet Cosmochimica Acta*, 51 (1987), 1561-1571.

8. K. Sarveswara Rao, P.K. Paramguru, R.P. Das and H.S. Ray, "The role of galvanic interaction during ammonia leaching of multimetal sulphides," *Miner. Engineering*, 11 (1998), 1011-1024.

9. K. Sarveswara Rao et al., "Study of leaching of multimetal sulphides through an interdisciplinary approach," *Min. Proc. Ext. Met. Rec*, 7 (1991), 209-234.

10. Chen Xia, "Associated sulfide minerals in thiosulfate leaching of gold: problems and solutions" (Ph.D. thesis, Queen's University, 2008), 64-66.

11. D. Feng, "Preg-robbing phenomena in the thiosulphate leaching of gold ores," *Minerals Engineering*, 14 (2001), 1387-1402.

12. P.L. Breuer and M.I. Jeffrey, "The reduction of copper(II) and the oxidation of thiosulphate and oxy-sulphur anions in gold leaching solutions," *Hydrometallurgy*, 70 (2003), 163-173.

13. D. Feng and J.S.J. Van Deventer, "Leaching behavior of sulphides in ammoniacal thiosulphate systems," *Hydrometallurgy*, 63 (2002), 189-200.

Rare Metal Technology 2015
Edited by: Neale R. Neelameggham, Shafiq Alam, Harald Oosterhof, Animesh Jha, David Dreisinger, and Shijie Wang
TMS (The Minerals, Metals & Materials Society), 2015

HYDROMETALLURGICAL EXTRACTION OF PRECIOUS, RARE AND BASE METALS USING AN OXIDIZING ACID CHLORIDE HEAP LEACH

[1]David Dreisinger, [2]Niels Verbaan, [2]Charlotte Forstner, [1]Ralph Fitch
[1]Trimetals Mining Inc., 2696 S. Colorado Blvd., Denver, CO, 80222, USA
[2]SGS Canada Inc., 185 Concession Street, Lakefield, ON K0L 2H0, Canada

Keywords: Silver, Indium, Gold, Gallium, Copper, Lead, Zinc, Malku Khota, Bolivia, Acid Chloride Leaching

Abstract

A process has been developed to extract gold, silver, indium, gallium, lead, copper, zinc, and other metals from a large resource using an oxidizing acid chloride heap leaching technology. The ore is crushed and/or ground to improve reactivity and then contacted with a solution containing NaCl-HCl-NaOCl. The value metals are leached into solution as chloro-complexes. The leachate is treated in a series of separation and recovery processes to produce separate saleable or disposable products. The paper will provide a summary of process development and application to a silver-indium ore from Bolivia. The possible application of the process to other ore types will be presented for discussion.

Introduction

Malku Khota is situated in the eastern part of the Bolivian Altiplano at elevations between approximately 3,800 to 4,580 meters above sea level. The project is located 98 km east-southeast of Oruro, and 85 km south of Cochabamba in a relatively remote area.

Figure 1. Project Location in Bolivia

The Malku Khota deposit has major values in Ag, In, Ga, Cu, Zn and Pb. Armitage et al (2011) summarized the measured and indicated resources and inferred resources for the project as shown below. Small amounts of gold are also found in the deposit.

Table 1. Resource Statement for Malku Khota (Armitage et al, 2011).

Resource		Measured	Indicated	M+I	Inferred
Ag (g/t)	Tonnes	30,989,448	224,001,987	254,991,434	230,013,794
	Grade	33.4	27.3	28.1	18.9
	Ozs	33,319,487	196,960,598	230,280,085	140,027,216
In (g/t)	Grade	6.1	5.8	5.8	4.1
	Tonnes	188	1,293	1,481	935
Ga (g/t)	Grade	4.5	4.3	4.3	4.3
	Tonnes	139	943	1,082	1,001
Cu (%)	Grade	0.02	0.02	0.02	0.02
	Lbs	13,947,823	106,366,881	120,314,704	102,081,329
Pb(%)	Grade	0.07	0.07	0.07	0.07
	Lbs	48,665,367	404,649,086	453,314,453	362,157,607
Zn(%)	Grade	0.02	0.05	0.04	0.05
	Lbs	16,239,090	230,573,723	246,812,812	240,292,377

Notes: The resource cut-off grade of 10 g/t Ag equivalent is based only on the values of Ag ($16 USD/oz) and In ($550USD/kg). Estimated metal content does not include any consideration of mining, mineral processing, or metallurgical recoveries.

The preliminary economic assessment by Armitage et al (2011) presented the base case development plan for the Malku Khota project at a scale of 40,000 t/d of ore treatment by an acid chloride heap leach process. The acid chloride process was successfully tested in order to extract the metals of value in the ore and produce final or intermediate products for sale. An alternative process involving cyanide heap leaching was also examined. As cyanide leaching was not able to recover indium and gallium the preliminary economic assessment was focused on the acid chloride leach process. The development of the acid chloride metallurgical process is described below.

Metallurgical Process Development

Mineralogy

The mineralization at Malku Khota is confined primarily to relatively shallow water to aeolian sandstone sequences and is believed to be sedimentary exhalative (SEDEX) in origin. The SEDEX related disseminated silver mineralization is associated with silver, lead, zinc, and barite. A later hydrothermal event introduced gold and bismuth and additional silver, base metals, indium and gallium. Earlier mineralization may have been partially redistributed by the hydrothermal event.

The majority of the disseminated silver mineralization at Malku Khota is hosted within the sandstones in the Malku Khota and Upper Wara Wara Units. The silver minerals identified in outcrop, tunnels and drill holes are mostly in the form of sulfides (acanthite), complex sulphosalts, oxides, iodides, and bromides often including iron, lead, and antimony in their structures. Ore minerals include acanthite, jamesonite group minerals (including owyheeite), fizyelite (Pb,Ag,Sb sulfide), Pb,Sb,Ag,Cd sulphosalt, silver, cerargyrite minerals, silver oxide, pyrite, galena, sphalerite, tetrahedrite and tennantite (one sample), copper sulfide (only in a few samples), lollingite (one grain), native bismuth, bismuthinite, arsenopyrite, greenockite, barite, valentinite (Sb oxide), stibioconite (Sb oxide), bindheimite (Pb,Sb oxide), scorodite (iron arsenate), massicote (Pb oxide), plattnerite (Pb oxide), tripuhyite (Fe,Sb oxide), eriochalcite (Cu chloride), stetefeldite (AgSb hydrate), stibnite, iron oxides, zinc oxide. Sulfides and oxides have a very small grain size and are commonly intermixed. Only under the SEM or with X-ray diffraction could they be determined. Acanthite is distributed in cracks in the sulphosalt, and in microfractures in quartz and barite (Hansley 2011). Metal oxides were clearly derived from the sulfides, the latter occurring as remnants in oxides.

Specific indium minerals have not been found but there is mounting evidence that the indium occurs within the minerals jamesonite and acanthite.

Secondary minerals within the sandstones include specular hematite, hematite, crandallite, kaolinite, illite, strontianite, chlorite, carbonate cement, calcite, rutile, primary and secondary monazite, primary and secondary tourmaline, ferroan carbonate, muscovite, plumbogummite ($PbAlPO_4$), anglesite, and anatase.

Three areas of strata-bound silver mineralization have been identified, the Limosna, Wara Wara and Sucre zones. This mineralization occurs within a strike length of approximately 4 kilometers and widths of 20 to 200 meters but is open along strike and to depth.

The deposit contains a large number of minerals with values in precious, rare and base metals. The balance of the gangue mineralization consists primarily of sandstone. The mineralogy of an ore deposit typically guides the metallurgical development team as to possible routes for extraction of valuable metals. The varied mineralogy (predominantly oxides with some sulfides) and lower grade nature of the deposit focused the development effort on heap or vat leach metal extraction with a lixiviant with the ability to extract the full range of metals. The acid chloride system was selected based on the ability of chloride to complex with the key metal values in the ore including silver and indium.

Leaching Studies

The leaching of value metals from Malku Khota ore can be represented by simplified reactions for elements, oxides and sulfides of various value metals. The acid chloride system can be represented as a mixture of acid (HCl), chloride source (NaCl) and oxidant (NaOCl (bleach)). The chloride salts of In, Ag, Au, Cu, Zn, Pb and Ga are water soluble in the following reactions.

$In_2O_3 + 6HCl = 2InCl_3 + 3H_2O$
$Ag_2O + 2HCl + 2NaCl = 2NaAgCl_2 + H_2O$
$Ag_2S + 2HCl + NaCl + NaOCl = 2NaAgCl_2 + S + H_2O$
$Au + 3HCl + 1.5NaOCl = NaAuCl_4 + 1.5H_2O + 0.5NaCl$
$CuO + 2HCl = CuCl_2 + H_2O$
$CuS + 2HCl + NaOCl = CuCl_2 + S + NaCl + H_2O$
$ZnO + 2HCl = ZnCl_2 + H_2O$
$ZnS + 2HCl + NaOCl = ZnCl_2 + S + NaCl + H_2O$
$PbO + 2HCl = PbCl_2 + H_2O$
$PbS + 2HCl + NaOCl = PbCl_2 + S + NaCl + H_2O$
$Ga_2O_3 + 6HCl = 3GaCl_3 + 3H_2O$

The consumption of acid in the leaching process is dictated by the acid consumed in the above reactions. The presence of acid consuming minerals of iron, aluminum, calcium, magnesium, manganese, antimony, arsenic and other metals is undesirable as demonstrated by the following simplified reactions (for example). These reactions all consume acid.

$Fe_2O_3 + 6HCl = 2FeCl_3 + 3H_2O$
$FeO(OH) + 3HCl = FeCl_3 + 2H_2O$
$Al_2O_3 + 6HCl = 2AlCl_3 + 3H_2O$
$AlO(OH) + 3HCl = AlCl_3 + 3H_2O$
$CaCO_3 + 2HCl = CaCl_2 + CO_2 + H_2O$
$MgCO_3 + 2HCl = MgCl_2 + CO_2 + H_2O$
$MnO + 2HCl = MnCl_2 + H_2O$
$Sb_2O_3 + 2HCl = 2SbOCl + H_2O$
$As_2O_3 + 3H_2O + 2NaOCl = 2H_3AsO_4 + 2NaCl$

The consumption of acid by these reactions is largely unavoidable and dependent on the minerals present in the raw material, the acid concentration employed in the process, the leach time and the temperature of leaching.

The result of treating Malku Khota ore with acid chloride leach solution would be expected to be a mixed solution containing remaining free acid, salt and dissolved metal-chloride salts of the value metals along with gangue mineral chloride salts. Note that gold is present in Malku Khota deposit at low levels. The acid chloride leach solution would be expected to extract gold as the auric-chloride species.

The acid chloride leaching studies for Malku Khota included;

- Agitated acid-chloride leaching of various samples;
- Bottle roll acid-chloride leaching of various samples;
- Column acid-chloride leaching
- Diagnostic leaching (soak tests).

The agitated leaching tests were well suited to testing finely ground material for a stirred reactor leach design. However for a heap leach process, these tests had limited utility. Bottle roll tests on coarser material for extended periods produced too much abrasion of the ore particles. Column leach testing is an excellent technique and was used in the development program. A type of diagnostic leaching process was pioneered in order to allow testing of many samples under a variety of crush sizes and solution conditions. This facilitated obtaining extraction versus time results quickly and inexpensively. The diagnostic leaching procedure was referred to as a soak test. Crushed ore and leach solution were added to a plastic bucket and covered. The bucket contents were periodically agitated by a manual swirling action with oxidizing ORP maintained by adding supplemental oxidant as required. Samples of solution were withdrawn to follow metal extraction versus time. Final solids were assayed to close the material balance. The soak test results were believed to be a good indicator of ore leaching performance for a strong acid leaching solution such as that used in this program.

A large number of samples of Malku Khota ore were leached using the soak test procedure. Table 2 shows a series of 11 samples tested in some of the early development work.

Table 2. Chemical Analysis of Series of Samples from Malku Khota for Acid Chloride Leaching.

	Chemical Analysis (g/t)										
	10-6	10-7	10-8	10-9	10-10	10-11	10-12	10-13	10-15	10-16	10-17
In	3	10	14	8.3	69	28	15	7.5	5.9	9.8	14
Ga	1.9	3	1.9	1.7	2.7	1.8	3.7	2.3	3.8	4	3.4
Au	< 0.02	< 0.02	< 0.02	< 0.02	< 0.02	< 0.02	< 0.02	< 0.02	< 0.02	< 0.02	< 0.02
Ag	115	41.8	32	21	71.2	35.3	38.5	38	106	41.8	220
Al	5900	8100	5000	6100	4900	4300	13000	12000	10000	9900	12000
As	17	22	150	33	200	99	94	22	71	19	980
Ba	59	1200	2800	840	5800	1400	1000	2500	3400	13000	230
Be	0.2	0.4	0.1	0.2	0.1	0.6	0.3	0.1	0.3	0.3	0.4
Bi	< 0.6	< 0.6	< 0.6	< 0.6	< 0.6	< 0.6	< 0.6	< 0.6	19	28	55
Ca	80	100	64	81	99	230	50	86	160	46	57
Cd	0.3	0.5	< 0.2	< 0.2	0.2	0.3	1.4	51	< 0.2	1.5	11
Co	1	1.9	0.6	1	< 0.6	1	0.7	3.8	0.7	0.7	1.8
Cr	78	170	63	79	85	82	89	69	75	88	150
Cu	20	32	55	15	48	31	50	43	480	410	2600
Fe	5300	11000	9800	5500	8200	5400	8000	5200	34000	23000	7700
K	600	230	520	190	420	350	1000	810	3100	170	4000
Li	17	16	13	15	9	12	16	12	45	40	25
Mg	80	76	37	39	48	77	64	150	340	60	180
Mn	160	160	18	37	16	47	16	41	21	67	30
Mo	7	12	1.8	2.3	2.4	1.7	1.5	2.7	2.7	2.3	3.2
Na	36	48	18	31	27	44	30	71	39	15	27
Ni	4	9	4	5	3	4	3	6	4	5	7
P	210	< 200	< 200	< 200	< 200	270	200	< 200	630	210	220
Pb	730	600	420	240	650	910	930	9600	1200	150	610
Sb	48	41	27	38	50	38	45	48	930	170	1000
Se	< 10	< 10	< 10	< 10	< 10	< 10	< 10	< 10	< 10	< 10	< 10
Sn	3	13	4	5	7	3	4	4	4	< 2	2
Sr	82	150	120	150	160	120	210	150	700	280	73
Ti	350	590	290	350	240	290	1000	450	1300	500	370
Tl	< 0.4	< 0.4	< 0.4	< 0.4	< 0.4	< 0.4	< 0.4	0.4	< 0.4	< 0.4	< 0.4
U	1.1	1.5	0.5	0.7	0.7	1.3	0.7	0.7	0.5	0.5	1
V	4	3	4	4	3	3	9	9	15	8	7
Y	1.2	1.7	0.9	1.2	0.7	1.3	1.5	1.4	1.9	1.1	2
Zn	220	350	86	130	76	190	300	4200	30	280	2890

The soak tests were carried out using a ratio of 0.1 kg of solid to 1 kg of solution containing 2.7 M NaCl and 100 g/L HCl with 1 g of NaOCl/L. The reagents (acid and hypochlorite) were supplied in excess to ensure sufficient reactant to measure the ore leaching response. The temperature was ambient (15-25 °C) over the 365 day leaching cycle. Ore was crushed or ground to sizes ranging from 270 mesh to 1 inch. Samples were taken more frequently and earlier for the finer samples.

There was a wide variety of results but generally extraction levels for key elements were excellent. Figures 2-5 illustrate some of the results of extraction versus time for various samples and particle sizes.

Final silver extractions were greater than 80% for all tests finer than 115 mesh. Indium extractions were all higher than 80%, except for coarse sizes of 10-16 and 10-17. There was a greater range of extractions for lead, but lower extracting samples tended to correspond with samples containing the highest lead levels in the feed and/or the coarsest crush size. The highest gallium extraction was only 64% with some samples realizing no extraction (10-6). Copper extractions range from 49% in 10-13 to 100% in 10-10. Zinc extractions were generally greater than 80%, with the lowest extractions coming from sample 10-15 at ~40%.

Acid consumption was calculated by representing all metal units in the leach solution as their molar equivalent chloride units. For example, iron and aluminum were calculated using a molar equivalent of 3 chloride units corresponding to the $FeCl_3$ and $AlCl_3$ complexes respectively. This assumes that every unit of metal that leached into the liquor reacted with chloride to do so and that no precipitation of leached metals (containing chloride) has occurred. The total of all mmol/L chloride equivalents was multiplied by the volume of liquor, multiplied by the molar weight of hydrochloric acid (HCl) and then divided by the weight of the feed sample for the test. Acid consumptions for the individual soak tests ranged from 10 to 77 kg hydrochloric acid per tonne of ore. Acid consumption is highly associated with the iron and aluminum content in the leach liquor as these two metals make up the bulk of the metal leached. The acid concentration in the leach solution decreased slightly during the test.

The leach solutions from the metal extraction studies consisted of a mixture of dissolved metals of value (Ag, In, Ga, Au, Cu, Pb, Zn), residual acid (HCl), background salt (NaCl) and dissolved aluminum and iron chlorides ($AlCl_3$ and $FeCl_3$).

A solution treatment process was developed in order to maximize recovery of value metals into commercial products, recycle excess acid and regenerate excess acid associated with iron chlorides. The development of this process is described in the following section.

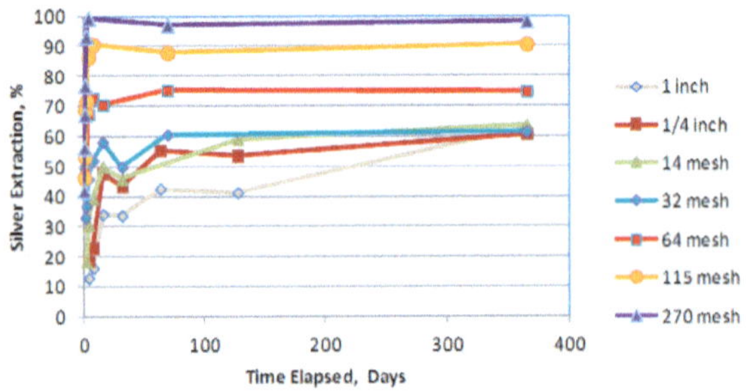

Figure 2. Silver Extraction Versus Time for Sample 10-10

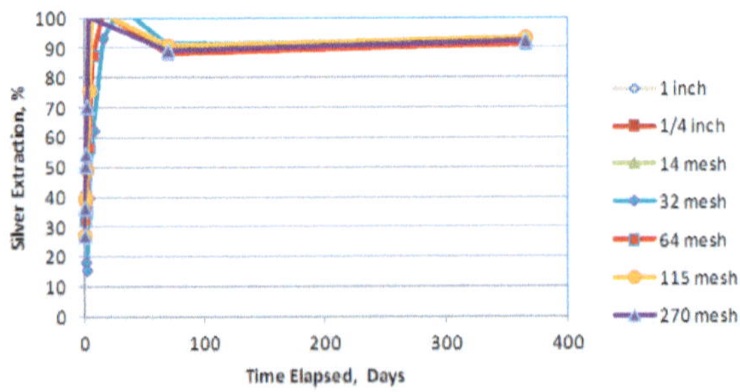

Figure 3. Silver Recovery Versus Time for Sample 10-13

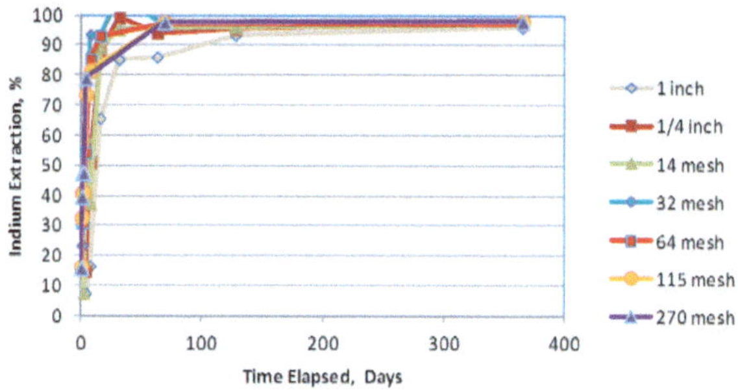

Figure 4. Indium Extraction Versus Time for Sample 10-12

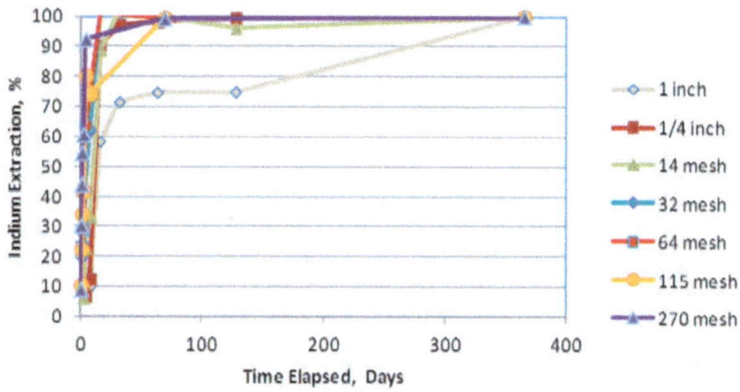

Figure 5. Indium Extraction Versus Time for Sample 10-10

Solution Treatment Studies

Early work on the solution treatment process indicated that an acid/salt separation could be accomplished using the Eco-Tec Recoflo process for acid recovery. This process is commonly used in recovery of acid from spent pickle liquor in the steel industry to separate acid (eg. HCl) from metal chloride salts (eg. $FeCl_2$) formed in the pickling process. The concept in use of the Eco-Tec system was to recover a low acid salt solution containing the value metals for processing and a high acid solution to recycle to the leaching process (Sheedy, 1997).

In later studies, attention turned to minimization of acid use in the leach process and solvent extraction of iron chlorides in order to separate iron from the leachate and enable the recycling of the acid (HCl) that was associated with the leaching of iron. Iron chloride salt(s) (e.g., $FeCl_3$) is generally responsible for a significant component of the consumption of acid (e.g., $Fe_2O_3 + 6HCl = 2FeCl_3 + 3H_2O$) in the process. The recovery of a ferric chloride solution would lead to the possibility of high temperature ferric chloride hydrolysis to form iron oxide and hydrochloric acid for recycle (the reverse of the reaction above.

The raffinate solution from the iron removal process could then be treated for recovery of Ag, Au and Cu by cementation, In and Ga by alkali precipitation and Zn and Pb precipitation as sulfides. The cements could then be further refined to pure products or toll processed through existing refineries. The In/Ga co-precipitate could be refined using conventional In recovery processes (for example similar precipitates are recovered from zinc hydrometallurgical refinery solutions). The final product in this case would be pure In and Ga for sale. The ZnS and PbS products are common compounds in mineral concentrates (eg. Sphalerite and galena) for treatment at zinc and lead smelters and refineries and could be processed by these routes.

Solvent extraction of iron

The solvent extraction process for iron recovery was designed to be "reagent free". I.e., no additional chemical consumption was to be used to minimize the cost. The process involves loading a mixed iron chloride-acid species (e.g., iron chloride-hydrochloric acid species such as for example $HFeCl_4$) onto a solvating extractant to form a loaded solvent. The solvating extractant may be, for example, tri-butyl-phosphate (TBP) or tri-alkyl phosphinate (Cyanex 923) or di-butyl-butyl phosphonate (DBBP). These solvating extractants were found to be highly selective in the extraction of the $HFeCl_4$ species from mixed salt solutions. The minor co-extraction of other species may be controlled by scrubbing of the solvent with water. Finally, water stripping can be used to form a $FeCl_3$ – HCl strip solution.

Tables 3 and 4 show the results of locked cycle testing of iron extraction using 100% TBP organic. Loading was performed with three countercurrent stages and an A/O of 3.65. Scrubbing was performed with one stage and an A/O of 2. The scrub solution containing iron (Table 3) was reused for a number of cycles before refreshing. Stripping was performed with four countercurrent stages and an O/A of 1.2. The loading, scrubbing and stripping were performed as sequential locked cycle tests.

The deportment analysis (Table 4) showed that the iron solvent extraction process resulted in 99.9% removal of iron from the PLS with negligible loss of the pay metals. Gold and gallium were not analyzed in solution but are expected to deport to the raffinate.

Table 3. Solvent Extraction of PLS Solution with 100% TBP Solution.

Stream	Chemical Analysis (mg/L)								
	In	Ag	Fe	Al	Ca	Cu	Na	Pb	Zn
PLS	22.7	75.3	13,300	959	259	409	95,900	5,940	523
Raffinate	23.5	98.5	9.7	921	170	366	110,000	6,430	672
Scrub Feed			88,100						
Scrub Product	1.83	0.47	79,300	16	116	7	2620	55	35
Strip Solution	0.18	<0.03	48,300	<8	369	<1	104	<20	<20
Loaded Organic			41,700		148	<0.1	235		15
Scrubbed Organic			50,600		188	<0.1	14		1
Stripped Organic			4,260		<2	<0.1	<2		0.1

Table 4. Deportment of the Various Elements through the Solvent Extraction Process

Stream	Deportment (%)								
	In	Ag	Fe	Al	Ca	Cu	Na	Pb	Zn
Raffinate	99.4	100	0.1	99.7	63.4	99.8	99.9	99.9	99
Scrub Liquor	0.4	0	23.8	0.1	2.1	0.1	0.1	0	0.2
Strip Liquor	0.2	0	76.2	0.2	34.5	0.1	0	<0.1	<0.7

Iron Chloride Hydrolysis

The hydrolysis of the strip solution allows recovery of HCl and formation of synthetic hematite.

$$2FeCl_3 + 2H_2O = Fe_2O_3 + 6HCl$$

There are a variety of engineering solutions to the hydrolysis system including spray roasting, pyrohydrolysis or heating to decomposition. Ideally, the engineered solution would result in (1) a filterable hematite product, (2) a weak acid solution for recycle to stripping and (3) a strong acid solution for recycle to the ore leach lixiviant makeup.

Initial indications were favourable toward using heating of the strip solution to form hematite. Progressive heating of ferric chloride solutions showed that the solutions were very concentrated (over 500 g/L Fe) with a very high boiling point (+180 C) with near complete recovery of acid to condensate.

Acid Neutralization

The raffinate from the iron solvent extraction process may still contain small amounts of acid. The neutralization of this acid can be accomplished with limestone.

$$2HCl + CaCO_3 = CaCl_2 + H_2O + CO_2(g)$$

The neutralization of acid is necessary to aid in the further downstream steps (eg. cementation on metallic iron or hydrolytic precipitation).

Silver and Gold Cementation on Copper

The three most noble metals remaining in the solvent extraction raffinate are (in order) gold, silver and copper. It is possible to cement these elements together using metallic iron. However, this then combines the precious metals with copper which may complicate the further refining of the gold and silver product. Copper is the least noble of the three metals so copper can be used to reduce the precious metals.

$$2NaAgCl_2 + Cu = 2Ag + CuCl_2 + 2NaCl$$
$$2NaAuCl_4 + 3Cu = 2Au + 3CuCl_2 + 2NaCl$$

In the presence of excess copper, cupric chloride will reduce to form soluble cuprous chloride, increasing the copper consumption.

$$CuCl_2 + Cu = 2CuCl$$

Table 5 shows a typical result for cementation. Copper powder was added to 187% of stoichiometric. The ORP of the final solution was 203 mV versus Ag/AgCl reference electrode. Ag recovery to the cement was 99.2% with negligible losses of other pay metals. A portion of the Sb was removed. This may have come off as part of the cement product. At an ORP of 203 mV, stibnine is not thermodynamically favoured to form. Arsenic levels in solution were near constant indicating that there was no precipitation or volatilization (as arsine) of arsenic. Gold was not present in the feed solution for cementation but would be expected to cement with silver.

Table 5. Cementation Test for Silver Recovery

Stream	Analysis (mg/L or %)								
	Ag	Cu	In	Pb	Ga	Zn	Fe	As	Sb
Fe Raff	90.7	358	10.1	6270	3.39	355	15.1	112	90
Soln	0.68	911	11.3	6550	3.75	381	15.8	111	56
Solid	25.35								

Copper Cementation

Copper cementation can be accomplished on metallic iron to recover copper values from solution. In practice, the bulk of this cement could be recycled back to the silver and gold cementation process to eliminate any need to procure scrap copper to the site.

$2CuCl + Fe = 2Cu + FeCl_2$

Table 6 summarizes the results of the copper cementation testing. The results showed removal of copper to 1.7 mg/L corresponding to 99.8% recovery. Once again the Sb and possibly minor As was removed from solution. The final ORP was -450 mV versus Ag/AgCl.

Table 6. Copper Cementation with Iron Results

Soln	Chemical Analysis (mg/L)									
	Cu	Fe	Pb	In	Ga	Al	As	Ag	Sb	Zn
Feed	991	15.8	6550	11.3	3.75	954	111	0.68	56	381
Filt.	1.7	1030	6330	10.8	2.92	981	100	<0.08	9	369

Indium and Gallium Precipitation

Indium and gallium can be removed from solution using pH adjustment with limestone or lime. Other trivalent metals will also precipitate at around the same pH point so the precipitate will be low grade and in need of further refining.

$InCl_3 + 1.5CaCO_3 = In(OH)_3 + 1.5CaCl_2 + 1.5\ CO_2(g)$
$AlCl_3 + 1.5CaCO_3 = Al(OH)_3 + 1.5CaCl_2 + 1.5\ CO_2(g)$
$FeCl_3 + 1.5CaCO_3 = Fe(OH)_3 + 1.5CaCl_2 + 1.5\ CO_2(g)$
$CrCl_3 + 1.5CaCO_3 = Cr(OH)_3 + 1.5CaCl_2 + 1.5\ CO_2(g)$
$GaCl_3 + 1.5CaCO_3 = Ga(OH)_3 + 1.5CaCl_2 + 1.5\ CO_2(g)$

Initial testing indicated that indium and gallium could be removed from solution at pH of 5.25.

Table 7 shows the results of a bulk precipitation at this pH value. The indium and gallium values were quantitatively precipitated to yield 0.227% In and 0.0478% Ga precipitate. The precipitate was contaminated with aluminum and lead, presumably as hydroxides. There was also some iron in the precipitate along with some arsenic and minor amounts of other elements.

It is possible that two – stage precipitation would yield a higher In/Ga grade and a cleaner precipitate for refining. Nevertheless, this sort of precipitate can be refined to produce the rare metals of indium and gallium using conventional methods.

Table 7. Precipitation of Indium and Gallium. Analysis in mg/L for solution and g/t for solid precipitate.

	Feed	Filtrate	Precipitate
Fe	1030	798	33700
Pb	6330	6020	77800
Ag	0.08	<0.03	4.6
In	10.8	<0.05	2270
Ga	2.92	<0.05	478
Al	981	3	177,000
As	100	12	14,200
Cu	1.7	1	110
Sb	9	0.06	538
Zn	369	3.05	4770

Lead and Zinc Precipitation as Sulfides

Lead and zinc can be sequentially precipitated as sulfides by addition of a source of sulfide (eg. NaSH). The initial selective precipitation of lead is accomplished by ORP control during NaSH addition.

$PbCl_2 + NaSH + NaOH = PbS + 2NaCl + H_2O$
$ZnCl_2 + NaSH + NaOH = ZnS + 2NaCl + H_2O$

The results indicate the tremendous potential of the sulfide process to precipitate highly pure zinc and lead sulfides (+80 and +50% of Pb and Zn metal content respectively). Conveniently, the sulfide precipitate also removed trace amounts of silver, arsenic, copper and antimony.

Table 8. Precipitation of Lead and Zinc with Sodium Hydrosulfide. Solution analysis in mg/L and solid analysis in %.

	Feed	PbS Filtrate	ZnS Filtrate	PbS	ZnS
Fe	798	740	734	80.8	2.18
Pb	6020	39	<2		
Ag	<0.03	<0.03	<0.08		
Al	3	<0.8	<0.8	0.02	
As	12	<3	<3		
Cu	1	0.41	<0.1		
Sb	0.6	<1	<1		
Zn	305	317	4.45	0.01	53.2
S				12.6	29.8

The solution remaining after zinc and lead removal can be recycled back to leach. The solution is essentially a near neutral sodium chloride solution containing minor amounts of iron as ferrous chloride. If warranted, a separate cleanup step (oxidation and precipitation) of iron could be performed, before recycling back to leach.

The simplified flowsheet for the fully integrated process is shown in Figure 6.

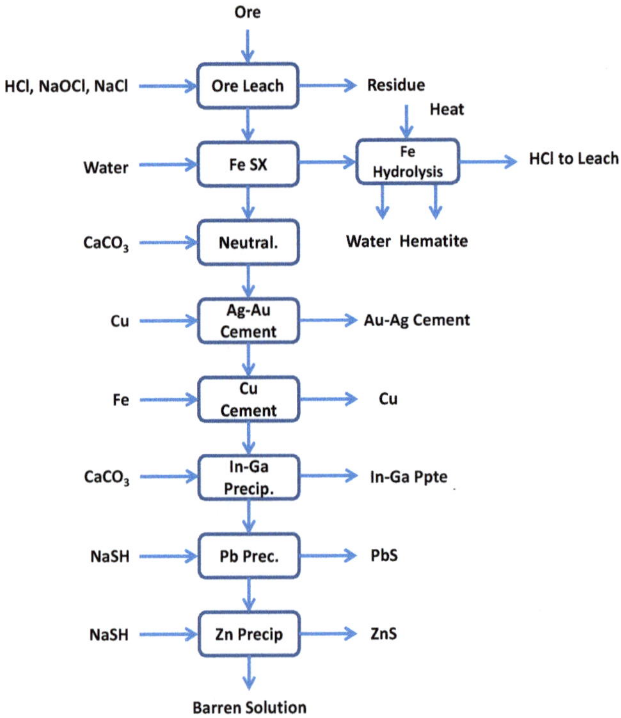

Figure 6. Simplified Process for the Acid-Chloride Leaching Process

Conclusions

The acid chloride leaching process has been developed to unlock the unique potential of the Malku Khota deposit in Bolivia with significant values in precious, rare and base metals. The soak test was developed to characterize the ore response and determine the impact of crush/grind size, temperature, reagent strength and time on the leaching of the pay metals. A novel process (Dreisinger et al, 2013) for treatment of the leachate was developed that included the recycling of acid associated with iron by solvent extraction and hydrolysis and then production of separate products of value including;

1) Ag-Ag cement
2) Cu cement
3) In-Ga hydroxide precipitate
4) PbS precipitate
5) ZnS precipitate

The Malku Khota project is no longer available for development by TriMetals Mining Inc. However, the acid chloride leach process could be used for treatment of alternate ore deposits. The company is actively searching for a suitable deposit to take advantage of the unique characteristics of the process for extracting value from complex deposits containing precious, rare and base metals.

References

Armitage, A., Desautels, P., Zurowski, G., Pennstrom, W., Malbran, F., Fitch, R., "Preliminary Economic Assessment Update", May 10, 2011. (www.sedar.com)

Dreisinger, D., Fitch, R., Verbaan, C., "Method for Recovering Indium, Silver, Gold and Rare, Precious and Base Metals from Complex Oxide and Sulfide Ores", US Patent 8,585,991 B2, Nov 19, 2013.

Sheedy, M., "Acid Recovery and Purification Using Absorption Resin Technology", Paper presented at the 126th TMS Annual Meeting, Orlando, Florida, USA, 9 February 1997

RECOVERY OF PLATINUM GROUP METALS USING PEROVSKITE-TYPE OXIDE

Kota NAGAI, Hiroki KUMAKURA, Shota YANAI, Takashi NAGAI

Chiba Institute of Technology; 4-6-1 Tudanuma, Narashino, 275-0016 JAPAN

Keywords: PGM, Perovskite-type oxide, Recycling

Abstract

A large amount of platinum group metals (PGMs) are used in automobile catalyst. It is important to recover the metals from wasted catalyst. Nomura et al [1] suggested a recovery process that PGM oxides were absorbed to perovskite-type oxide based on $LaScO_3$. In this study, the PGM oxide absorption properties of various perovskite-type oxides are investigated to develop a new recovery process. This process has the potential to separate each element in the process if the perovskite-type oxide is selected properly. Some of perovskite-type oxides were synthesized with metallic platinum, palladium or rhodium at high temperature in air atmosphere. And the PGM contents in the oxide are determined by chemical analysis. Then a new recovery process of PGMs from wasted automobile catalyst, in which PGMs can be separate to each element, was designed.

Introduction

A large amount of Platinum (Pt), Palladium (Pd), and Rhodium (Rh) are used in automobile catalysis.[1] Resource of their metals is limited and mining of them has high environmental load. Therefore, the recycling of PGMs from wasted automobile catalysis is very important.

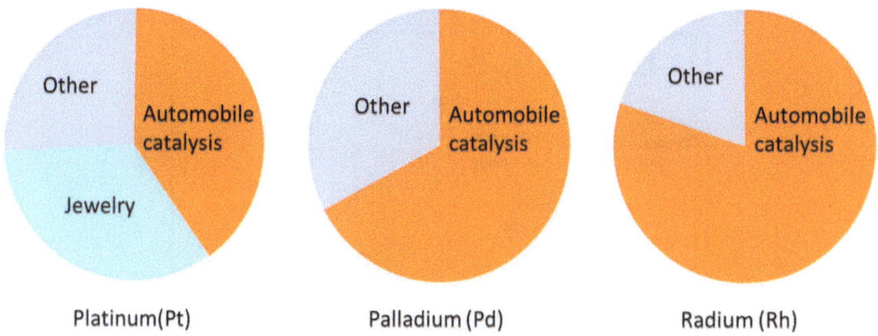

Fig. 1 Demand of Pt, Pd and Rh in 2012.

The Rose process (Fig. 2) was developed as a process to recover PGM from wasted automobile catalysis.[2] In this process, the automobile catalysis is first crushed mechanically, and then extracting agent and flux are melted at high temperature. The mixture separates into a metal phase, (copper (Cu) alloy containing PGM) and slag. The PGM is enriched by oxidized Cu to Cu oxide. The PGM are then recovered and separated to each element in a wet process. However, it is hard to dissolve to acid because PGM are stable chemically and strong oxidation agent such as aqua regia is used. In addition, to separate PGM to each element, solvent extraction process is used.

Nomura et al.[1] reported that $La_{0.7}Sr_{0.2}Ba_{0.1}ScO_{3-d}$ (LSBS), which is a perovskite-type oxide based on $LaScO_3$ absorbed PGM oxides and proposed a new recovery process. (Fig. 3) In this case, the PGM absorbed to perovskite-type oxide can be dissolved to acid easily, because the PGM is absorbed as PGM oxide. Then the PGM can be recovered from the solution by traditional methods.

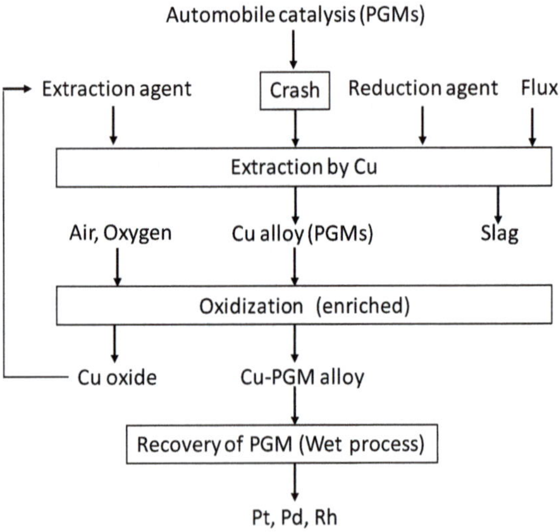

Fig.2 Flow of Rose process.

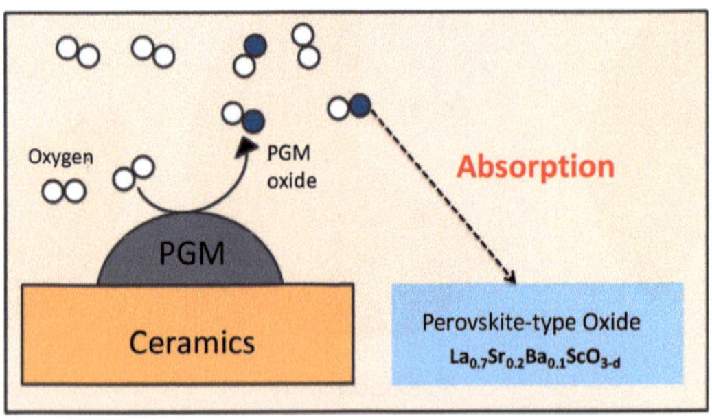

Fig. 3 New Process proposed by Nomura et al.

It is possible to absorb Pt, Pd, and Rh by some perovskite-type oxide respectively, since there is compatibility between a perovskite-type oxide and PGM. (Fig. 4) In a new recycle process, PGM are recovered and separated at once. It is not necessary to use strong oxidization agent to dissolve PGM to acid, because the PGM are absorbed by the perovskite-type oxide as PGM oxide.

In this study, absorption properties of many perovskite-type oxides are investigated to find the perovskite-type oxide which absorbs only Pt, Pd, or Rh.

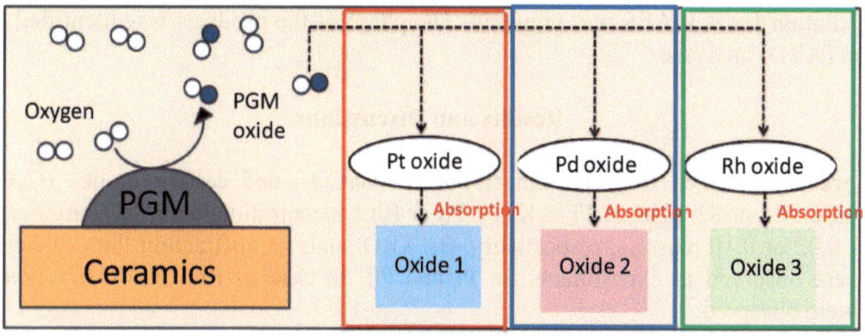

Fig. 4 A new recovery process of PGM from wasted automobile catalysis.

Experimental procedure

Perovskite-type oxides were prepared by mixing two reagent grade powders. The desiccated mixture was charged in a Pt crucible and heated in an electric resistant furnace in case of investigation of Pt absorption property. (Fig. 5)

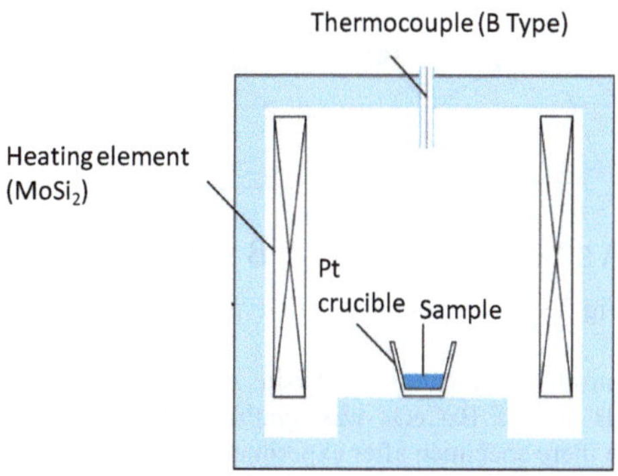

Fig 5 Schematic configuration of the apparatus for investigation of Pt absorption property.

In case of investigation of Pd or Rh absorption properties, the mixture was heated on a Pd plate or Rh plate.(Fig 6)

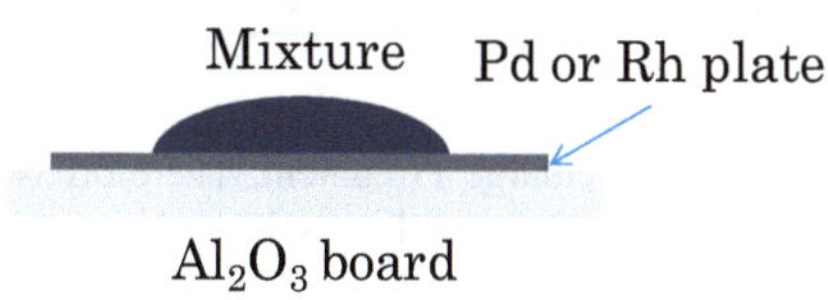

Fig 6 Schematic configuration of the apparatus for investigation of Pd or Rh absorption property.

The mixture was heated at 1473 – 1873 K for 4 – 8 h in air atmosphere. Pt, Pd or Rh concentrations of the specimen after experiment were determined by inductively coupled plasma atomic spectroscopy (ICP-AES). By about 0.1 - 0.5 g specimen was dissolved to acid, the aqueous solution for ICP-AES was prepared. The phase of the products was identified by X-ray diffraction (XRD) analysis.

Results and Discussions

$BaCeO_3$ was synthesized from barium carbonate ($BaCO_3$) and cerium oxide (CeO_2) in Pt crucible or, on Pd or Rh plate at 1773 K. Pt, Pd or Rh concentration in $BaCeO_3$ after experiment was 0.23, 0.02 or 0.10 mass %, respectively. By XRD analysis, diffraction lines of $BaCeO_3$ and $BaCO_3$ were observed in experiments for Pt and Pd. In case of Rh, only diffraction lines of $BaCeO_3$ were observed.

There are two lattice sites which are named to A site and B site in the perovskite structure. (Fig. 7), the Another experiment was performed to determine which site absorbs PGM.

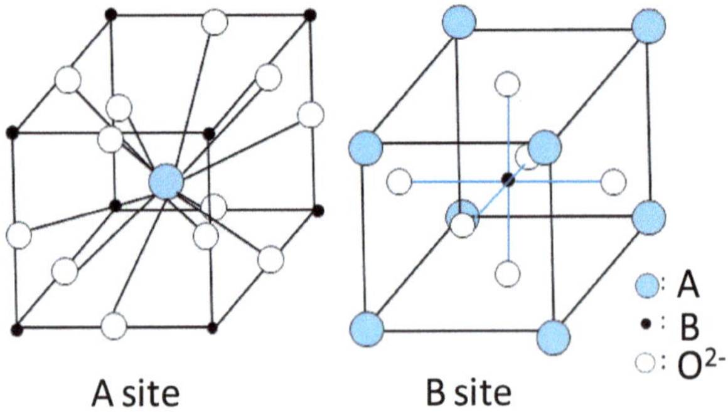

Fig 7 Structure of perovskite-type oxide (ABO_3)

$Ba_{0.95}CeO_3$, which is introduced vacancy to A site of $BaCeO_3$, and $BaCe_{0.95}O_3$, which is introduced vacancy to B site of $BaCeO_3$ was synthesized in Pt crucible at 1573 K in air atmosphere. Pt content in there specimen after experiment is listed in Table 1.

Table 1 Pt content in $BaCeO_3$

Oxide	Time	Pt content C_{Pt} (mass%)
$BaCeO_3$	4	0.15
$Ba_{0.95}CeO_3$	4	0.08
$BaCe_{0.95}O_3$	4	0.48
$BaCe_{0.95}O_3$	8	0.9[illegible]

By introducing vacancy to B site (Ce site), the absorption properties are improved. And vacancy at A site (Ba site) is not effective to enlarge Pt content. Therefore, Pt is absorbed at B site in provskite-type oxide. This is good agreement of the results by Nomura et al.[2]

It is necessary to investigate the PGM oxide absorption property of many provskite-type oxides, and to discover the provskite-type oxide which has selectivity of PGM oxide for a new recovery process.

Conclusions

Some of perovskite-type oxides were synthesized with metallic platinum, palladium or rhodium at high temperature in air atmosphere. And the PGM contents in the oxide are determined by chemical analysis.$BaCeO_3$ absorbed Pt (0.23 mass%) and Rh (0.10 mass%)

Pt is absorbed to A site in perovskite-type oxide.

References

[1] Katsuhiro Nomura and Hiroyuki Kageyama; "Recycling Technology for Platinum Group Metals by Using Perovskite-type Oxides" Materia Japan, 52(2), 58-63, 2013
[2] Platinum 2013, Johnson Matthey, 2013

Rare Metal Technology 2015
Edited by: Neale R. Neelameggham, Shafiq Alam, Harald Oosterhof, Animesh Jha, David Dreisinger, and Shijie Wang
TMS (The Minerals, Metals & Materials Society), 2015

RESEARCH ON PROCESS OF HYDROMETALLURGICAL EXTRACTING Au, Ag, AND Pd FROM DECOPPERIZED ANODE SLIME

Yongbin Yang, Wei Yin, Tao Jiang, Bin Xu, Qian Li

School of Minerals Processing and Bioengineering, Central South University, Changsha, Hunan, P R China, 410083

Keywords: Decopperized Anode Slime; Pretreatment; Leaching Method; Au; Ag; Pd

Abstract

Copper anode slime is an increasingly significant resource for precious metals such as Au, Ag and Pd. The extraction of Au, Ag and Pd from decopperized anode slime by different pretreatments and different lixiviants including cyanide, chloride, thiourea and thiosulfate have been investigated. When oxidizing roasting pretreatment was carried out under the conditions of roasting time 120min and temperature 625 ℃, the removal ratios of S and Sb could reach 82.18% and 98.52%, respectively. And the following alkaline leaching of calcine could result in the dissolution of 46.40% Cu and 98.48% Pb. The optima leaching ratios of Au, Ag and Pd reached 99.48%, 97.69%, 94.79% by cyanide leaching, and 95.01%, 79.84%, 96.59% by chloride leaching, and about 95.21%, 95.66%, 48.79% by thiourea/thiosulfate leaching.

1 Introduction

Copper anode slime is an important byproduct in the process of copper electrolysis refining, which contains a large number of valuable metals, especially gold, silver, platinum, palladium, copper, tin, etc. that made the double value of anode slime. Generally, the conventional method for the treatment of copper anode slime is pyrometallurgical process. However, low precious metals recovery, environmental pollution, complex process and high cost are main disadvantages of this method [1]. Hydrometallurgical processes as a flexible and high efficiency method have been widely researched to recover valuable metals from anode slime in recently years [2].

Due to a large number of base metals are contained in decopperized anode slime, therefore the pretreatments are conducted before the comprehensive recovery of precious metals from the anode slime [3]. The purposes of pretreatment are to remove the base metals of Cu, Sb, Pb etc. and to enrich precious metals. There are many reports about pretreatment researches [4, 5], the method of pressure oxidation leaching has some advantages in reagent cost, environmental protection, etc., but corrosion and high risk of equipments are main disadvantages of this method [6]. Although sulfation-roasting is efficient, it has the disadvantages of high production cost and environmental pollution [7]. Compared to above mentioned methods, oxidizing roasting, mechanical activation and alkaline leaching are able to dramatically improve the extraction of Au, Ag and Pd from decopperized anode slime [8, 9].

Cyanide leaching, as the primary method of extracting of Au, Ag and Pd at present, is characterized by low production cost, simple equipment and high recoveries of precious metals [3]. Among the non-cyanide leaching, chloride leaching has been widely applied in industry in the past, and it has the advantages of high efficiency and good economic benefit [10]. The thiourea leaching has also been widely studies in the past three decades, and it is highly effective and not sensitive to base metal impurities [11]. In addition, the thiosulfate leaching is attractive in virtue of non-toxic and low cost of reagent [12].

Base on the above, the combined process of pretreatments followed by leaching of precious metals has been employed to treat decopperized anode slime in this work. The pretreatments for removing base metals completely and leaching methods for extracting of Au, Ag and Pd have been investigated.

2 Materials and Pretreatments

2.1 Materials

The decopperized anode slime used in experiments was provided by one of copper smelters in China. Before experiments, the original copper anode slime was processed by preliminary oxidation roasting and sulfuric acid leaching to recover the copper. The decopperized anode slime was 87.95% less than 31μm, and its chemical compositions were shown in Table I.

Table I. Chemical compositions of the decopperized anode slime

Components	Au	Ag	Pd	Sn	Cu	S	Sb	Pb	Ni	As	Fe
Wt.%	103.21*	4124.72*	87.00*	48.00	0.88	9.41	8.00	9.81	1.34	0.28	0.56

*unit g/t

2.2 Physico-chemical Characterization

The specific surface area (S_A) was determined by the DBT-127 apparatus, which could measure the corresponding resistance by using a certain amount of air going through a certain thickness of the compacted powder layer.

X-ray diffractometry was accomplished by using a diffractometer D/max-2500 (Rigaku D, Japan) equipped with a CuKα source operating at 40 kV and 250 mA. Data were collected every 1 s and the detector was moved at a rate of 8°min^{-1}. The diffractogram of the decopperized anode slime, shown in Figure 1, indicates the presence of stannic oxide (SnO_2) and lead compounds ($PbSO_4$, Pb_3O_4, Ag_2PbO_2). And the phase results of Au and Ag in decopperized anode slime were analyzed by the method of chemical phase analysis and listed in Table II.

It could be seen from Table 2 that 83.52% of gold was exposed, and the others was mainly encapsulated in sulfides and oxides. As much as 87.88% of silver was metallic silver, and the rest mainly existed in the form of sulfide, encapsulated in sulfides and oxides. On the other hand, only 0.1% of gold and 0.73% of silver were encapsulated in silicates.

The main chemical reagents, such as sodium hydroxide, hydrochloric acid, sodium thiosulfate, hydrogen peroxide, ammonia solution, copper(II) sulfate pentahydrate, ferric sulfate, etc. that used in this study were all analytical grade.

Table II. The results of chemical phases analysis of Au and Ag in decopperized anode slime

Au	Phase	Exposed gold	Encapsulated in sulfides		Encapsulated in oxides		Encapsulated in silicates		Total
	Content g/t	86.20	10.02		6.89		0.10		103.2
	Distribution%	83.52	9.71		6.67		0.10		100.0
Ag	Phase	silver oxide	metallic silver	silver chloride	silver sulfide	Encapsulated in sulfides	Encapsulated in oxides	Encapsulated in silicates	Total
	Content g/t	4.40	3624.93	7.33	208.34	208.67	41.05	30.00	4124.7
	Distribution%	0.11	87.88	0.18	5.05	5.06	1.00	0.73	100.0

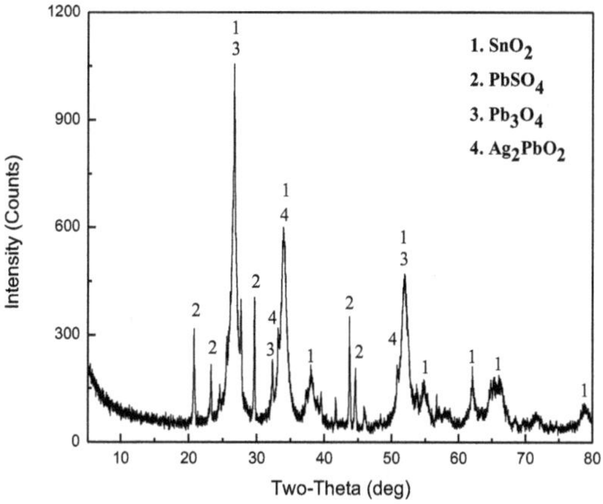

Figure 1. X-ray diffraction pattern of decopperized anode slime

2.3 Pretreatments

2.3.1 Oxidizing Roasting In order to facilitate Sb, Cu, As, S and Pb oxidized to corresponding sulfates and oxides, suitable air linear velocity in hearth was necessary. The flue gas was treated by lye absorption before it was released.

A certain amount of decopperized anode slime was put into the porcelain boat, and then sent to the tube furnace when the setting temperature of the oxidizing roasting was reached. Finally, took it out to cool naturally after keeping the scheduled time (2 hours) in tube furnace at constant temperature.

2.3.2 Mechanical Activation Mechanical activation of minerals was performed with different methods including ball-milling (RK/ZQMΦ160×60, Wuhan Rock & Crush Grand Equipment Manufacture Co., Ltd, China) and high pressure roller mill (Modified Roller Mill Φ250×120, Central South University Molybdenum Processing Plant, China). The former under the following conditions: grinding media filling proportion of 20%, grinding time of 0 to 30 min as specified and mill speed of 112 rpm. However, the latter conditions: pressure of roller mill 3 MPa, and roller mill times 3.

2.3.3 Alkaline Leaching The alkaline leaching was investigated using a 500mL three-necked flask in which 150mL of alkaline leaching solution (1-3.5M NaOH) and 30g of roasting slag were added. The leaching was performed for 120 min at 90℃, using a stirring rate of $6.67 s^{-1}$ in the reactor.

3 Leaching of Au, Ag and Pd

3.1 Cyanide Leaching

The decopperized anode slime was directly leached in this experiment, and the leaching test apparatus include a 400 mL glass reactor, digital stirrer and pHS-3c acidity meter. The stirred leaching was performed for time 48 hours at pH 11 and stirring rate $8.33 s^{-1}$.

3.2 Chloride Leaching

Palladium oxide in decopperized anode slime was difficult to extract by chloride leaching, so it should be reduced to palladium by Formic acid. Optimal operating parameters were established by conditional experiments as formic acid 60g/L, liquid-to-solid ratio 3, temperature 95℃ and time 60min.

Hydrogen peroxide and hydrochloric acid in this experiment were used as the oxidizing agent and chlorinating agent, respectively [10]. The chloride leaching was investigated using a 500mL three-necked flask in which 120mL of alkaline leaching solution (1.7M HCl+10%H_2O_2) and 30g of residue were added. The leaching was performed for 120 min at 90℃, using a stirring rate of $5s^{-1}$ in the reactor. The best flowsheet of the whole process was given in Figure 2.

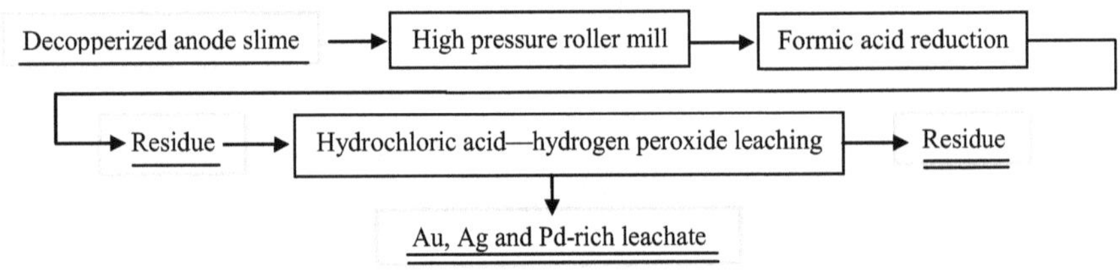

Figure 2. Flowsheet of the whole process of chloride leaching

3.3 Thiourea Leaching

The leaching was conducted in a 400 mL glass reactor into which 150 mL of leaching solution (40 g/L $CS(NH_2)_2$ + 1.5% $Fe_2(SO_4)_3$) and 30 g of pretreatment residue were added.

3.4 Thiosulfate Leaching

The leaching was performed by using a 400 mL glass reactor in which 150 mL of leaching solution (0.53M $Na_2S_2O_3 \cdot 5H_2O$ +5 g/L $CuSO_4 \cdot 5H_2O$) and 30 g of pretreatment residue were added, and the best flowsheet of the whole process was given in Figure 3.

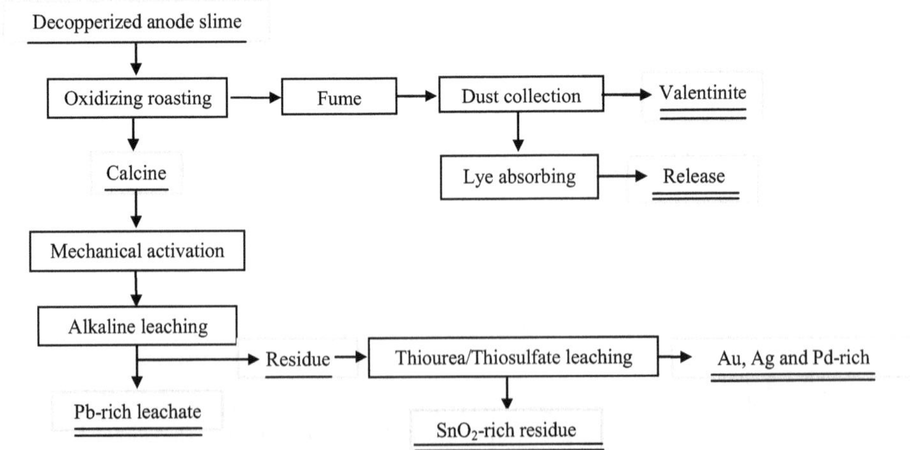

Figure 3. Flowsheet of the whole process of thiourea/thiosulfate leaching

4 Results and Discussion

4.1 Effect of Pretreatments

4.1.1 The Effect of Roasting Temperature on Removal Ratio of S and Sb The dependence of removal ratio of S and Sb on roasting temperature varying from 400℃ to 700℃ was illustrated in Figure 4. When the temperature changed from 400℃ to 625℃, the removal ratio of Sb reached 98.52% with that of S increased from 76.60% to 82.18%. According to the results of Figure 4, 625℃ was chosen as the optimal temperature.

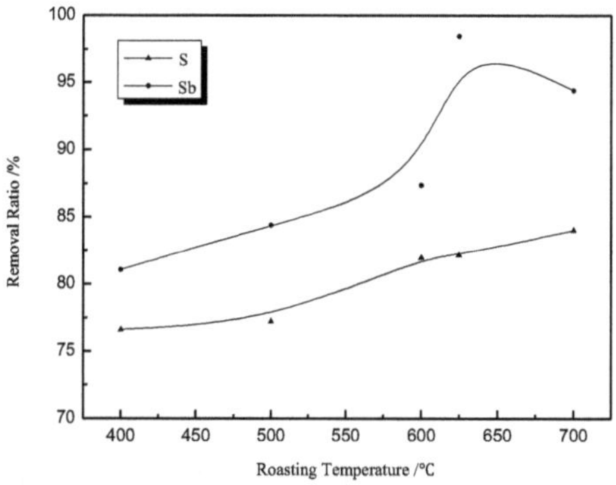

Figure 4. The effect of roasting temperature on the removal ratios of S and Sb

4.1.2 Effect of Different Mechanical Activation Methods on S_A The results were shown in Table III, and it could be seen that the specific surface area (S_A) of material which was treated through high pressure roller mill (HPRM) increased from 3362.36 $cm^2 \cdot g^{-1}$ to 4221.94 $cm^2 \cdot g^{-1}$, while that of ball-milling increased to 3733.91 $cm^2 \cdot g^{-1}$. Then the following alkaline leaching and thiosulfate leaching were implemented to the mechanically activated materials, and the leaching ratio of Au and Pd were represented in Table IV, showed that high pressure roller mill (HPRM) was the optimal mechanical activation method.

Table III. The results of different activation methods on S_A

Sample	Decopperized anode slime	Activated materials	
		By HPRM	By ball-milling
S_A /$cm^2 \cdot g^{-1}$	3362.36	4221.94	3733.91

Table IV. The results of different mechanical activation methods on leaching ratio of Au and Pd

Mechanical activation	The content in residue /$g \cdot t^{-1}$		Leaching ratio /%	
	Au	Pd	Au	Pd
By HPRM	23.50	30.00	77.87	68.83
By ball-milling	42	41	57.56	54.29

4.1.3 The Effect of NaOH Concentration on Leaching Ratio of Lead The plot in Figure 5 describes the effect of NaOH concentration on leaching ratio of Pb in roasting slag. Clearly, the leaching ratio of Pb was improved rapidly due to the increase of concentration of NaOH in the beginning, but it was gradually stabilized and reached to 98.48% when the concentration of NaOH was higher than 4mol/L. Meanwhile the leaching ratio of Cu was improved 46.40%.

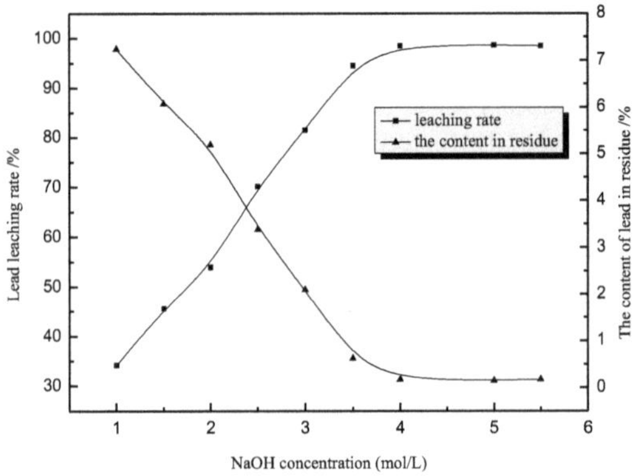

Figure 5. The effect of NaOH concentration on Pb leaching ratio

4.2 Cyanide Leaching

Table V showed that the extraction of Au from decopperized anode slime depends on sodium cyanide concentration. The Au recovery was 92.66% under the sodium cyanide concentration of 10kg/t and increase to 99.48% under the sodium cyanide concentration of 30kg/t. At the same times, the leaching ratios of Ag and Pd were up to 97.69% and 94.79%, respectively.

The results demonstrated that high recovery of Au, Ag and Pd were achieved by cyanide leaching. On the other hand, sodium cyanide is a kind of highly toxic substance, and its use is very restrictive.

Table V. The effect of sodium cyanide concentration on Au leaching ratio

Sodium cyanide concentration /$kg \cdot t^{-1}$	The content in residue /$g \cdot t^{-1}$	Leaching ratio (%)
10	8.1	92.66
20	4.0	96.82
30	0.6	99.48

4.3 Chloride Leaching

The direct treatment of chloride leaching to decopperized anode slime only enables to achieve the lower recoveries of Au, Ag and Pd. The results summarized in Table VI further showed that high pressure roller mill pretreatment was helpful to improve the leaching ratio of precious metals (shown in Figure 2). The leaching ratio of Au increased from 48.97% to 94.65%, and that of Pd increased from 72.37% to 94.99%, while that of Ag increased only from 12.67% to 66.52%, which demonstrated that chloride leaching has the limitation for the recovery of Ag. Therefore, the measures to strengthen recovery of Ag were necessary. Non-oxidative chloride leaching (NOCL) [12] was performed by conditional tests as hydrochloric acid 1.5 mol/L, ammonium chloride 300 g/L, liquid-to-solid ratio 4, temperature 40 °C and time 90 min. The extraction of Ag in the whole process achieved 79.84%.

Table VI. The results of different chloride leaching methods on leaching ratio of Au, Ag and Pd

Method	Leaching ratio /%		
	Au	Ag	Pd
Chloride leaching	48.97	12.67	72.37
HPRM—Chloride leaching	94.65	66.52	94.99
HPRM—Chloride leaching—NOCL	95.01	79.84	96.59

4.4 Thiourea Leaching

The pretreatment of decopperized anode slime by alkaline leaching was carried out and then the residue was performed by thiourea leaching. The effect of thiourea concentration on recovery Au was shown in Table VII. It indicated that the leaching ratio of Au first increased and then decreased with enhancing the thiourea concentration. When the thiourea concentration changed from 0.35mol/L to 0.53mol/L, the leaching ratio of Au increased from 50.03% to 62.22%, and then decreased to 55.94% under the thiourea concentration of 0.70mol/L. According to the results of above, thiourea concentration of 0.53mol/L was regarded as the optimal concentration of thiourea leaching.

Table VII. The effect of thiourea concentration on Au leaching rate

Thiourea concentration / $mol \cdot L^{-1}$	The content in residue / $g \cdot t^{-1}$	Leaching ratio /%
0.35mol/L	57.00	50.03
0.53 mol/L	43.00	62.22
0.70mol/L	50.30	55.94

The whole process of thiourea leaching was accomplished according to the Figure 3, and the results were summarized in Table VIII. It showed that 95.21% of Au, 95.66% of Ag and 48.79% of Pd were extracted by thiourea. So it was necessary to choose an appropriate lixiviant to extract the rest of precious metals in residue.

Table VIII. The results of the whole process of thiourea leaching

Elements	Residue components / $g \cdot t^{-1}$	Leaching ratio /%
Au	5.80	95.21
Ag	203.00	95.66
Pd	54.00	48.79

4.5 Thiosulfate Leaching

Decopperized anode slime was directly leached in the thiosulfate solution under the normal temperature and pressure, and the low leaching ratios of Au and Ag were 34.13% and 69.21%, respectively. Therefore, it's very important to appropriate pretreatments to improve the recoveries of precious metals. The whole process of thiosulfate leaching was investigated according to the Figure 3, and the results were shown in Table IX, from which it could be seen that 95.04% of Au, 94.96% of Ag and 46.67% of Pd were dissolved from the pretreatments residue.

Table IX. The results of the whole process of thiosulfate leaching

Elements	Residue components / $g \cdot t^{-1}$	Leaching ratio /%
Au	6.40	95.04
Ag	260.00	94.96
Pd	58.00	46.67

5 Conclusions

The combined process of pretreatments followed by leaching of precious metals were employed to treat decopperized anode slime, and they enable to extract not only precious metals, including Au, Ag and Pd, but also other metals (Cu and Pb), furthermore,SnO_2 was concentrated in final residue.
The decopperized anode slime was directly leached by sodium cyanide, and over 99.48% of Au, 97.69% of Ag and 94.79% of Pd could be solubilized in the cyanide solution. As a highly toxic substance, cyanide's use is very strictly restricted.
The direct treatment of chloride leaching to decopperized anode slime was only able to achieve the low recoveries of Au, Ag and Pd, while the pretreatment of high pressure roller mill which could increase the specific surface area was helpful to improve the leaching ratios of Au, Ag and Pd to 94.65%, 94.99% and 66.52%, respectively. Moreover, non-oxidative chloride leaching could have obtained the extraction of Au, Ag and Pd in the whole process of 95.01%, 79.84% and 96.59%, respectively.
The whole process of thiourea/ thiosulfate leaching achieved the high leaching ratios of Au and Ag. In the step of oxidizing roasting, decopperized anode slime was completely oxidized and 82.18% of S and 98.52% of Sb were oxidized to the corresponding oxides. And 46.40% of Cu and 98.48% of Pb were dissolved into alkaline solution in the step of alkaline leaching. Furthermore, precious metals were enriched further in alkaline residue. The optima leaching ratios of Au, Ag and Pd reached about 95.21%, 95.66% and 48.79% by thiourea/ thiosulfate leaching. However, there were still as much as 51.21% of Pd remained in the thiourea/ thiosulfate leaching residue. Therefore, it was necessary to choose an appropriate intensive leaching to extract the rest of precious metals in residue.

Acknowledgement

Financial assistance and support from the project of NSFC (Grant No. 51074182) and China Postdoctoral Science Foundation (Grant No. 2014M550422) are gratefully acknowledged.

References

1. Yang, M, "Advancement in processes and technology of comprehensive utilization of lead anode slime," *Precious Metals*, 19 (3) (1998), 55-60.
2. Talip, H., Bu, D., Cafer, C, "Optimization of removal of lead from bearing-lead anode slime," *Journal of Industrial and Engineering Chemistry*, 16 (2010), 355–358.
3. Xue, G., et al, *Gold and silver hydrometallurgy and analytical test methods* (Bei Jing, Science Press, 2009), 647.
4. Hait, J., Jana, R.K., Kumar, V., Sanyal, S.K, "Some studies on sulfuric acid leaching of anode slime with additives," *Ind. Eng. Chem. Res.*, 41, (2002), 6593-6599.
5. Liang, J., Liu S., "Research progress on technology of dealing with copper anode slime," *GOLD*, 12(29) (2008), 32-38.
6. Li, C., Wei, C., Xu, H., Li, M., Li, X, "Oxidative pressure leaching of sphalerite concentrate with high indium and iron content in sulfuric acid medium," *Hydrometallurgy*, 102(1-4) (2010a), 91-94.
7. Hait, J., Jana, R.K., Sanyal, S.K, "Processing of copper electrorefining anode slime," *Miner. Process. Ext. Metall*, 118 (2010), 240-252.
8. Ficeriová, J., Baláž, P., Villachica, C.L, "Thiosulfate leaching of silver, gold and bismuth from a complex sulfide concentrates," *Hydrometallurgy*, 77 (2005), 35-39.
9. Xin, B., Zhang, J., Feng, Y., Guo, S., Ren, B, "Recovery process of palladium from waste palladium catalyst," *Modern Chemical Industry*, 30(2) (2010), 253-255.
10. Liu et al., 1995, "Study on gold extraction from copper anode mud by thiourea process," *GOLD*, 16(6), 37-42.
11. Ficeriová, J., Hashemzadehfini, M., Abkhoshk, E., Shahraki, B, "Effect of mechanical activation on thiosulfate leaching of gold from a complex sulfide concentrate," *Trans. Nonferrous Met. Soc. China*, 21 (2011), 2744-2751.
12. B.Xu, Y.B.Yang, Q.Li, G.H.Li, T.Jiang, "Fluidized roasting-stage leaching of a silver and gold bearing polymetallic sulfide concentrate," Hydrometallurgy, 147-148 (2014), 79-82

Rare Metal Technology 2015

Rare Earth Metals

Rare Metal Technology 2015
Edited by: Neale R. Neelameggham, Shafiq Alam, Harald Oosterhof, Animesh Jha, David Dreisinger, and Shijie Wang
TMS (The Minerals, Metals & Materials Society), 2015

STATUS OF SEPARATION AND PURIFICATION OF RARE EARTH ELEMENTS FROM KOREAN ORE

Joon Soo Kim[1†], Hyun Soo Kim[1], Myong Jun Kim[1], Jin-Young Lee[2], Jyothi Rajesh Kumar[2*]

[1]*Department of Energy and Resources Engineering*
Chonnam National University (CNU), Gwangju 500-757, Korea

[2]*Mineral Resources Research Division*
Korea Institute of Geoscience and Mineral Resources (KIGAM), Daejeon 305-350, Korea

Abstract

Rare earth materials have an ever growing variety of applications in the modern and hi-tech technology. They apply to many an industry with crucial materials and they provide numerous a customers with benefits. Rare earth materials utilization increasing day-by-day in the worldwide, at the same time the resources is very limited in many countries like Korea.

The present general article divided four sorts such as 1) mineral processing of Korean ore deals the gravity, magnetic separation and froth flotation, 2) decomposition and leaching of concentrates included sodium hydroxide fritting, sodium hydroxide hot digestion, hydrochloric acid leaching, 3) separation and purification of rare earth elements in mixed rare earth chloride solution consists reduction and oxidation method, solvent extraction, 4) preparation of rare earth oxide compound from purified elemental rare earth chloride solution having oxalate salt precipitation by reaction crystallization. The last part consists of a summary of the present study.

†Presenting author: jskim1186@jnu.ac.kr, *Corresponding author: rkumarphd@kigam.re.kr
Phone: 82-42-868-3313; Fax: 82-+42-868-3418

Introduction

The field of rare earth elements (REEs) is charming. An important research and development effort continue globally to explore and establish ways and want to put the REMs to use, in the service of humankind [1]. The REEs are in an ever growing variety of applications in the modern technology. Rise in demand for rare earth materials due to high-tech industries. They provide many an industry with crucial materials and they provide numerous a customer with benefits. Rare earth primary products are mainly used as raw materials for high-purity individual rare earth chemicals, and in making of petroleum and environmental protection catalysts, misch metal and polishing powders. From these beginnings and over many years, industrial applications of REMs have developed in metallurgy, magnets, ceramics and electronics, chemical, optical, medical, nuclear technologies, permanent magnetic, catalyst special alloy and abrasive materials (Fig. 1). Main use of rare earth materials is phosphor, the current scenario of the securities for rare earth mineral and functional materials at strategic aspect is mainly objective; necessity of stable supply quantities for related industries of rare earth materials.

2.5 decades ago the rare earth materials demand worldwide are about 33,000 tons. Now in the new millennium (year 2011) the demand reaches ~105,000 tons. For example fields like magnets (5 to 20%), phosphors (4 to 8%), ceramics (5 to 7%) are increased whereas others are slightly decreased also. Ceramics consists of mixed REOs (rare earth oxide's), phosphors having Eu, Y, Tb metals, magnets manufactured by Nd, Pr and Dy metals, glass/catalysts are made with Ce and La.

Although the demand for rare metals will further increase in future, the problem of balancing between medium and heavy rare metals with light ones will keep on make a trouble as before. The rare earth ore contains various metals in a nearly constant composition. Since rare earth cost level is likely to go up when, as a result of increasing demand for some specific rare metals the remaining ones are unfertilized, the rare earth manufacturers will find it difficult to maintain stability in the supply of large quantities. There is a need to develop a demand for utilizing rare metals or preferably a balanced demand for the available quantities of various rare metals.

Discussion

There is still another problem of distinguishing between the resources producing countries and resources consuming countries. Except America, which possesses a vast reserve, major rare earth consuming countries like Europe, Korea and Japan; as they don't have natural resources, it is difficult for them to guarantee either a stable price or a stable volume of rare earth supply. Therefore, it might be necessary in order to promote mutual exchange of information and market exploration between the producing countries and to establish international collaboration at every stage from the starting material to final product.

Properties of rare earth materials deal mainly uneven distribution of rare earth minerals in the world and preponderance of preparation technology. Difficulty of separation and purification technology for rare earth elements, because of similar properties as well as rare earth materials have a unique characteristic (magnetic, optics) because of electron configuration properties. Rare earth materials have a role of comparable to vitamin for high tech. industries and utilization of the additives for variable materials. Based on the report written by US geological survey reported on year 2011, the deposits of rare earth minerals available in China 48% tREO (total rare earth oxide), USA 11% tREO, India 3% tREO and Australia 1% tREO other countries 20% tREO. Production of tREO worldwide is about 125 thousand tons per year whereas consumption is about 111 thousand tons per year. 95% production is provided by China. The price trend of rare earth materials is presented in Fig. 2.

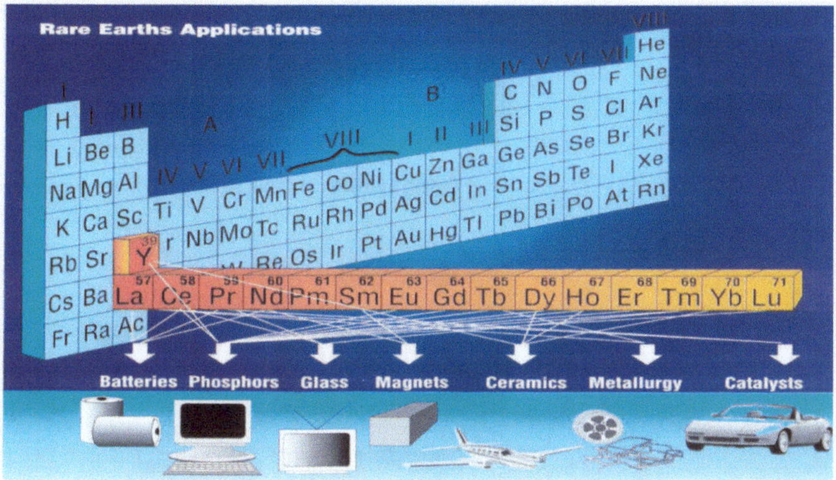

Fig. 1. Applications of rare earths in diversifying fields

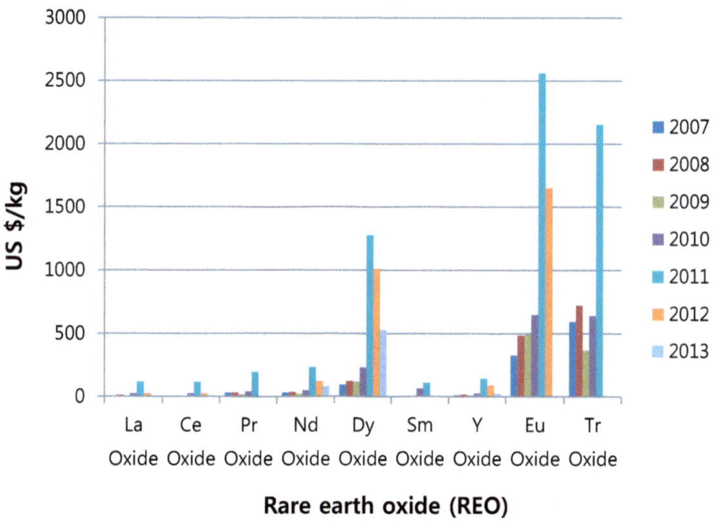

Fig. 2. The price trend of rare earth materials (REO basis : US $/kg (Oxide 99% up), Y&Eu Oxide 99.99%, Rare Earths & Yttrium: Market Outlook to 2015 [Roskill Information Services Ltd (2011)], [Asian metal(2013)]) [2]

Preparation technologies of rare earth materials fundamental and basic step are mineral dressing and metallurgical processing's the rare earth ore. The flowsheet begins with the crushing and grinding of the rare earth ore further mineral processing at this stage gangue minerals were separated then concentration of rare earth mineral adding acids or alkali's decomposition and leaching of the ore processed. By-products are and impurities separated the leach liquor further treated with proper extractants to extract the concern rare earth metal. The final step is to prepare the rare earth compounds with the addition of the reagents (Fig. 3).

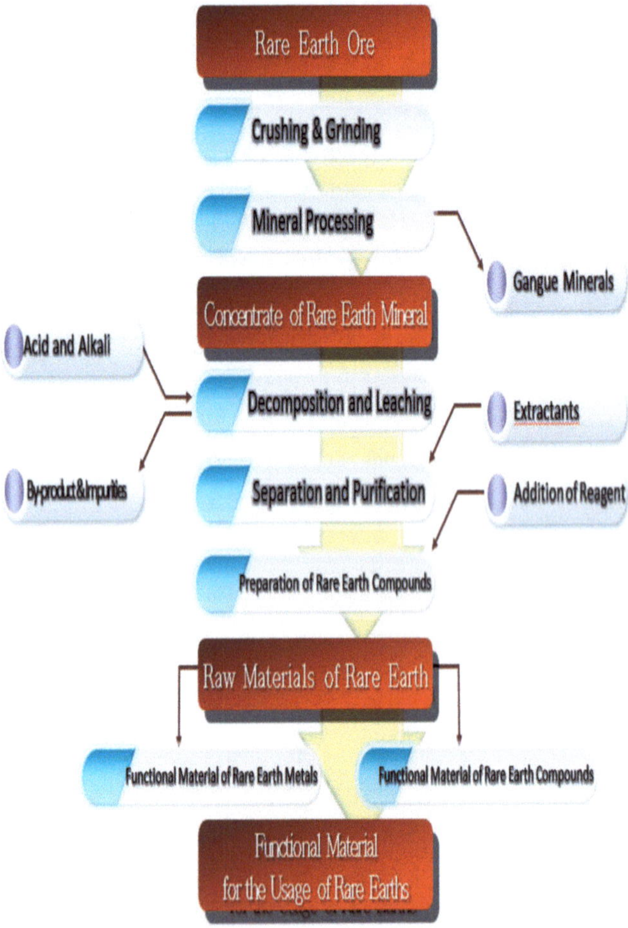

Fig. 3. Flowsheet of the rare earth ore processing for the functional materials for the use of rare earths

Variables and properties of rare earth ores of monazite, bastnasite, xenotime and ion adsorption clay are reproduced below. Monazite ore consists of light rare earth mineral and phosphate mineral, deposits mainly located in countries such as Australia, India, Brazil, Egypt and Malaysia etc and deposits of Korean sand mineral is 45,000 ton, 0.05% tREO. Bastnasite ore having light rare earth mineral and fluorocarbonate minerals, deposits located in China, USA, Africa, Brazil etc and mineral processing is difficult than monazite ore. Xenotime ore having phosphate mineral with a rich amount of yttrium elements and deposits located in Malaysia, China, Indonesia and Australia etc. Ion adsorption clay located only in China country, Jiangxi province having a high content of heavy rare earth elements.

Mineral processing of Korean ore

Mineral processing of rare earth ore deals the four basic techniques such as gravity separation (Jig and shaking table separator) its responsible for density difference, magnetic separation (Dry and wet magnetic separator)- it shows magnetic properties difference, electrostatic separation (Electrostatic separator) will shows the conductor and non-conductor difference, Final method is froth floatation separation (Froth flotator) will separate the ore material based on hydrophilic and hydrophobic properties difference. Mineral processing of Korean rare earth ore presented in Fig. 4. The total recovery rate of TREO is about 35% and grade is 50%.

Decomposition and leaching of concentrates

Decomposition and leaching process selective factor mainly depends on the following properties: type of properties of concentrate minerals, shape of final rare earth product, recovery and utilization of non-rare earth elements, consideration of operational condition and environmental contamination and finally economic aspects. Decomposition and leaching rate is ~92% (Fig. 5).

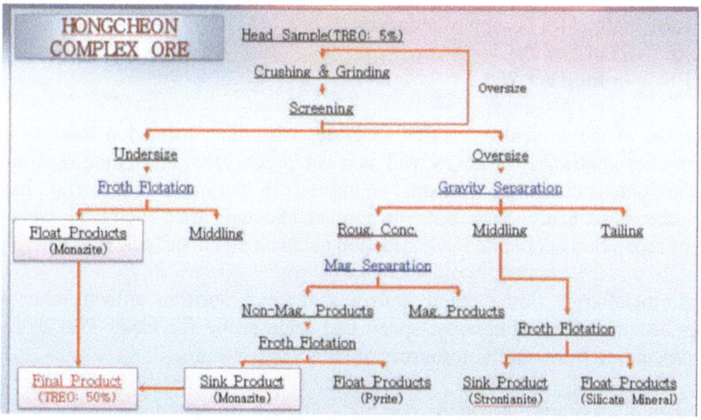

Fig. 4. Mineral processing of the Korean rare earth ore (Hongcheon Complex Ore)

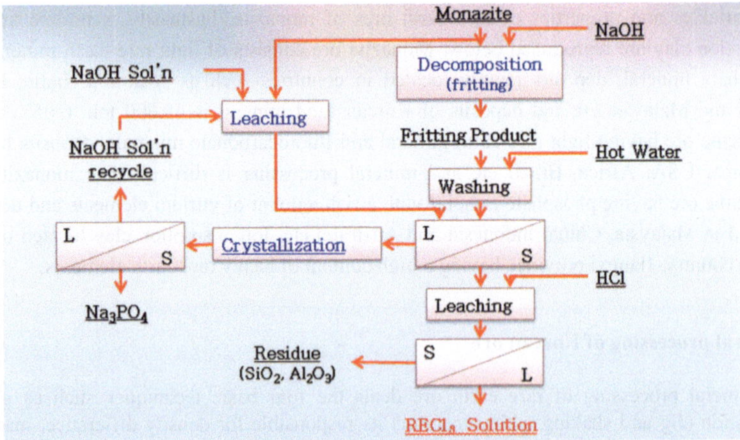

Fig. 5. Decomposition and leaching process of concentrate of monazite

Separation and purification of rare earth elements

Conventional separation methods of rare earth elements are fixed as fractional crystallization, fractional precipitation and selective oxidation/reduction methods. The first one can work on solubility difference of double salt compounds for rare earth elements; whereas, second one is solubility difference of double salt sulfate oxalate precipitation for rare earth elements. The third method is based on the oxidation and reduction process. For example, following equations represent the third methodology:

Oxidation method for cerium separation:

$$2Ce(OH)_3 + 1/2O_2 \rightarrow 2Ce(OH)_4 \qquad (1)$$

Zinc reduction method for europium recovery:

$$2EuCl_3 + Zn \rightarrow 2EuCl_2 + ZnCl_2 \qquad (2)$$

$$EuCl_2 + H_2SO_4 \rightarrow EuSO_4 + 2HCl \qquad (3)$$

An outline of the solvent extraction (SX) describes that, metal ion transportation from phase to another immiscible phase as well as it is a process that allows the separation of two or more components due to their unequal solubilities in two immiscible liquid phases. In SX processes, there are many cases both in applied and industrial chemistry were the main objective of separation is obtained by extraction using a reagent called extractant [3].

In hydrometallurgy, this method enjoys a favored position among other separation techniques because of its simplicity, speed and wide scope for metal ions extraction and recovery processing from natural resources or secondary resources.

By utilizing apparatus, no more complicated than a separatory funnel and requiring several minutes at most to perform. Extraction procedures offer much to the metallurgist. The solvent extraction process applied in metallurgy, which can be described simply by an equation [3, 4]:

$$M + \overline{E} = \overline{ME} \quad (4)$$

Where, M = Metal, $\overline{E}$ = Organic phase, $\overline{ME}$ = Loaded organic phase with metal

The distribution of metals from one phase to the other phase represents as "*D*" means distribution ratio *D* = metal concentration in organic phase [$M_{(Org)}$] / metal concentration in the aqueous phase [$M_{(Aq)}$], for the extraction process, whereas scrubbing and stripping processes the *D* is reverse (metal concentration in organic phase / metal concentration in aqueous phase).

$$D = M_{(Org)} / M_{(Aq)} \quad (5)$$

The separation factors (β) of the two or more metal ions calculate by using the desired metal ion distribution ratio / other metal ion distribution ratio.

$$\beta = \frac{D_A}{D_B} = \frac{C_{A(o)}/C_{A(a)}}{C_{B(o)}/C_{B(a)}} \quad (6)$$

The extractant requires following properties such as, good selectivity, low density, high extraction capacity, low viscosity and high ignition temperature, stable during storage or contact with acids/alkalies and very low or negligible solubility in the aqueous phase. For most commonly rare earths extraction processing's PC 88A (2-Ethylhexyl phosphonic acid mono-2-ethylhexyl ester) used as an extractant system (Fig. 6).

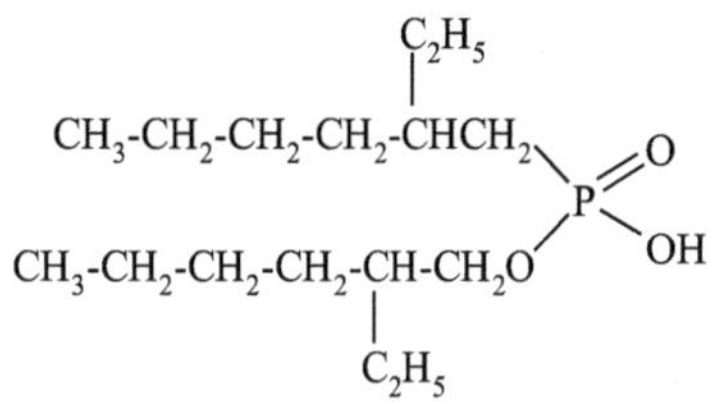

Fig. 6. The structure of 2-ethylhexyl phosphonic acid mono-2-ethylhexyl ester (PC 88A)

The overall REEs extraction by using PC 88A as an extractant system the following equation is the representative form:

$$REE^{n+} + nH_2A_2 = RE\,(HA_2)_n + nH^+ \quad (7)$$

Where REE = Rare earth element, H_2A_2 = dimeric form of PC 88A, *n* = number

The process design of solvent extraction requires multi-stage extraction process using mixer-settlers and counter current extraction (CCE) is called reflux extraction. Karr column, pulsed column and mixer settler are used as an apparatus for the solvent extraction process. Mixed rare earth chloride solution processing for rare earth metals procedure set forth in Fig. 7.

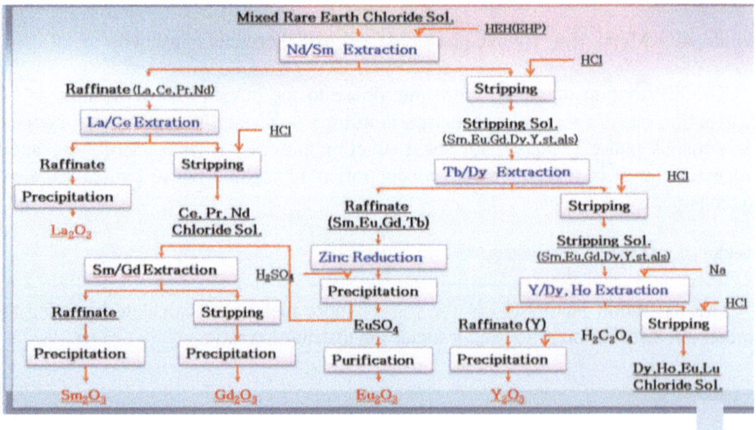

Fig. 7. Flowsheet for the separation of rare earth elements in Korea

An outline of the ion exchange method is the distribution of the ions (cation or anion) between a solution and a solid exchanger and it has been studied pas ~100 years but the theoretical aspects studied since a few decades ago [5]. Preparations of high purify materials this method can apply, but the productivity is low. The reaction equation for exchange method for cation and anion of resin is representing the following equation:

$$3RH + RE^{+3} = R_3RE + 3H^+ \quad (8)$$

Where, $R = -SO_3^-$

Preparation of rare earth oxide compound and metal

One of the most significant preparation technologies of rare earth materials is reaction crystallization method with oxalic acid. This method is originally, pH adjustment of rare earth solution obtained by purified process. The following reaction equations are the crystallized precipitation of rare earth solution with oxalic acid.

$$2RE^{3+} + 3C_2O_4^{-2} = RE_2(C_2O_4)_3\downarrow \quad (9)$$

$$2RE_2(C_2O_4)_3 + 3O_2 = 2RE_2O_3 + 12CO_2 \quad (10)$$

The experimental parameters is required to optimize for reaction crystallization method is process control of particle size, size distribution and shape according to acidity, oxalic acid concentration, temperature and agitation speed. The crystallization rate is ~99%.

Other important preparation technologies of rare earth materials are metal preparation of rare earth is the fused salt electrolysis (Fig. 8). The electrolysis of chloride (or fluoride) process is represented by the following equations. The efficiency of fused salt electrolysis is ~30%.

$$2RECl_3 \rightarrow 2RE^{3+} + 6Cl^- \quad (11)$$
$$2RE^{3+} + 6e^- \rightarrow 2RE_{(Cathode)} \quad (12)$$
$$6Cl^- \rightarrow 3Cl_2 + 6e^-_{(Anode)} \quad (13)$$

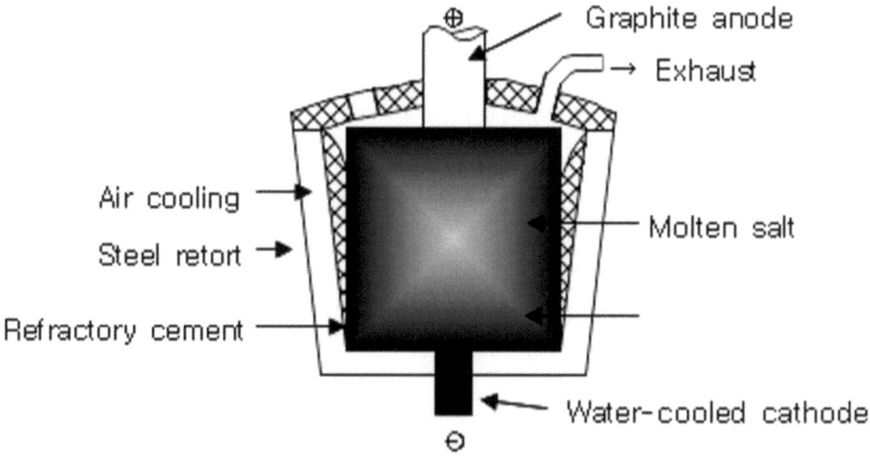

Fig. 8. Diagram of fused salt electrolysis process

Summary

The present general article summarizes about the status of separation and purification of rare earth elements from Korean ore by mineral and metallurgical routes. The applications of the rare earth materials required 33,000 tons in last millennium (year 1990) whereas in new millennium it raised to 105,000 tons means 3.18 times more; it concludes that the materials utilization growing rapidly. Korean ore mineral processing touched to 50% of TREO final product. Rare earth chloride leach liquors were prepared by decomposition and hydrochloric acid treatment. From the mixed rare earth chloride solutions various rare earth oxides were prepared by precipitation methodology. Fused salt electrolysis applied to the final metal preparation in Korea.

Acknowledgements

We would like to acknowledge the financial support from the R & D Convergence Program of MSIP (Ministry of Science, ICT and Future Planning) and NST (National Research Council of Science and Technology) of Republic of Korea (Grant B551179-11-01-00)

References

[1] Guptha, C.K., Krishnamurthy, N, (2004): Extractive metallurgy of rare earths, CRC Press, Boca Raton, USA.

[2] Roskill Information Services Ltd, (2011).

[3] Kumar, J.R., Lee, J.Y., (2010) Liquid-liquid extraction of tetravalent platinum, Verlag Dr. Muller Aktiengesellschaft & Co. KG, Saarbrucken, Germany (ISBN: 978-3-639-24889-0)

[4] Ritcey, G. M., Ashbrook, A. W. (1984) Solvent Extraction Principles and Applications to Process Metallurgy Part I, Elsevier, Amsterdam.

[5] Nachod, F.A., (1949), Ion exchange theory and application, Academic Press INC. Publishers, NY, USA

Rare Metal Technology 2015
Edited by: Neale R. Neelameggham, Shafiq Alam, Harald Oosterhof, Animesh Jha, David Dreisinger, and Shijie Wang
TMS (The Minerals, Metals & Materials Society), 2015

OPTIMIZATION OF RARE EARTH LEACHING

Grant Wallace[1], Sean Dudley[1], William Gleason[1], Courtney Young[1], Larry Twidwell[1], Jerome Downey[1], Hsing-Hsin Huang[1], Rod James[1], Ed Rosenberg[2]
[1]Montana Tech of The University of Montana; 1300 W. Park; Butte, MT, 59701, USA
[2]The University of Montana; 32 Campus Dr.; Missoula, MT, 59812, USA

Keywords: Earths, Leaching, Statistical Optimization

Abstract
The use of applied chemistry in the production and optimization of leach solutions from Rare Earth Element (REE) ores and concentrates is being investigated. Ore and concentrate samples were characterized using scanning electron microscopy/mineral liberation analysis (SEM/MLA), X-ray diffraction (XRD), and Atomic emission inductively-coupled plasma spectroscopy (ICP-AES). Multiple leach tests were performed to analyze the effects of temperature, residence time, and reagent concentration on the leaching of REEs. Analysis of leach solutions was carried out using ICP-AES. Modeling and statistical analysis of extraction behavior was carried out using DesignExpert 9. Modeling data for cerium extraction indicates that extraction is greatly influenced by temperature and reagent concentration, while leaching time plays a much less important role. Experimental design techniques are being utilized to optimize REE recovery. Results, conclusions, and directions for future studies will also be discussed.

Introduction
In recent years, the United States has begun to investigate the development of domestic sources or rare earth elements (REEs). Also known as the lanthanides, these elements have become highly valuable resources as they have become integral to numerous applications, such as the production of high-powered magnets, computer components, and defense technologies.

The fact that REEs are given the title "rare" is somewhat misleading. REEs are frequently found within the earth's crust, however, they are not often found in concentrations that make them economical to mine and process. In addition to this, the similarities in the chemical and physical properties of REEs make identifying and separating REEs a difficult task.

This study investigates the leaching of ore and concentrate samples originating from a proposed REE mine near Bear Lodge, WY. Three ore samples and three concentrate samples were provided. The concentrates have been labeled RE1, RE2, and RE3, while the ores have been labeled as RE4, RE5, and RE6. There is limited published data on direct REE leaching of ores or concentrates and little available on statistically optimized leaching processes. The goal of this study is to optimize REE extraction from the Bear Lodge samples via leaching with hydrochloric acid (HCl) as a precursor to the development of a cost-effective exctraction process and to demonstrate the use of statistical analysis software to the industry.

Characterization
To determine mineralogy for the various REE samples, each sample was examined using SEM-MLA. The characterization work was done using a LEO 1430VP SEM outfitted with two Ametek Apollo-40 EDS detectors. The MLA software utilized the X-ray Backscattered Electron (XBSE) method. This method relies on using variations in the gray-scale of backscattered

electrons to differentiate mineral phases. The X-rays gathered from these mineral phases are compared to a mineral X-ray database for identification.

Rare earth carbonates, primarily bastnasite and ancylite, were the overall largest source of REEs in the samples. Other REE-bearing minerals include cerianite, and monazite. The principle minerals composing each of the samples are presented in **Table 1.**

Table 1. Modal Mineralogy of REE Samples From MLA

Sample (Wt%)	RE1	RE2	RE3	RE4	RE5	RE6
Ancylite [1]	0.58	0.45	2.28	0.11	0.02	16.68
Bastnasite [2]	5.31	5.79	24.83	2.09	0.14	0.05
Biotite [3]	12.34	11.89	0.00	3.42	9.46	5.84
Calcite [4]	0.00	0.00	0.04	0.73	0.02	39.06
Cerianite [5]	5.88	5.31	0.01	2.19	0.48	0.02
Iron Oxides [6]	17.37	21.27	0.11	34.67	13.16	20.44
Monazite [7]	8.14	5.29	22.30	1.51	0.91	1.21

[1] $Sr(Ce,La)(CO_3)_2(OH)\cdot H2O$; [2] $(Ce,La)(CO_3)F$; [3] $K(Mg,Fe)_3(AlSi_3O_{10})(OH)_2$; [4] $CaCO_3$; [5] $(Ce,Th)O_2$; [6] FeO; [7] $(Ce,La)PO_4$

The REEminerals present are primarily sources of light REEs with the exception of trace amounts of xenotime, YPO_4, which is often associated with heavy REEs.

XRD was used in conjunction with SEM-MLA in the mineralogical characterization of the six Bear Lodge samples. A small amount of each sample was finely ground (-100micron) using a mortar and pestle. The sample was loosely packed onto a microscope slide. A Rigaku Ultima IV X-ray Diffractometer using Cu-Kα radiation at 40kVand 40mA was used to analyze the samples. The XRD pattern for sample RE1 is shown in **Figure 1.**

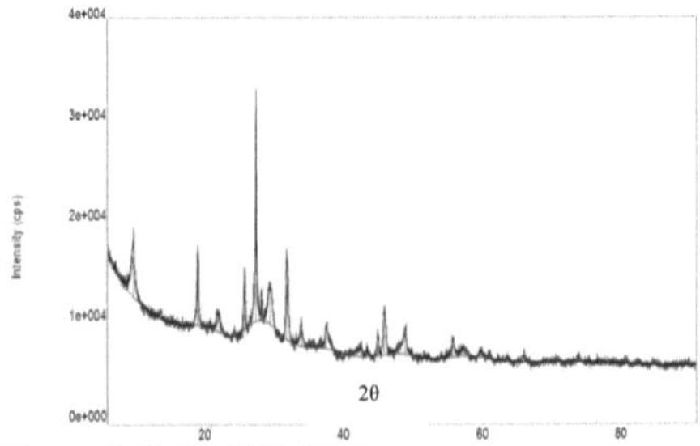

Figure 1. RE1 XRD Pattern

Analysis of the XRD spectra is done using PDXL software. From the analyses, the existence of the major phases, identified via MLA analysis, are confirmed.

Elemental analysis of the REE samples is determined by inductively-coupled plasma/atomic emission spectroscopy (ICP-AES) through the use of lithium tetraborate (LiB_4) fusions. 1.0g of LiB_4 was blended with 0.1g of sample/standard. Half of the LiB_4 was added to a graphite crucible, followed by the REE sample. These were blended together before the remaining half of the LiB_4 was added. The samples were placed in a furnace and fused at 1000°C for 15 minutes. Each LiB_4 bead was placed in 50mL of ICP blank (10%HCl, 5%HNO_3) and allowed to digest for 24 hours while being stirred using a shaker table. Samples were diluted with ICP blank solution by a factor of two and analyzed. **Table 2** shows the resulting ICP data for several key REEs. Samples were analyzed using an ICP Thermo-Scientific iCAP 6000.

Table 2. ICP-AES Values For REE Concentrations (wt %)

Sample	Ce	La	Nd	Eu	Gd	Dy	Pr
RE1	10.049	9.626	4.446	0.130	0.331	0.261	1.282
RE2	10.414	10.573	4.620	0.134	0.347	0.278	1.337
RE3	19.253	20.709	7.011	0.210	0.661	0.495	2.133
RE4	2.666	2.806	1.100	0.036	0.105	0.084	0.341
RE5	0.930	0.820	0.468	0.018	0.047	0.031	0.139
RE6	4.019	4.207	1.598	0.041	0.112	0.088	0.486

The results from the fusions were used to determine the amount of REE extraction experienced during the leaching experiments. Extraction is measured, in this study, as a ratio of the amount of REE leached into solution divided by the concentration of REE present in the sample according to the LiB_4 fusions.

Leaching Scope Testing

Leaching experimental design and statistical modeling are done using DesignExpert 9. This program was capable of producing statistically valid experimental design matrices, as well as modeling and analysis of the resulting data. Variables for the DesignExpert9 statistical analysis were temperature, time, and HCl concentration (g HCl/0.5g sample). To produce an effective design matrix, upper and lower boundaries for the three variables needed to be defined. Upper and lower boundaries for HCl concentration were defined using data produced by the MLA analysis of the RER samples in conjunction with stoichiometric calculations based on reactions involving the principle REE-bearing minerals and HCl. The minimum stoichiometric requirement of HCl was used as the lower boundary for the design matrix and the upper boundary was determined by multiplying this minimum requirement by a factor of 50. Upper and lower boundaries for temperature were set at 25°C and 90°C to prevent the leach solution from boiling. The upper and lower boundaries for time were set at 30 and 90 minutes based on the concept of mimicking industrial time constraints. The reactions for each of the REE minerals are as follows:

Bastnasite

$$CeCO_3F + 3HCl_{(aq)} = CeCl_3 + H_2O + CO_{2(g)} + HF_{(aq)}$$

Cerianite

$$CeO_2 + 4HCl_{(aq)} = CeCl_{4(aq)} + 2H_2O$$

Monazite

$$CePO_4 + 3HCl_{(aq)} = CeCl_{3(aq)} + H_3PO_{4(aq)}$$

Xenotime

$$YPO_4 + 3HCl_{(aq)} = YCl_{3(aq)} + H_3PO_{4(aq)}$$

The scoping trials consisted of ten individual tests. Five of these tests were carried out on RE1 to measure the effect on REE concentrates. Another five tests were carried out on RE4 to measure the effect on ore samples. For each scoping test, 37.35% by weight HCl solution was used as the leaching agent. **Table 3** shows the scoping test matrices for RE1 and RE4.

Table 3. Scoping Test Design Matrix

Test no.	Temp (°C)	g HCl/0.5g Sample
1	25	0.2
2	90	0.2
3	25	10.0
4	90	10.0
5	60	5.1

Solutions were analyzed using ICP-AES. The iCAP 6000 was calibrated for REEs using multi-element REE standards at 10, 50, and 100 ppm concentrations.

Various elements responded differently to the different sets of conditions. Enough variation was present between the different scoping tests to support that 0.2g and 10g HCl would serve as suitable high and low boundaries for a statistical design matrix. The amount of variation indicates that it is possible to affect REE extraction by changing the given parameters (temperature, time, and concentration). Statistical design matrices are appropriate for extraction optimization.

Design Matrix Testing

An experimental design matrix for RE1 was prepared using DesignExpert 9. The matrix consisted of 15 individual experiments. The three variables analyzed by the design matrix are reagent concentration (gHCl/0.5g sample), temperature (°C), and leaching time (min). The design matrix used to analyze the extraction of REEs from RE1 is shown in **Table 4.**

Table 4. DesignExpert 9 Matrix

Std	gHCl/0.5g solid	Temp (deg C)	Time (min)
1	10.0	90.0	30.0
2	10.0	25.0	120.0
3	0.2	90.0	120.0
4	0.2	25.0	30.0
5	0.2	57.5	75.0
6	10.0	57.5	75.0
7	5.1	25.0	75.0
8	5.1	90.0	75.0
9	5.1	57.5	30.0
10	5.1	57.5	120.0
11	5.1	57.5	75.0
12	5.1	57.5	75.0
13	5.1	57.5	75.0
14	5.1	57.5	75.0
15	5.1	57.5	75.0

In addition to Ce, the matrix also included La, Eu, Nd, Pr, Th, Y, Gd, Fe, and Ca extraction values as responses that are capable of being individually analyzed and modeled.

The experiments carried out at 25°C were agitated using a shaker table. Elevated temperature experiments were carried out on a hot plate and agitated using a magnetic stirrer. Experiments 5, 6, and 11-15 were performed simultaneously using a Cole-Parmer multi-stage hot plate.

Modeling Results

A model for Ce is produced based on a square root transform of the data. ICP data is modeled using a modified quadratic relationship. The program models the effect of the three variables individually and shows the interactions between variables. The model can be expressed as an equation which is used to fit the data.

$$\sqrt{Ce\ Extraction} = -0.1674 + 0.0968A + 0.0095B + 0.000289C - 0.00068AB + 0.00041AC - 7.034 \times 10^{-5}BC - 0.00814A^2 + 4.497 \times 10^{-5}B^2 + 7.157 \times 10^{-6}ABC$$

$$A = reagent\ concentration\ (gHCl/0.5g\ solids)$$
$$B = Temperature\ (^\circ\mathrm{C})$$
$$C = Time\ (\mathrm{minutes})$$

A series of statistical diagnostics were carried out to determine how well the experimental extraction data fit the mathematical model. Examples of these diagnostics are provided.

A graph of the data as predicted by the model vs. experimentally determined values for Ce is shown in **Figure 2**.

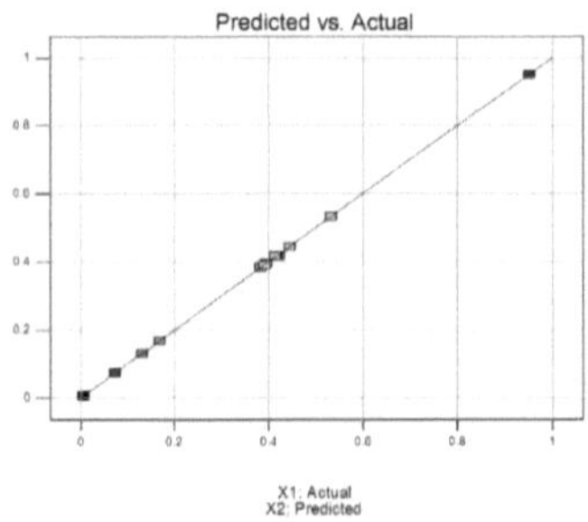

Figure 2. Predicted vs Actual Plot (Ce)

The individual points on the graph in **Figure 2** represent the experimental data while the line represents the values predicted by the model equation. The greater the amount of deviation from the line a data set experiences, the greater the possibility that the model is not acceptable.

DesignExpert 9 was able to provide further graphical analyses. The data was expressed in the form of three-dimensional response curves that are shown in **Figures 3** and **4.** Each response curve shows the effect on REE extraction from the variation of two variables, while the third variable is held constant.

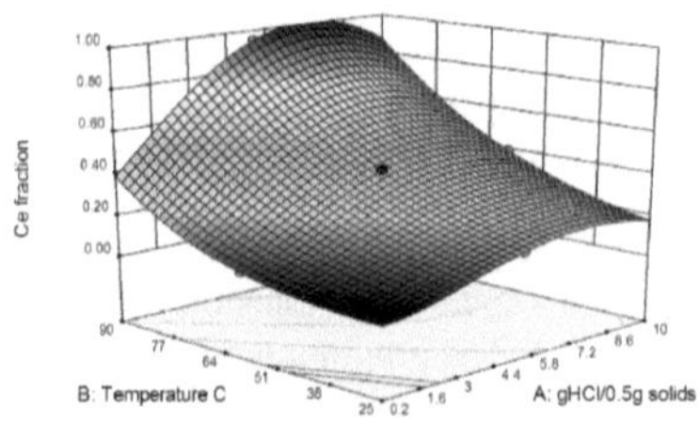

Figure 3. Ce Response Surface Plot (75min)

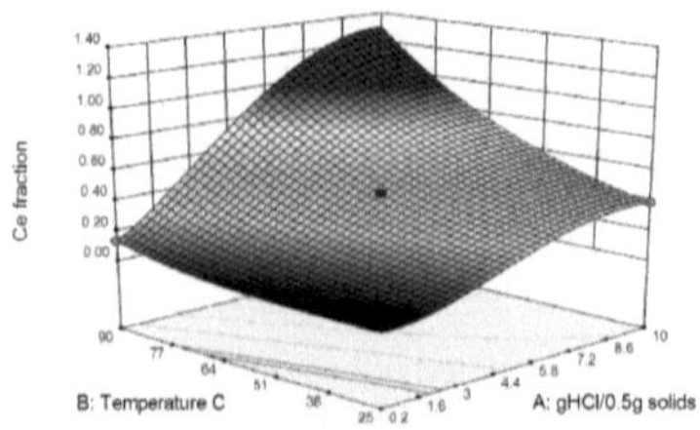

Figure 4. Ce Response Surface Plot (120min)

In the examples presented here, Ce extraction is plotted on the z-axis while temperature and reagent concentration are plotted on the x and y-axes. In **Figures 3** and **4**, time is held constant at

75 and 120 minutes respectively. The point in the center of each graph is a representation of the midpoint conditions and the Ce response associated with those parameters.

Discussion

From the model, it is possible to observe that thermodynamics (temperature) and reaction mechanics (reagent concentration) have the greatest effect on Ce extraction. Kinetics (time), does not have as great of an effect; however, it is still significant when considered in conjunction with the effects of temperature and reagent concentration. From the response surface plots, maximum Ce recovery occurs at high reagent concentration and temperature. The variation in the response surface plots also indicates that optimization of Ce extraction is possible.

The Predicted vs Actual plot shown in **Figure 2** validates the model used to predict the behavior of Ce extraction under the experimental design parameters. The alignment of the data points along the predicted vs. experimentally determined line demonstrates that the model fits the data very well. Additional diagnostics not presented here further support the use of this model as a means of predicting leaching behavior.

Optimization parameters were determined based on the DesignExpert 9 model. A range of values for reagent concentration and temperature were determined that could produce Ce extraction values that are greater than 0.75 and should equate to approximate recovery. According to the design matrix data, Ce extraction could exceed 0.75 provided HCl concentrations exceed 5g HCl/0.5g RE1 and temperatures remain above 75°C. Time must be held at 120 minutes.

The results show that it is possible to statistically optimize the extraction of individual REEs from the Bear Lodge samples.

Future Work

Due to difficulties associated with characterizing REEs, an error analysis of the sample assays (LiB_4 fusions), will be carried out. In addition, a sample of each of the concentrates and ores will be sent for analysis using ICP-MS. Mercury porosimetry and particle size analysis via Coulter Counter will also be carried out on each of the Bear Lodge samples. DSC calorimetry data, previously gathered, will also be analyzed and interpreted for enhanced characterization of the Bear Lodge samples. Finally, X-ray fluorescence will be conducted to provide additional support to the MLA and XRD data.

Improvements will be made to the data models by adding an additional five experiments to the design matrix. Response surface analyses will be conducted for each of the elements observed in the RE1 design matrix and the remaining Bear Lodge samples. Once the design matrices have been produced, it will then be possible to compile all of the data into a historical analysis matrix that will be able to model the data using material type as a fourth variable. This will allow for a comparison to be made between ores and concentrates. Optimization conditions will be determined using DesginExpert 9 and these conditions will be used in larger-scale experiments.

Acknowledgements

The authors would like to thank the Office of Naval Research for providing support and funding for this project through research grant N000141210592: Technical Progress Report on REE Selective Processing by Leaching and Chelating SPCs. The authors would also like to acknowledge the support of Rare Element Resources who supplied the ore and concentrate samples used in this project.

References

Chi, R., Tian, J., Zhu, G., Wu, Y., Li, S., Wang, C., et al. (2006). Kinetics of rare earth leaching from a manganese-removed weathered rare-earth mud in hydrochloric acid solutions. *Separation Science and Technology,* (41), 1099-1113.

Chi, R., Zhang, X., Zhu, G., Zhou, Z. A., Wu, Y., Wang, C., et al. (2004). Recovery of rare earth from bastnasite by ammonium chloride roasting with fluorine deactivation. *Minerals Engineering,* (17), 1037-1043.

de Vasconcellos, M., da Rocha, S., Pedreira, W., Queiroz, C., & Abrao, A. (2006). Enrichment of yttrium from rare earth concentrate by ammonium carbonate leaching and peroxide precipitation. *Journal of Alloys and Compounds,* (418), 200-203.

Feng, X., Long, Z., Cui, D., Wang, L., Huang, X., & Zhang, G. (2013). Kinetics of rare earth leaching from roasted ore of bastnaesite with sulfuric acid. *Transactions of Nonferrous Metals Society of China, 23*(3), 849-854.

Kandil, A., Aly, M., Moussa, A., Kamel, A., Gouda, M., & Kouraim, M. (2010). Column leaching of lanthanides from abu tartur phosphate ore with kinetic study. *Journal of Rare Earths, 28*(4), 576.

Kang, Z. C., Ting, Z. L., & Eyring, L. (1992). The preparation and characterization of hydroxycarbonate colloidal particles of individual and mixed rare earth elements. *Journal of Alloys and Compounds,* (181), 477-482.

Kim, W., Bae, I., Chae, S., & Shin, H. (2009). Mechanochemical decomposition of monazite to assist the extraction of rare earth elements. *Journal of Alloys and Compounds,* (486), 610-614.

Li, M., Zhang, X., Liu, Z., Hu, Y., Wang, M., Liu, J., et al. (2013). Kinetics of leaching fluoride from mixed rare earth concentrate with hydrochloric acid and aluminum chloride. *Hydrometallurgy, 140*(0), 71-76.

Parhi, P. K., Park, K. H., Nam, C. W., Park, J. T., & Barik, S. P. (2013). Extraction of rare earth metals from deep sea nodule using H_2SO_4 solution. *International Journal of Mineral Processing, 1*(19), 89-92.

Tian, J., Yin, J., Chi, R., Rao, G., Jiang, M., & Ouyang, K. (2010). Kinetics on leaching rare earth from the weathered crust elution-deposited ores with ammonium sulfate solution. *Hydrometallurgy,* (101), 166-170.

B. Kronholm, C. Anderson, P. R. Taylor. "A Review of Hydrometallurgy Applied to Rare Earths." *JOM* 2013, 65, 1321.

Parhi, P. K., K. H. Park, C. W. Nam, J. T. Park, and S. P. Barik. "Extraction of rare earth metals from deep sea nodule using H2SO4 solution." *International Journal of Mineral Processing* 119 (2013): 89-92. Web. 8 Sept. 2013

Rare Metal Technology 2015
Edited by: Neale R. Neelameggham, Shafiq Alam, Harald Oosterhof, Animesh Jha, David Dreisinger, and Shijie Wang
TMS (The Minerals, Metals & Materials Society), 2015

NUMERICAL SIMULATION OF THE MASS TRANSFER FOR RARE-EARTH CONCENTRATE IN LEACHING PROCESS

Tingyao Liu[1], Yong Sheng[1], Teng Yang[1], Bao Wang[1,2], Lihui Han[3]; Qing Liu[1]

[1]State Key Laboratory of Advanced Metallurgy, University of Science and Technology Beijing; 30 Xueyuan Road, Haidian District, Beijing, P. R. China
[2]Department of Engineering, University of Leicester; University Road, Leicester, United Kingdom
[3]School of Metallurgical and Ecological Engineering, University of Science and Technology Beijing

Keywords: Rare-Earth, Simulation, Mass Transfer, Leaching

Abstract

This paper mainly investigates the mass transfer phenomena of rare-earth concentrate from leaching process in Rushton stirred tank reactor with rotate speed of 100-600 rpm. In the research, a certain amount of sphere particles with 50-450 μm in diameter is treated as the mineral particles. Comparison of laboratory experiment data with numerical simulation is performed. The results show that the critical impeller speed for complete dispersion is 300 rpm. And the maximum value of turbulence kinetic energy appeared 15 mm away from paddle, where the mixing time is decreased 55% and 80%, respectively, when mineral particles are injected into stirred tank through one or four filling ports. Moreover, with mixing process going on, the numbers of particles near the paddle region accounts for as much as 58% of the total amount. Meanwhile, following the particle diameter increasing, the particle concentration of the upper part of stirred tank reactor is increased.

1 Introduction

Leaching is an important process of extracting in hydro-metallurgy of RE (rare earths) and many leaching methods have been used to extract RE, such as acid, alkali and salt leaching. These methods normally used stirred tank as the reaction equipment, for example, extracting RE from Baotou RE concentrate with stirred tank by L Mei in 2014 [1], and Feng X.L uses stirred tank to leach RE from roasted ore of bastnaesite with sulfuric acid in 2013 [2].The stirred tank has many advantages as leaching equipment: (1) It provides beneficial hydrodynamic condition for mineral grains to transfer; (2) It can effectively accelerate the chemical reaction; (3) It reduces pollution to protect environment because the stirred tank is airtight container. In early days, some researchers analyzed experiment and empirical data to determine stirred tank working condition [3, 4], but in recent year, more and more scholars have studied kinetics of fluid and transfer principle of particles to research leaching process [5].

The current paper is to study the mass transfer of RE concentrate in stirred tank by the numerical simulation. Firstly, comparing the experiment data to determine turbulence modeling; secondly, selecting drag formulation to accurately predict the dispersed mineral grains holdup distribution; finally, analyzing the particles number and position to reflect leaching results.

2 Mathematical Model

As is well known, flow in stirred tank is moving rapidly due to the interaction between fast rotating impellers and fluid, in the meantime, mineral grains bear the force mainly from gravity and turbulent flow, so choosing accurate turbulence modeling and force balance equation is the key to obtaining accurate simulation result.

2.1 The Turbulence Model

Because the rotation speed is normally above 100 rpm, and Reynolds number can surpass 1.4×10^4 in the leaching process, so the fluid moving can be described by Reynolds-averaged continuity and momentum equations [6,7].

Continuity equation:

$$\frac{\partial(\rho u_j)}{\partial x_j} = 0 \qquad (1)$$

Where, ρ is density (kg/m^3), u_j is velocity component in x_j direction (m/s), x_j is direction along different axis (m).

Momentum equation:

$$\rho\frac{\partial u_i u_j}{\partial x_j} = -\frac{\partial P}{\partial x_i} + \frac{\partial}{\partial x_j}(\mu\frac{\partial u_i}{\partial x_j} - \rho\overline{u_i' u_j'}) \qquad (2)$$

Where, P is the pressure (Pa), μ is kinematic viscosity (Pa·s), and $\rho\overline{u_i' u_j'}$ is Reynolds stress tensor. The Reynolds stress tensor is hardly obtaining accurate value due to turbulence complexity and randomness, a common method employs the Boussinesq hypothesis to relate the Reynolds stresses to the mean velocity gradients:

$$\rho\overline{u_i' u_j'} = u_t\left(\frac{\partial u_i}{\partial x_j} + \frac{\partial u_j}{\partial x_i}\right) - \frac{2}{3}\left(\rho k - \mu_t\frac{\partial u_k}{\partial x_k}\right)\delta_{ij} \qquad (3)$$

Where, u_t is turbulence viscosity (Pa·s), k is turbulence kinetic energy (m^2·s^2), δ_{ij} is Kronecker denotation.

Equation 3 shows that the governing equations are closed by the Boussinesq hypothesis, however, the hypothesis has a defect that turbulence is considered isotropy. Although this hypothesis can apply to most of turbulence environment, highly swirling and rotating flows like in stirred tank have obvious anisotropy. If the Boussinesq hypothesis is still used that may cause calculation error.

So researchers abandon the Boussinesq hypothesis and solve transport equations for the Reynolds stresses, together with an equation for the dissipation rate for close the governing equations, this method is called Reynolds Stress Model (RSM) [8].

Reynolds Stress Transport Equation:

$$\frac{\partial}{\partial \mathrm{t}}\left(\rho\overline{u_i u_j}\right)+\frac{\partial}{\partial \mathrm{k}}\left(\rho U_k\overline{u_i u_j}\right)=-\frac{\partial}{\partial x_k}\left[\rho\overline{u_i u_j u_k}+\overline{p\left(\delta_{kj}u_i+\delta_{ik}u_j\right)}\right]+\frac{\partial}{\partial \mathrm{k}}\left(\mu\frac{\partial}{\partial x_k}\overline{u_i u_j}\right)-$$
$$\rho\left(\overline{u_i u_j}\frac{\partial U_j}{\partial x_k}+\overline{u_j u_k}\frac{\partial U_i}{\partial x_k}\right)-\rho\beta\left(\mathrm{g}_i\overline{u_j\theta}+\mathrm{g}_j\overline{u_i\theta}\right)+p\left(\frac{\partial u_j}{\partial x_i}+\frac{\partial u_i}{\partial x_j}\right)-2\mu\overline{\frac{\partial u_j}{\partial x_i}\frac{\partial u_i}{\partial x_j}}-$$
$$2\rho\Omega_k\left(\overline{u_j u_m}\varepsilon_{ikm}+\overline{u_i u_m}\varepsilon_{jkm}\right) \tag{4}$$

Where, Ω is rate-of-rotation tensor; ε_{ij} is the dissipation tensor.

2.2 The Force Model

In order to analyze the force situation of mineral grains conveniently, the particle of RE concentrate is treated as spheres. As the mineral grains is immersed by solution, the particle moving through a fluid will experience a net force of the fluid on the particle, For objects symmetrical perpendicular to the upstream flow, this force will be in the direction of the free stream—a drag D_F, If the object is not symmetrical, there may also be a force normal to the free stream—a lift L_F [9].

Figure 1 shows free-body diagram of particle in flow. The RE particle mainly bear three kinds of force: The weight of the particle, W; The additional acceleration force such as buoyancy force, F_A; The drag force of the flow on the particle (Lift force is counteracted because of the particle is symmetrical), D_F

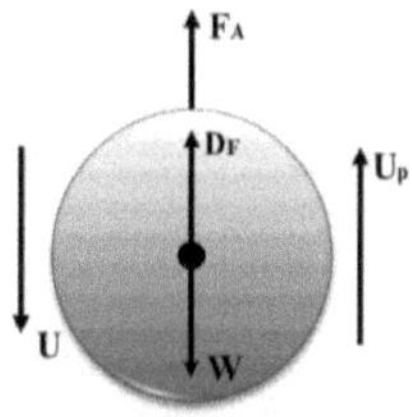

Fig.1 Free-body diagram of particle

On the basis of Newton's second law, this force balance equation can be described as follows:

$$\frac{dU_{\mathrm{p}}}{dt}=D_{\mathrm{F}}\left(\overrightarrow{U_{\mathrm{p}}}-\vec{U}\right)+\vec{g}\frac{(\rho_{\mathrm{p}}-\rho)}{\rho_{\mathrm{p}}}+\overrightarrow{F_{\mathrm{A}}} \tag{5}$$

Where, U_{p} is the RE particle velocity (m/s), U is the flow velocity (m/s), ρ_{p} is the RE particle density (Kg/m^3), ρ is the solution density (Kg/m^3) and D_{F} can be obtained as follows:

$$D_{\mathrm{F}}=\frac{18\mu}{\rho_{\mathrm{p}}D_{\mathrm{p}}^2}\frac{C_{\mathrm{D}}Re}{24} \tag{6}$$

Where, D_{p} is the RE particle diameter (m), Re is the relative Reynolds number, C_{D} is drag coefficient, as Barnea and Mizrahi model could exactly calculate the dispersed phase holdup distributions, the model employed in this paper is shown as follows [10]:

$$C_D = \left(1 + a_d^{\frac{1}{3}}\right)\left(0.63 + \frac{4.8}{\sqrt{Re_d}}\right)^2 \quad (7)$$

Where, a_d is the dispersed phase holdup; Re_d is the particle Reynolds number.

3 Experimental and Simulation Condition

The current study uses the standard six-blade Rushton stirred tank as leaching equipment. Figure 2 shows it is consisted by a cylindrical and flat-bottomed tank. In order to measure conveniently and safely, the solution is replaced by water and the liquid height is equal to the tank diameter. In the meantime, the flow speed data is measured by Laser-Doppler to compare with simulation results.

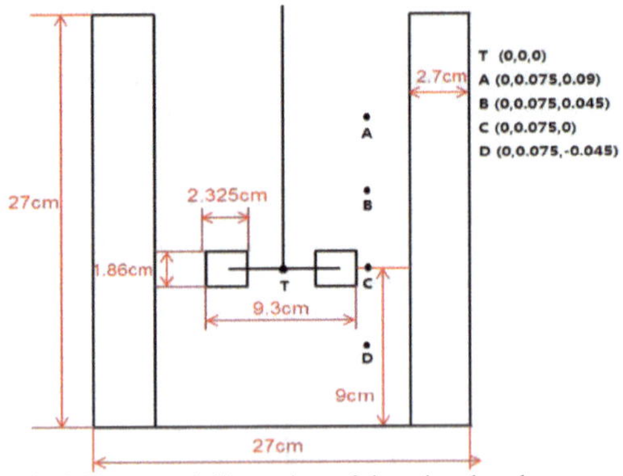

Fig.2 Shape and dimension of the stirred tank

Generally, the specific gravity of RE concentrates ranges from 4.3 to 5.9, and the particle diameter in leaching process is smaller than 300μm. Because this paper mainly focus on the mass transfer of particle in leaching process, the particle density and diameter are considered constant, and two hypothesis are adopted in this paper: (1) The particle is sphere and Non-deformation; (2) The fluid in the tank is an isothermal process with no mass transfer between chemical reactions and phases. The physical properties of particle are listed in Table Ⅰ.

Table I. Physical properties of particle

Parameter	Value
Diameter	150 (μm)
Density	4.8×10^3 (Kg/m^3)
Deformation rate	— (%)

4 Results and discussion

The mass transfer of RE particle is influenced by different factors in the tank, such as flow field, the mixing time and particle diameter. Through discussions and analysis of these factors, the current paper gets optimized solution about improve the efficiency of leaching process in the stirred tank.

4.1 Flow Field

The numerical simulation is used to simulate the fluid flow and particle motion. The case simulated here is with 200rpm rotate speed. In Figure 3, z/w is axial coordinate/ impeller blade width; U_r/U_{tip} is mean radial velocity/ impeller tip velocity. In Figure 4, $U\theta/U_{tip}$ is mean average tangential velocity/ impeller tip velocity. Fig.3 and Fig.4 indicate that employing Reynolds Stress Model can accurately predict the flow velocity in the tank. As is shown in Fig.3, mean radial velocity sharply increase when the flow is close to impeller and the values are approximately equal to zero when 2z/w is greater or less than ± 2.0. Fig.4 has the same situation, it indicates that sufficient stirring region is limited, therefore, increasing the rotate speed is hardly to improve the ability of stirring, and the flow filed is uniformity in the one-stage stirred tank.

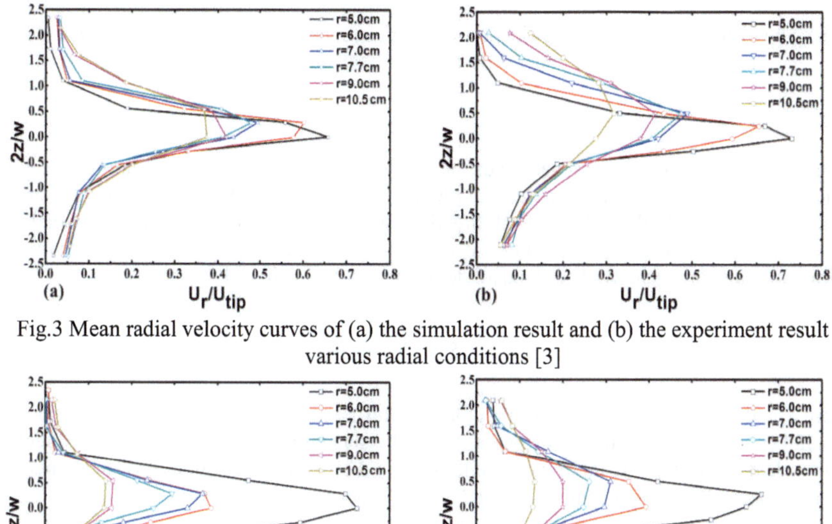

Fig.3 Mean radial velocity curves of (a) the simulation result and (b) the experiment result at various radial conditions [3]

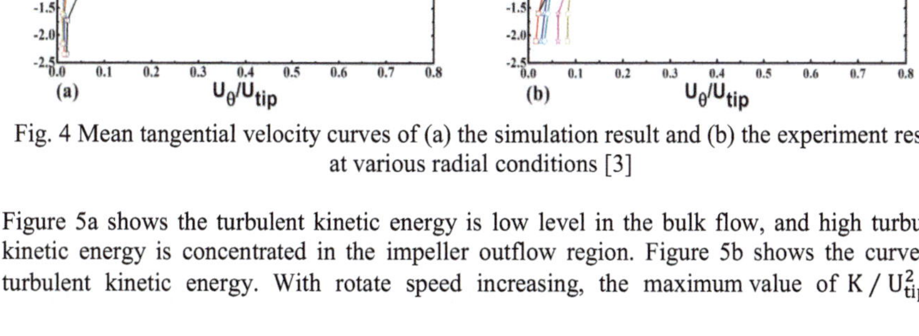

Fig. 4 Mean tangential velocity curves of (a) the simulation result and (b) the experiment result at various radial conditions [3]

Figure 5a shows the turbulent kinetic energy is low level in the bulk flow, and high turbulent kinetic energy is concentrated in the impeller outflow region. Figure 5b shows the curves of turbulent kinetic energy. With rotate speed increasing, the maximum value of K / U_{tip}^2 is significant changed. When rotate speed is 200rpm, the maximum value is less than 0.07, however, rotate speed is 1000rpm, the maximum value reached 1.7 and the position remained maintain 15 mm away from paddle, that means this place have optimum kinetic condition in the flow.

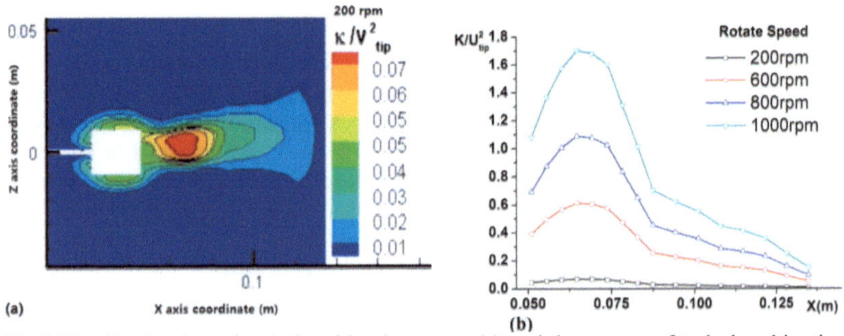

Fig.5 The distribution of turbulent kinetic energy (a) and the curves of turbulent kinetic energy under different rotational speed

4.2 Mixing Time

Mixing time is an important parameter indicator for leaching in the stirred tank, through counting the number and distribution of particles, the mixing time is obtained. Figure 6a shows the tank is divided into six floods along the vertical direction, and Figure 6b shows the tank is divided into four regions in the horizontal direction, at the same time, standard deviation is employed to indicate mixing homogeneity. The standard deviation can be calculated from equation 8:

$$S = \sqrt{\frac{\sum_i^N (X_i - X)^2}{N}} \tag{8}$$

Where X_i and X are particle number of i region for realistic and ideal condition, respectively, N is the number of divide region. In this paper, when the standard deviation of horizontal direction (S_H) reached the ninety percent of ideal value (V_i) and the standard deviation of vertical direction (S_V) is dynamically stable, this time is called mixing time.

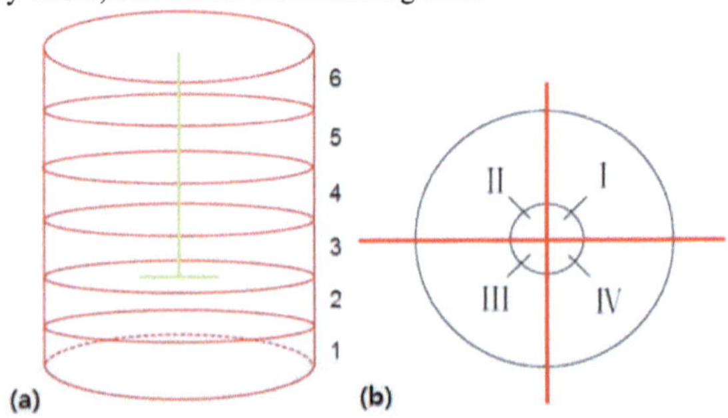

Fig.6 Division of the stirred tank in the vertical direction (a) and horizontal direction (b)

4.2.1 Effect of Injection Port Position. The injection port position influence the mixing time of particles because the flow field is non-uniform in the tank. At the beginning of the simulation, four injection ports (A~D) are setting up and five thousand RE particles are injected into the tank through each port. Fig.2 shows the coordinate values of these ports in detail. Figure 7b indicates that the track time is approximately 8s for port A and another is approximately 15s when S_V reaches dynamically stable. As is shown in Figure 7a, the track time of S_H reaching the V_i line is 18s for port C and 40s for port A, the mixing time can reduce above 55% if particles are injected into the tank from port C, since port C is located in the region where it has highest level of turbulent kinetic energy.

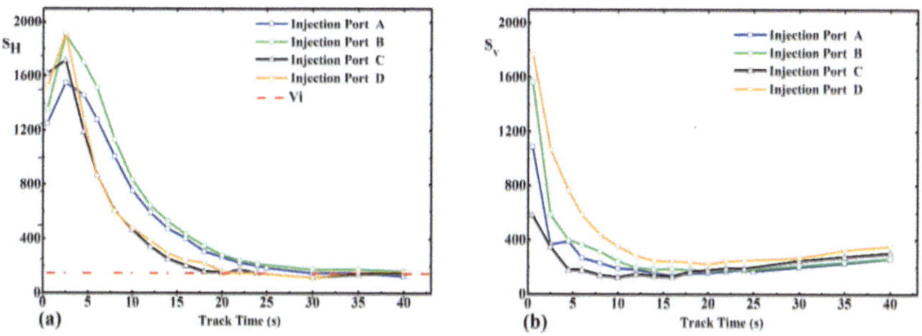

Fig.7 Standard deviation of (a) horizontal direction and (b) vertical direction under different position of injection ports

This paper also considers effect of multi-injection ports. The distance of agitating shaft and Z-axis coordinate values of these ports are equal to port C, multi-injection ports are symmetrically distributed. Figure 8b shows the results of different number ports are similar. Figure 8a shows multi-injection port is superior to single-injection port because symmetrical injection port can make particles have well distribution at the beginning, and mixing time of multi-injection ports are only 8s, comparing other single-injection ports, saving the mixing time up to 80%. In conclusion, the study indicates that changing the number and position of injection port can effectively reduce S_H to shorten the mixing time in leaching process.

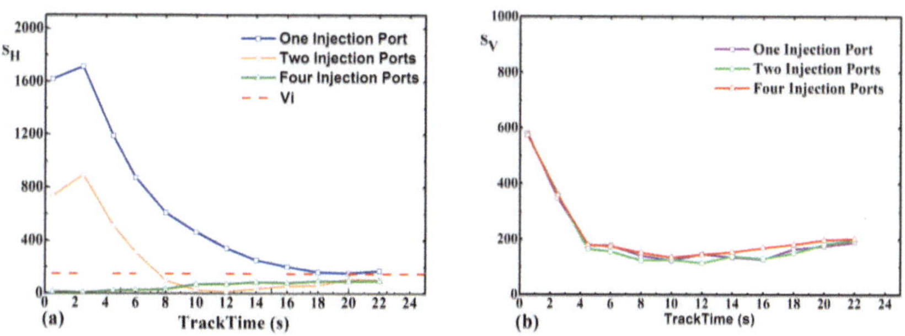

Fig.8 Standard deviation of (a) horizontal direction and (b) vertical direction under different the number of injection ports

<u>4.2.2 Effect of Rotate Speed.</u> It is well known that increasing the rotate speed can obtain well mixing results. Figure 9 shows that the curve is declined slowly with 100rpm, when rotate speed is 200rpm or 300rpm, S_V can quickly reach dynamically stable in 13s, and with the rotate speed increasing, the distribution of particles become disordered. So the rotate speed has critical value for complete dispersion and the critical impeller speed is approximately 300rpm in this case.

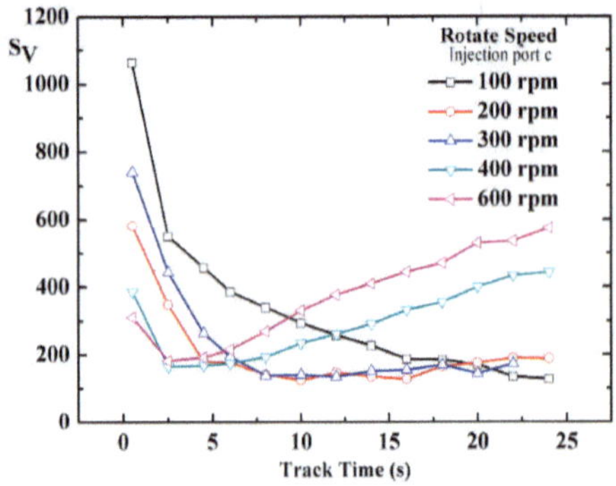

Fig.9 Profiles of standard deviation of vertical direction with different rotate speeds

<u>4.3 Particle Diameter</u>

RE particle has a wide range of diameters due to different production processes. In the current simulation, five hundred particles with five different diameters are injected into the tank. Figure 10 shows the distribution of particles at the mixing time, and Fig.6a shows the paddle exists in floor 3. Fig.10 shows the percentage of particles approach the maximum near the paddle. Especially, the percentage is as much as 58% when diameter of particle is 450μm, for the flow turbulence intensity of near the paddle is higher than other regions, and more particles are drifted into this place, therefore, using multi-stage stirred tank to make the flow homogeneity and the distribution of particles are reasonable. Fig.10 also indicates that following the particle diameter increasing, the particle concentration of the upper part of stirred tank reactor is increased, that means large particles will lead to distribution become confused.

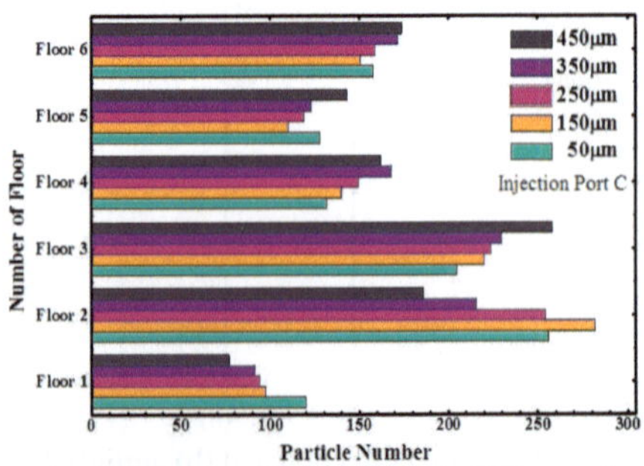

Fig.10 The distribution of particle with different diameters

5 Summary

A good mass transfer of RE particle phenomenon is beneficial to leaching process, mixing time and degree are important indicators for efficiency and quality of production. The study results show the RSM can accurately simulate the flow field in the tank, and maximum value of turbulence kinetic energy appeared 15 mm away from paddle, hence setting up multi-injection ports in this region can effectively reduce mixing time. The study also indicates critical impeller speed for complete dispersion in leaching process, and good homogeneity of distribution under critical impeller speed. Meanwhile, multi-stage stirred tank can improve mass transfer of particles, and small diameter particle not only add surface area for reaction, but also decrease partial segregation of particles distribution.

Acknowledgement

The authors are grateful to Ministry of Science and Technology of the People's Republic of China for providing National High Technology Research and Development Program of China (No.2013AA065703).

References

1. LI Mei et al. "Extracting rare earth from Baotou rare earth concentrate with salt leaching method," *Journal of Central South University (Science and Technology)*, 45(6) (2014), 1759-1765.

2. Feng X et al. "Kinetics of rare earth leaching from roasted ore of bastnasite with sulfuric acid," *Transactions of Nonferrous Metals Society of China,* 23(2) (2013), 849-854.

3. Wu H, Patterson G K. "Laser-Doppler measurements of turbulent-flow parameters in a stirred mixer," *Chemical Engineering Science*, 44(10) (1989), 2207-2221.

4. Jaworski Z, Nienow A W, and Dyster K N. "An LDA study of the turbulent flow field in a baffled vessel agitated by an axial, down-pumping hydrofoil impeller," *The Canadian Journal of Chemical Engineering*, 74(1)(1996), 3-15.

5. Sun X et al. "A Lagrangian–Lagrangian coupled method for three-dimensional solid–liquid flows involving free surfaces in a rotating cylindrical tank," *Chemical Engineering Journal*, 246(2014), 122-141.

6. J. O. Hinze. *Turbulence* (New York, NY: McGraw-Hill Publishing Co. 1975).

7. Xin Feng et al. "Numerical simulation of solid–liquid turbulent flow in a stirred tank with a two-phase explicit algebraic stress model," *Chemical Engineering Science*, 82(2012), 272-284.

8. Resende P R, Pinho F T, and Cruz D O. "A Reynolds stress model for turbulent flows of viscoelastic fluids," *Journal of Turbulence*, 14(12) (2013), 1-36.

9. Munson et al., *Fundamentals of Fluid Mechanics* (Hoboken, NY: John Wiley & Sons, Inc. 2009), 493-510.

10. Barnea E, Mizrahi. "A generalized approach to the fluid dynamics of particulate systems part 2," *The Canadian Journal of Chemical Engineering*, 53(5) (1975), 461-468.

Rare Metal Technology 2015
Edited by: Neale R. Neelameggham, Shafiq Alam, Harald Oosterhof, Animesh Jha, David Dreisinger, and Shijie Wang
TMS (The Minerals, Metals & Materials Society), 2015

APATITE CONCENTRATE, A POTENTIAL NEW SOURCE OF RARE EARTH ELEMENTS

Tianming Sun[1], Mark William Kennedy[2], Gabriella Tranell[2] and Ragnhild E. Aune[2]

[1]Department of Materials Science and Engineering, Royal Institute of Technology (KTH), Stockholm, SWEDEN
[2]Department of Materials Science and Engineering, Norwegian University of Science and Technology (NTNU), Trondheim, NORWAY

Keywords: Apatite, Rare earth elements, REE

Abstract

Rare Earth Elements (REEs) are critical for the production of many advanced materials and products. Mining and primary production of REEs are generally associated with significant environmental impacts. By recovery of REEs from secondary sources, through recycling or as a byproduct, the adverse environmental impacts may be partly avoided.

In the present study apatite concentrate originating from iron ore beneficiation done by LKAB in Kiruna, Sweden, is explored as a possible source for REEs. The composition and distribution of the REEs within the concentrates are studied using ICP-MS, XRD and SEM-EDS/EPMA, as well as the melting point and heat of fusion using DTA-TGA-MS. The behavior of the material during melting has also been studied through wetting and slag saturation experiments. The overall aim of the study is to evaluate the implications for future smelting experiments aimed at concentration and recovery of the REEs.

Introduction

Rare Earth Elements (REEs) are a group of 15 similar transition metals called "lanthanides", which also includes scandium and yttrium, due to their similar properties. All of these elements occur in nature except promethium (Pm). Rare earth elements are not in fact 'rare' on Earth. Some of them have similar or even higher abundances than commonly used materials such as copper. Although REEs occur in more than 200 types of minerals, in most of the minerals they can be recovered only as a coproduct or byproduct due to their low concentration[1]. Moreover, the concentration for different REEs in each mineral is quite different. There are three main type of minerals used commercially to produce rare earth metals: bastnasite, xenotime and monazite[1].

After huge rare earth resources were found in Inner Mongolia, China replaced the U.S. to become the largest REE supplier in the world during the late 1970's. Nowadays more than 90% of the REEs are exported from China[2]. Unfortunately, China has reduced the export of REEs, due to continued environmental damage and to encourage the development of alternative industries. This has created a potential for short-term gaps between supply and demand. On the other hand, with the increasing REE research intensity, the applications of these elements are being widely extended for both military and civilian uses. Since REEs are useful and important, and may face a future supply risk, both the E.U. and U.S. have listed REEs as critical strategic elements together with other elements, such as Indium (In) and Lithium (Li)[3, 4].

The development of secondary and sustainable resources would help to provide both strategically reliable sources and to bridge any supply gaps for REEs. China, along with Japan, the US and Europe has started research on the recycling of these elements. There are different groups of researchers using various methods trying to recycle REEs. Recycling of permanent RE-magnets has been a popular research topic. Hydro-metallurgical routes have been widely used by researchers[5-7], but due to the similarity to the processes used in primary REE extraction, they are expected to have similar environmental issues such as water pollution. Other methods are being explored including: (i) Electro-Slag-Refining (ESR) using molten salt to reduce the usage of harmful acids[8], (ii) liquid metal extraction[8], (iii) a glass slag method, which has been successful at the lab scale[9], and (iv) gas-phase extraction[10]. So far, there is no recycling method economically viable for commercial use. The main problems remain excessive energy use and toxic byproducts. None of the proposed methods, hydro- or pyro-metallurgical, use mineral byproducts (such as apatite concentrate) as raw materials.

REEcover, an European Commission funded project, is currently focused on the recovery of rare earth elements from magnetic waste fractions separated during Waste Electrical and Electronic (WEEE) recycling (a low volume and potentially high concentration secondary resource) and from apatite concentrate produced from iron ore tailings (a very high volume primary resource). This project has two main aims: (i) improve the reliability of supply of critical REEs, and (ii) strengthen the positions of Small and Medium Enterprises (SMEs) in REEs production and recovery throughout the value chain. Through cooperation between nearly 10 different industrial and academic partners, this project will determine the potential economic viability of REE recovery from the targeted sources. The REEcover project will study several pyro-metallurgical and hydro-metallurgical processing routes to produce either Rare Earth Oxides (REO) or Oxy-Carbides (REOC) for subsequent conversion to metal using electro-chemical means.

In the current study, apatite ($Ca_5(PO_4)_3(OH,F,Cl)$) concentrate produced by another REEcover work group from the Luleå University of Technology (LTU) using tailings originating from iron ore beneficiation done by LKAB in Kiruna, Sweden, has been utilized as the raw material. This concentrate has been analyzed and characterized for composition, structure and RE-elemental distribution by several different methods including ICP-MS[11], XRD and SEM-EDS/EPMA.

Preliminary experiments have been performed to study the behavior of the apatite during heating and melting, and to find suitable container materials for use during processing at larger scales. Wetting experiments were performed on a number of different substrates and gave a direct view of the melting and the reactivity of the apatite concentrate with different materials. DTA-TGA-MS experiments observed at which temperatures reactions such as calcining and melting took place, quantified the mass loss and heat flows, and allowed any potentially toxic product gases to be identified.

Due to the lack of accurate REE phase diagrams, a number of slag saturation/slow-cooling experiments were performed to attempt to concentrate the REE into new crystalline phases and to better understand the reactivity of the molten apatite to different refractory materials for use in later melting experiments. Pertinent results are combined in the plans for further work.

Experimental

Material Composition and Size

The general analyses of an apatite concentrate produced from Kirunavaara iron ore is shown in Table I. The particle size of the apatite concentrate after grinding was about 14-100 μm and the moisture content is 0.02%.

Table I. Apatite Concentrate Chemical Analysis Expressed as Simple Oxides[11]

Component	wt-%	Component	wt-%
SiO_2	2.04	MnO	0.31
Al_2O_3	0.23	Na_2O	0.16
CaO	51.10	P_2O_5	32.10
Fe_2O_3	1.77	TiO_2	0.07
K_2O	0.14	REE*	3517
MgO	1.45	Others	10.28

*Unit for REE is in mg/kg and is the mean value from two flotation tests

Applied Analytical Methods

X-ray diffraction (XRD) and scanning electron microscopy (SEM) are typically combined to identify the phases that exist in solid samples. In this work, the focus of the SEM/XRD work was to compare the phases (particularly those containing REEs), which existed before to those created during thermal treatment.

XRD

Phase determination was done by XRD using a powder diffractometer (Bruker D8 Advance A25) operating in Bragg Bretano mode, and utilizing Cu K-α radiation. A 2θ range of 5 to 70 degree and 30kV voltage was applied. Collected data were analyzed using the software DIFFRAC.EVA V3.2 from Bruker with the database PDF-4+ 2013 RDB.

SEM-EDS

The morphologies of the samples were studied by low voltage field emission SEM (Zeiss Supra 55 VP) combined with energy dispersive X-ray spectroscopy (EDS) and the accelerating voltage was set at 15kV.

EPMA

A JEOL JXA-8500F thermal field emission electron probe micro analyzer (EPMA), equipped with 5 wavelength dispersive X-ray spectrometers (WDS) and an energy dispersive X-ray spectrometer (EDS), was used to help identify the location of REEs in the different phases and provide overall elemental mapping using electron backscatter images. The accelerating voltage was 15 kV.

Heating Experiments

Three different types of heating experiments were applied to the apatite concentrate: (i) wetting, (ii) DTA-TGA-MS and (iii) slag saturation experiments.

Wetting Experiments

Several wetting experiments were done using a custom designed furnace. The schematic of the setup for this horizontal tube furnace is shown in Figure 1.

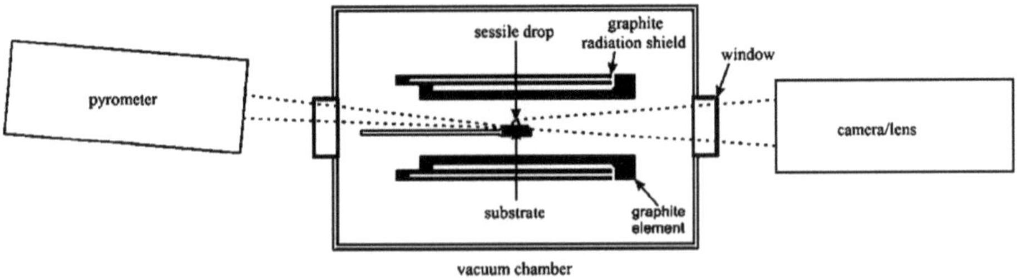

Figure 1. Wetting furnace schematic diagram

The sample holder sitting in the middle of the furnace is normally inserted from the left and the camera lens stays on the right. The apatite concentrate sample was pressed into a cylinder 4±0.2 mm in diameter and 5±1 mm in height to ensure the melted droplet did not fall off the substrate. The sample was then put on different substrates: (i) high purity graphite plate (ISO-88, Tanso), (ii) molybdenum foil, and (iii) tantalum foil. The foils are both 0.25±0.02 mm in thickness.

The wetting furnace used a graphite heating element and a graphite heat shield surrounding the chamber. This allows the sample to be heated to 1600°C in 1 minute under vacuum, or by using an inert or reducing gas. The maximum temperature of the furnace is 2400°C. A two-color pyrometer (Keller PZ40) was focused on the sample holder to measure the temperature in the furnace beginning from 900°C. A type 'C' thermocouple (tungsten-rhenium) was used to measure the sample temperature while <900°C.

The furnace chamber was evacuated and then argon (HiQ® Argon 6.0 supplied by AGA) was used with a flow rate of 0.5 L/min during the whole heating experiment. The oxygen content of the argon was further reduced by reaction with Mg chips at 400°C. An oxygen sensor (Cambridge Sensotec Rapidox 2100) was used to continually measure the partial pressure in the gas outlet and the oxygen partial pressure was maintained at $\sim 10^{-18}$ ppm throughout the entire experiment.

The apatite concentrate sample and substrate were quickly heated to 950°C at 300°C/min to reduce the opportunity for the substrate to oxidize. The material was then slowly heated at 10°C/min until the apatite concentrate was visually observed to have completely melted via the camera image. There was no defined peak temperature, because the true melting point was at this time unknown.

A Sony XCD-SX910CR digital video camera with a tele-centric lens (Navitar 1-50993D) was used to record images from the sample at 960×1280 pixels.

DTA-TGA-MS

Properties of the raw materials during heating were measured using a Setaram Setsys 2400 differential thermal and thermo gravimetric analysis (DTA-TGA) apparatus, also equipped with a Pfeiffer Quadrupole QMA 400 mass spectrometer (MS). This work was conducted in cooperation with the SINTEF Materials and Chemistry department. The maximum experiment temperature was 1700°C, so a glassy carbon furnace tube and type 'C' thermocouple (tungsten-rhenium) were used. From the results of the wetting experiments, molybdenum was chosen (due to the low reactivity) as the crucible for the DTA measurements of the apatite concentrate. The size of the crucibles used were 5±0.2 mm in outer diameter, 8±0.2 mm in height, 0.5±0.1 mm in wall thickness and 0.75±0.1 mm in bottom thickness. Sample weight was kept at 140.62±0.02 mg and the crucible without lid was held in a holder suspended in the furnace by a 0.5 mm tungsten wire. The furnace was first placed under vacuum and then backfilled and flushed with high purity argon (HiQ® Argon 6.0 supplied by AGA) at 20 mL/min during the whole experiment.

The first stage of heating used a 10°C/min heating rate up to 150°C, and was then held for 5 minutes to remove any moisture and equalize the temperature. The samples were then heated with the same heating rate until 1680°C and kept there for 30 minutes. The samples were subsequently cooled to room temperature at 50°C/min. A Faraday detector was used to measure the mass spectrum. Several potentially hazardous gaseous compounds: HF, CO, SO_2, PH_3, H_2S, HCl, Cl_2 and CO_2 were measured at 0.2 second intervals to ensure that they would not be created in future larger scale experiments.

Slag saturation experiments

Saturation experiments were conducted in an electrically heated high temperature muffle furnace (Nabertherm LHT 04/18). This furnace use pure alumina fiber as insulation materials and molybdenum disilicide as heating elements (Kanthal Super™ $MoSi_2$), which allowed the furnace has a theoretical maximum operating temperature of 1800°C and a full speed heating rate of 30°C/min. Although this furnace cannot achieve a completely oxygen free condition, different types of inert gas can be used such as Ar and N_2.

Apatite concentrate samples with a mass of 5±0.01 g were placed into Al_2O_3 (99.5%), SiO_2 (99.9%) and Al_2O_3-YSZ crucibles respectively. The dimensions of the crucibles were the same: 30±0.2 mm in outer diameter, 30±0.2 mm in height and 2±0.1 mm in thickness including the bottom. Then the crucibles with a lid made of the same material were placed in indentations in a high purity alumina (99.5%) plate with the thickness of (40±0.2 mm) to avoid damage to the furnace in the event of leakage.

These preliminary experiments were executed in an air atmosphere, and therefore no purging gas was utilized. Using the results from the DTA-TGA measurements, the temperature program was designed to allow for an initial drying, then calcination and subsequent melting periods: (i) fast heating of the sample at 10°C/min to 400°C then held for 15 minutes, (ii) continued heating at 10°C/min to 650°C then held for 30 minutes, (iii) slow heating at 5°C/min to 850°C then held for 30 minutes to complete calcination reactions; (iv) slow heating at 5°C/min up to 1650°C to achieve complete melting, then held for 4 hours to react with the crucible material, and finally (v) slow cooling at 5°C/min to room temperature. After cooling, the samples were taken out to determine weight loss and sectioned for characterization.

Results and Discussion

Initial Apatite Characterization

The XRD patterns identified that the predominant phases present in the apatite concentrate were fluorapatite, calcite and a silicate containing phase. SEM-EDS and EPMA, have identified three phases containing REEs, i.e. fluorapatite $Ca_5(PO_4)_3F$, allanite $(Ce,Nd)Al_2(SiO_4)(Si_2O_7)O(OH)$ and monazite $(Y,La,Ce,Nd)PO_4$ which is in good agreement with the results of the study previously published by Pålsson et al. of LTU[11].

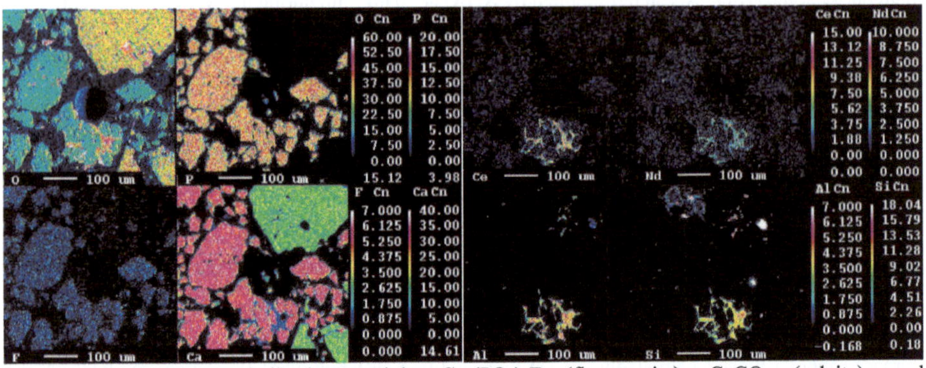

Figure 2. a) EPMA indicating mainly $Ca_5(PO_4)_3F$ (fluorapatite), $CaCO_3$ (calcite), and $(Ce,Nd)Al_2(SiO_4)(Si_2O_7)O(OH)$ (allanite).

Figure 2 a) shows an EPMA image of apatite concentrate showing mainly fluorapatite, calcite, and allanite. An SEM micrograph of the same area is shown in Figure 2 b). Other phases have been analyzed using EDS at the point locations indicated in red and are identified in Table II.

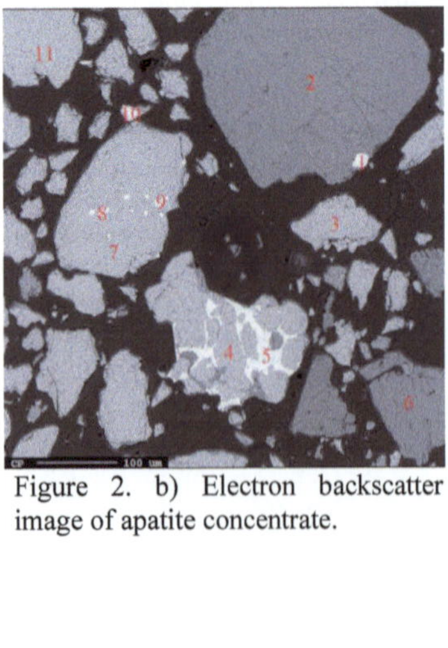

Figure 2. b) Electron backscatter image of apatite concentrate.

Table II. Phases Indicated in Figure 2 b) using EDS

	Fluorapatite	Allanite	Calcite	Iron-oxide
1				X
2			X	
3	X			
4	X			
5		X		
6			X	
7	X			
8	X			
9				X
10				X
11	X			

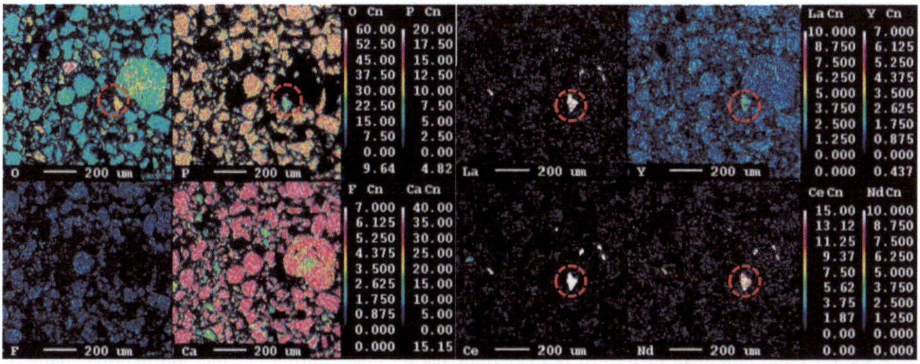

Figure 3. EPMA indicating (Y,La,Ce,Nd)PO_4 monazite phase (large bright particle in red circle and nearby small white dots).

In Figure 3) an EPMA image of a different area shows a large monazite (Y,La,Ce,Nd)PO_4 particle in the dotted red circle, as well as several nearby small monazite particles. The total percentage of La, Ce, Nd and Y is much higher (of the order of tens of percent) in the monazite and allanite phases than in the general apatite phase, raising hopes that it may be possible to melt, cool, precipitate and grow large separable crystals containing high concentrations of REEs for subsequent physical separation. The possibility to grow such crystals was the subject of slag saturation and slow cooling experiments.

Thermal Experiments

Wetting Experiments

Wetting experiments gave a visual indication of the melting of apatite concentrate, as well as the reactability of apatite with different substrates, the objective of which was to find suitable crucible materials for the subsequent DTA-TGA analyses. Moreover, the wetting experiment indicated a probable melting point of the sample and hence a target temperature for the subsequent thermal analyses.

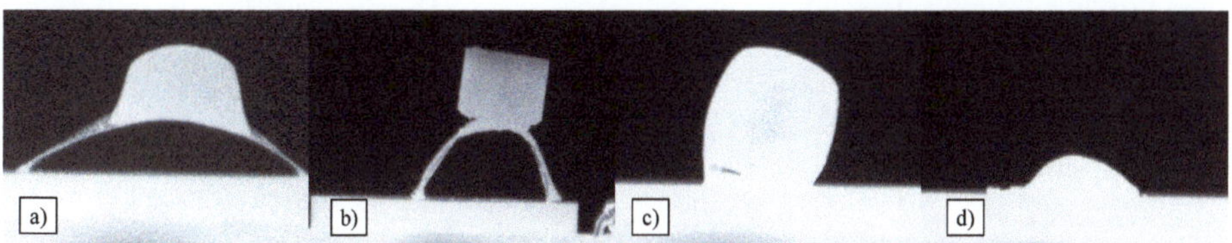

Figure 4. a) molybdenum foil as substrate at 1607°C; b) tantalum foil as substrate at 1313°C; c) graphite as substrate at 1601°C; d) same experiment as c) after 30 minutes at 1636°C.

Images were taken of the wetting experiments every 0.9 second. The images were then processed to create short movies of each experiment. From Figure 4 a) and b), it can be seen that both molybdenum and tantalum foil arched upward in different degrees. The tantalum foil bent more and the sample actually fell off before melting. Molybdenum was therefore selected as the crucible material for the subsequent DTA-TGA analyses.

According to literature, the melting points of various types of synthetic apatite should be of the order of 1608-1622°C[12]. All of the wetting experiments indicated a softening point (initial deformation) of apatite concentrate at ~1598±5°C. It should be noted that the concentrate being a 'natural' material was not pure apatite, and the presence of other species probably reduced the initial melting temperature. Before the sample completely melted (at the liquidus temperature), physical changes had already been observed inside the sample.

Figure 4 c) and d) show the experiment with apatite concentrate in direct contact with the graphite plate. The sample reacted quite significantly with the graphite. The samples were observed to release gas as shown in Figure 4 c) (grow and shrink rhythmically in volume) during melting. DTA-TGA was required to obtain a more complete picture of the reactions taking place during heating and melting. After 30 minutes, only a small amount of the melted sample remained on the substrate, as seen in Figure 4 d).

<u>DTA-TGA-MS</u>

Figure 5 shows the results of a representative DTA-TGA experiment for apatite in a molybdenum crucible. It can be observed from the figure that an endothermic reaction begins at 552°C and continues up to 962°C. Combined with an ~10% decrease in mass and the mass spectra results, this reaction appears to be the calcination of calcite, reaction (1), which has a normal ($\Delta G^0 = 0$) at 886°C. With this weight loss, the 'equivalent' calcite concentration was calculated as ~22% mass of the raw materials, i.e. all mass loss attributed to reaction (1). The reaction could begin earlier, due to the flushing with argon gas.

$$CaCO_3\ (s) \rightarrow CaO\ (s) + CO_2\ (g) \qquad (1)$$

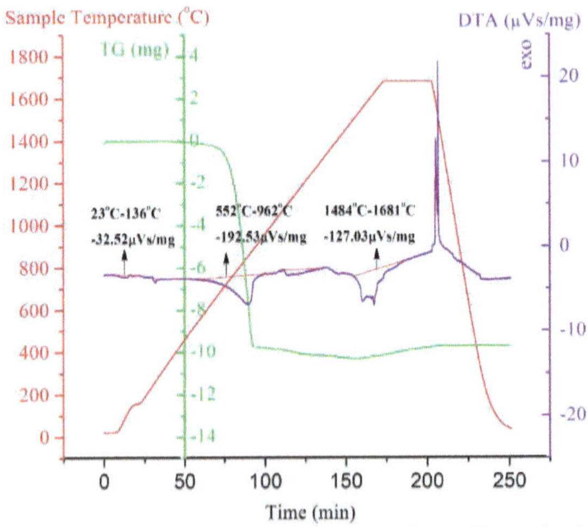

Figure 5. DTA temperature curve (red line and axis), TG curve (green line and axis) and heat flow curve vs time (purple curve and axis) for Apatite concentrate in a molybdenum crucible in flowing argon gas.

A second endothermic reaction started from ~1484°C and continued to 1681°C and represents the melting of the apatite concentrate. The heat flow went nearly to zero after 30 minutes indicating complete melting, similar to the wetting experimental results.

The behavior observed during the DTA-TGA experiments were then used to establish the temperature program used during the saturation experiments, having also established that toxic or damaging gas species (to the $MoSi_2$ heating elements) were not likely to be produced.

Saturation Experiments

Three preliminary saturation experiments have been performed with different crucible materials as the only variable. The results show a significant difference, as indicated in Figure 6.

Figure 6. Crucibles after saturation experiments. From left to right are YSZ, SiO_2 and Al_2O_3 crucibles. ZrO_2 and SiO_2 crucibles are embedded in part of the holding block, which was made of alumina.

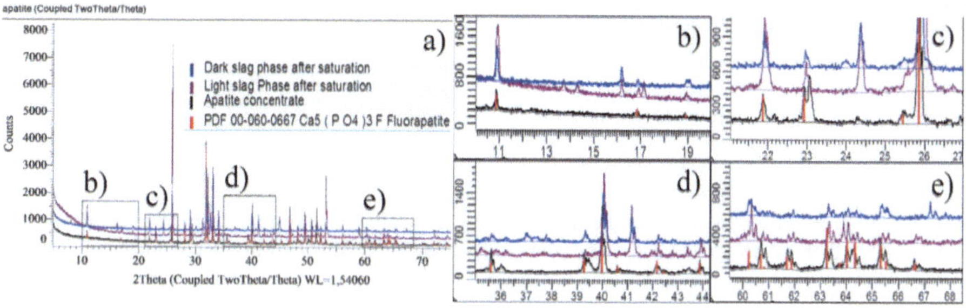

Figure 7. a) Comparison of XRD pattern from sample before and after the saturation experiment with in an Al_2O_3 crucible. b), c), d), e) are expanded images showing different 2θ ranges.

The experiment with a ZrO_2 crucible had an extensive reaction with the apatite essentially 'merging' with the original sample. The SiO_2 (99.9%) crucible not surprisingly produced a more amorphous glassy phase after reaction with the apatite. The transparent SiO_2 crucible became white due to the recrystallization during slow cooling. In addition, this product was quite brittle, which can be seen by the fractures in Figure 6. The last crucible in Figure 6 with Al_2O_3 (99.5%) was more robust as a crucible material and yielded a distinctly crystalline needle shaped product. Different phases appeared to be present, (i) a dark bottom phase and (ii) a yellowish top phase. There was also a recrystallized ring over the slag on the sides of the crucible.

During the saturation experiment using the Al_2O_3 crucible 0.36 g of the 5 g sample mass was lost, which is equivalent to 16 wt-% of calcite assuming all lost mass was CO_2 according to reaction (1). This is reasonable agreement with the 22% mass loss measured from the DTA-TGA-MS result considering the much smaller sample mass that was used previously (only ~146 mg).

The XRD results in Figure 7 and SEM-EDS results in Figure 8 both show changes in phases. Differences can be seen in some peaks in all of the 2θ ranges in Figure 7. While the slag had a visual color difference between the top and bottom portions, this was reflected in some differences in the XRD results. The 'dark' slag phase (blue curve) in Figure 7 still shows an apatite phase; however, the 'light' slag phase (purple curve) now indicates some new alumina containing phases. The calcite phase of the original apatite sample (black curve) in Figure 7 has disappeared in both the 'dark' and 'light' slag phases.

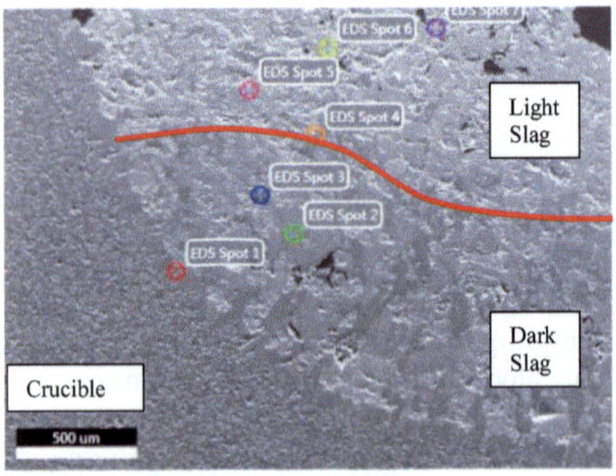

Figure 8. SEM-EDS analysis for slag saturation experiment using an Al_2O_3 crucible showing the crucible slag-interface and areas of dark and lighter coloured slag.

Table III. EDS Chemical Results for 'Spots' as indicated in Figure 8

Wt-%	O	Al	Si	P	Ca	F
Spot 1	45.24	48.27			6.94	
Spot 2	40.66	31.37	1.72		26.52	
Spot 3	46.07	47.15			6.78	
Spot 4	45.76	47.84			6.40	
Spot 5	38.77			19.19	41.99	0.04
Spot 6	38.97			19.00	41.93	0.09
Spot 7	38.15		0.41	19.13	42.16	0.15

From Figure 8, the different phases and crucible is easy to recognize. However, from EDS results in Table III, it was still difficult to detect REEs. Therefore, additional EPMA work must be conducted to determine the REE distribution.

Conclusions

This paper studied the high temperature behavior of apatite concentrate as a foundation for further experimental work. Three main REE (predominantly Y, La, Ce and Nd) containing phases where identified: (i) apatite, (ii) allanite and (iii) monazite. Ce and Nd were the predominant REEs in the allanite phase. Y, La, Ce, and Nd were all present in significant quantities in the monazite phase and Y was also widely dispersed in the apatite phase. If a procedure could be developed to grow phases (like monazite and allanite) then physical recovery of concentrated REEs might be possible. From the DTA-TGA-MS analysis, the melting point (liquidus) of this sample is ~1580°C. Calcination of calcite impurities (of the order of 20 wt-%) in the apatite with attendant release of CO_2 occurred over a temperature range of 552-

962°C in flowing argon. The mass spectrometry results did not indicate the formation of any toxic or corrosive compounds from the apatite during melting even under reducing conditions. The saturation experiment using an Al_2O_3 crucible showed promising results due to the formation of large new crystalline phases, which may facilitate physical upgrading of REEs. Al_2O_3 is also the most likely refractory material for use as a future crucible.

Future Work

Additional EPMA analyses must be performed to study the distribution of REEs in the recrystallized 'saturation' samples, as the EDS did not reveal the location of the REEs within the recrystallized phases. Further XRD analyses are also required to identify all of the new phases created during recrystallization.

Future experiments will focus on increasing the concentration of REE from apatite concentrate in the new crystalline phases. Meanwhile different saturation experiments will be conducted to identify the optimal conditions.

The potential for the removal of detrimental elements such as P by liquid metal extraction will also be studied.

Production of RECO by reaction with graphite will also be studied to see the possibility of produce RECO during the heat treatment.

Acknowledgments

The authors would like to thank the co-participants in the REEcover project for their assistance and the EU for funding. Thanks are given to: J.R. Tolchard for help during the XRD analyses, Y. Yu and M.P. Raanes for their advice and work with SEM-EDS and EPMA, T. Anzjøn and S. Bao for their help with the wetting and DTA-TGA-MS analyses and T. Schanche and D. Leroy for experimental assistance. The primary author also would like to give his thanks to the Chinese Scholarship Council (CSC) for funding his PhD research.

References

[1] C. K. Gupta and N. Krishnamurthy, *Extractive Metallurgy of Rare Earths*. (Boca Raton: CRC PRESS, 2005), 72-83.
[2] T.E. Graedel, J. Allwood, J.P. Birat, M. Buchert, C. Hageluken, B. K. Reck, and S. F. Sibley, "Recycling Rates of Metals," (UNEP 2011), 48.
[3] European Commission report, "Critical Raw Materials for the Eu," (Report of the Ad-hoc Working Group on defining critical raw materials, 2010), 33-41.
[4] U.S. DOE report, "Critical Materials Strategy," (2011), 113-119.
[5] J.W. Lyman, G.R. Palmer, and Murray, "Scrap Treatment Method for Rare Earth Transition Metal Alloys," U.S. Patent 5129945, (1992).
[6] T. Itakura, R. Sasai, and H. Itoh, "Resource Recovery from Nd–Fe–B Sintered Magnet by Hydrothermal Treatment," *Journal of Alloys and Compounds*, vol. 408–412, (2006), 1382-1385.
[7] Y. Baba, F. Kubota, N. kamiya, and M. Goto, "Recent Advances in Extraction and Separation of Rare-Earth Metals Using Ionic Liquids," *Journal of Chemical Engineering of Japan*, vol. 44, (2011), 679-685.

[8] T. W. Ellis, F. A. Schmidt, and L. L. Jones, "Methods and Opportunities in the Recycling of Rare Earth Based Materials," presented at the The Metallurgical Society (TMS) conference on high performance composites, Rosemont, IL (United States), (1994), 9.
[9] T. Saito, H. Sato, S. Ozawa, J. Yu, and T. Motegi, "The Extraction of Nd from Waste Nd–Fe–B Alloys by the Glass Slag Method," *Journal of Alloys and Compounds,* vol. 353, (2003), 189-193.
[10] K. Murase, K. Shinozaki, Y. Hirashima, K.-i. Machida, and Gin-yo. Adachi, "Rare Earth Separation Using a Chemical Vapour Transport Process Mediated by Vapour Complexes of the $LnCl_3$-$AlCl_3$ System," *Journal of Alloys and Compounds,* vol. 198, (1993), 31-38.
[11] B.I. Palsson, O. Martinsson, C. Wanhainen, and A. Fredriksson, "Unlocking Rare Earth Elements from European Apatite-Iron Ores," presented at the 1st European Rare Earth Resources Conference (ERES), Milos (Greece), (2014), 10.
[12] J.R.Pandya and B.J.Mehta, "The Melting Point of Synthetic Apatites," *Mineralogical Magazine,* vol. 37, (1969), 527-528.

Rare Metal Technology 2015
Edited by: Neale R. Neelameggham, Shafiq Alam, Harald Oosterhof, Animesh Jha, David Dreisinger, and Shijie Wang
TMS (The Minerals, Metals & Materials Society), 2015

RARE EARTH ELEMENTS GALLIUM AND YTTRIUM RECOVERY FROM (KC) KOREAN RED MUD SAMPLES BY SOLVENT EXTRACTION AND HEAVY METALS REMOVAL/STABILIZATION BY CARBONATION

Thenepalli Thriveni[1], Jyothi Rajesh Kumar[1], Ch. Ramakrishna[2], Yujin Jegal[1], Ji Whan Ahn[1]*

[1]Mineral Processing Division, Korea Institute of Geoscience and Mineral Resources (KIGAM), Gwahagno-124, Yuseongu-gu, Daejeon-305350.

[2]Department of Energy Resources Engineering, Inha University, Incheon.

Keywords: Rare earth elements, Yttrium recovery, solvent extraction, Korean red mud, heavy metals removal, carbonation

Abstract

In alumina production or alumina refining, bauxite is converted to alumina (aluminium oxide)

The rare-earth elements (REEs) are a group of 17 chemically similar metallic elements (15 lanthanides, including scandium and yttrium). Rare earth elements play a key role in many industrial applications, petroleum refining, clean energy, automobile, electronics, and defense applications. These rare earths are essential to the success of green technologies. China today controls more than 95 percent of the rare-earth supplies and nearly all rare-earth refining and after 2009 year rare earths crisis raises due to China's export restrictions and growing internal demand for rare earth elements. The extraction of gallium and yttrium from Korean red mud (KC red mud) by selective acid leaching was explored. The recovery of gallium and yttrium from a leach solution of the Korean red mud using solvent extraction was studied. A number of extractants were investigated for the extraction of those mentioned two rare earth elements and also we proposed a flow sheet of rare earth elements respectively. The toxic heavy metals presented in KC red mud were removed by CO_2 stabilization.

Introduction

The rare earths or rare-earth elements (REEs) are a group of 17 chemically similar metallic elements (15 lanthanides, including scandium and yttrium). Rare earth elements play an important role in many industrial, petroleum refining, clean energy, automobile, electronics etc. applications and in the military also, widely used for precision-guided munitions, communication systems, lasers, radar systems, avionics, night vision equipment and satellites. These rare earths are highly demanded and essential to the success of green technologies[1] (Fig.1). Figure.2 shows global demand of the rare earths.

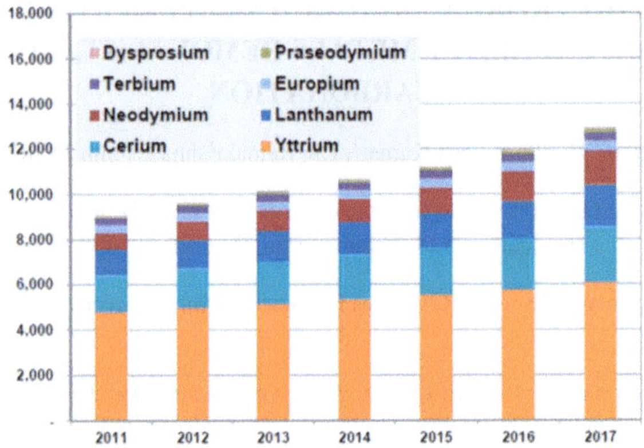

Figure.1. Rare earth metals demand for the cleantech industry, world markets 2011-2017 (Source: from Pike Research, 2011)

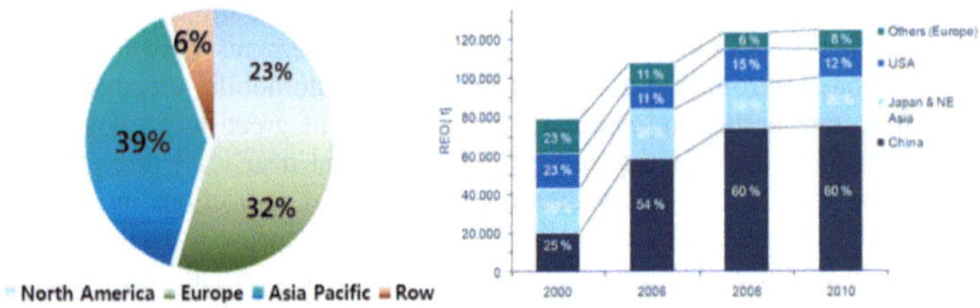

Figure.2. Share of rare earth metals demand by region, world markets 2017 (Source: Modified from Pike Research, 2011 & D. Kingsnorth, POLINARES working paperN.37, 2012)

China occupied about 90 - 95% of the world's rare earth oxides production especially two strongest magnets, samarium (Sm) and neodymium (Ne)[2].

Rare Earth Applications

The applications of rare earths are prominent in green technological areas. The use of REEs to combat global warming and improve energy efficiency has attracted significant attention. The use of several REEs in petroleum fluid cracking catalysts and automotive pollution control catalysts is well known. REE used in magnets reduces the weight of many pieces of equipment such as automobiles. Some REEs are used in in the reduction of carbon dioxide emissions, and have attracted noteworthy attention from public and governmental authorities.

Status of USA in Rare Earths Applications

According to the USGS survey, U.S. demand for rare earth elements are also projected to rise as per world demand continues to climb. Demand increases are also expected for rare earths in flat panel displays, hybrid vehicle engines, and defense from cruise missiles to missile guidance systems, smart bombs, night-vision technology and medical applications. The 2015 composition of U.S. and world demand is shown in Figure. 3.

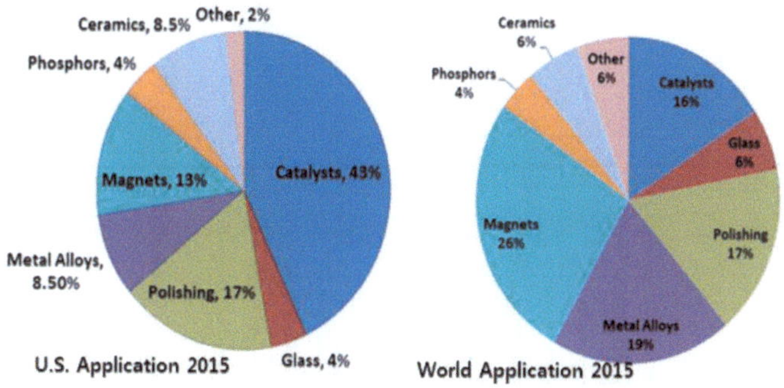

Figure.3. Rare earth metals demand by application U.S. and World, 2015 (Source: IMCOA, 2011)

South Korea

Situation in South Korea

South Korea has lack of rare earth deposits and imports from China. Recent china rare earth export limitation looking for alternative sources of rare earth elements.

Alternative sources for rare earth elements

Gallium and Yttrium were extracted from Korean red mud by solvent extraction. South Korea has found an undetermined amount of rare earth minerals in a deposit in the eastern

Gangwon province, state-run Korea Resources Corp (KORES). The South Korean state-run mining firm discovered veins containing rare earths while re-developing an iron ore mine.

Gallium Applications

Recently published article reviewed the importance of gallium in medical applications and its toxic effects. Over the past three decades, gallium compounds have gained importance in the fields of medicine and electronics[3] (Figure.4). In clinical medicine, radioactive gallium and stable gallium nitrate are used as diagnostic and therapeutic agents in cancer and disorders of calcium and bone metabolism. In addition, gallium compounds have displayed anti-inflammatory and immunosuppressive activity in animal models of human disease while more recent studies have shown that gallium compounds may function as antimicrobial agents against certain pathogens[3].

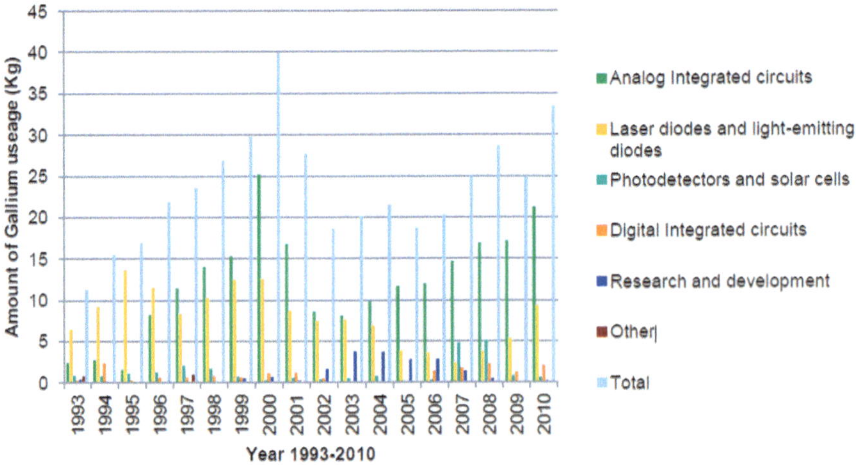

Figure.4. Principal Uses of Gallium in United States (1993-2010)

Extraction of Gallium from Korean Redmud

We are extracted Gallium from Korean redmud (KC red mud) by applying lime method[4] (Carbonation process). We followed the procedure method (Reference no.4) and here we are presenting the carbonation method. In this we are investigated some experimental parameters including temperature and pH.

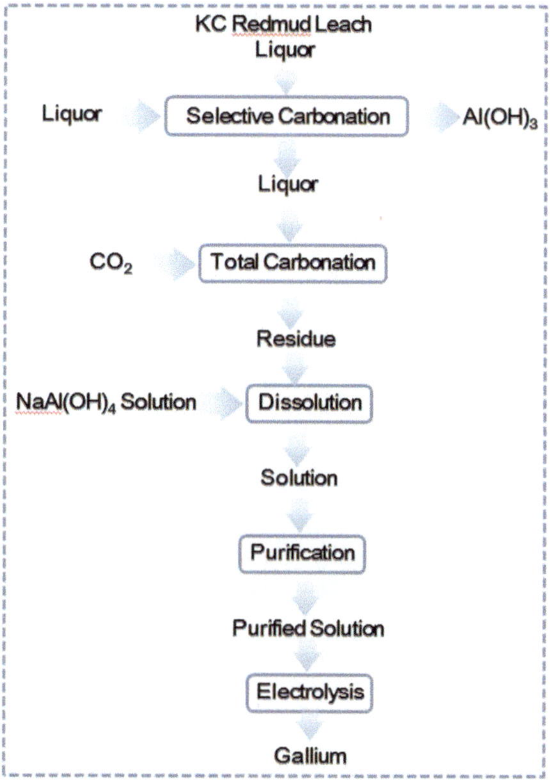

Figure 5. The flow sheet of carbonation method

By the carbonation method, we are successfully gallium was extracted 99% from Korean redmud.

Yttrium Global demand and its Applications

Yttrium oxide (Y_2O_3) is the critical enabler of energy efficient lamps, ubiquitous electronic devices, and other advanced materials used in defense, aerospace, energy infrastructure, and medical applications. Critical as it may be, the market is being hampered by a fog of uncertainty because of China's hegemony over global Y_2O_3 production[5].

Yttrium is quite used in ceramics, producing alloys, and optical glasses (Fig.6). It is also used in special catalysts and in opto-electronic devices such as lasers, filters for microwaves and europium "phosphorus" for the red color in television sets. Some of its isotopes are radioactive, being used in cancer radiotherapy.

Yttrium compounds have the following uses:

• Yttrium oxide is used to produce yttrium iron garnets which are useful microwave filters.Yttrium oxide is used in ceramic and glass formulations as it has a high melting point and imparts thermal shock resistance and low expansion characteristics to glass

• Yttrium oxide is widely used for making compounds such as YVO_4europium and YVO_4europium phosphors which are responsible for the red colour in television tubes.

• Yttrium iron ($Y_3Fe_5O_{12}$), yttrium aluminium ($Y_3Al_5O_{12}$) and yttrium gadolinium garnets possess interesting magnetic properties. Yttrium iron garnets are extremely efficient transmitters and transducers of acoustic energy.Yttrium aluminium garnet has a hardness of 8.5 and is finding application as a gemstone (synthetic diamond)

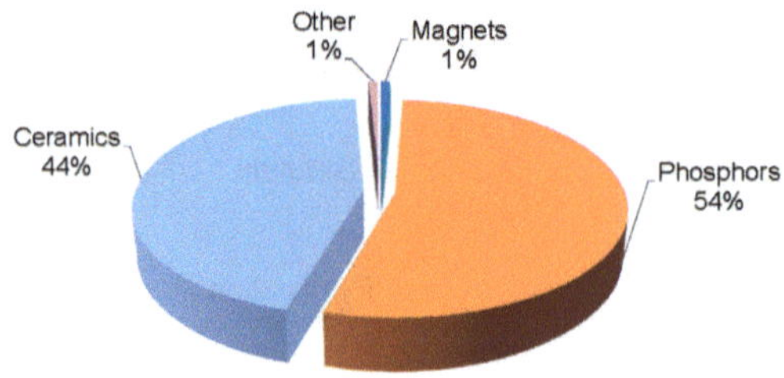

Figure 6.Yttrium Applications

Yttrium Extraction from Korean Redmud

Figure 7 showed the molecular structures of extractants. Figure 8 showed the flow sheet of yttrium extraction by using several kinds of extractants such as PC 88A, Cyanex 272 and Cyanex 302. Among these extractants PC 88A is more suitable for the yttrium extraction.

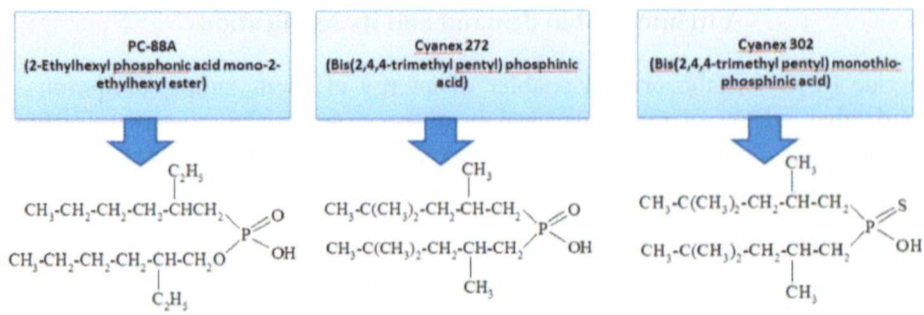

Figure. 7. Molecular structures of Extractants

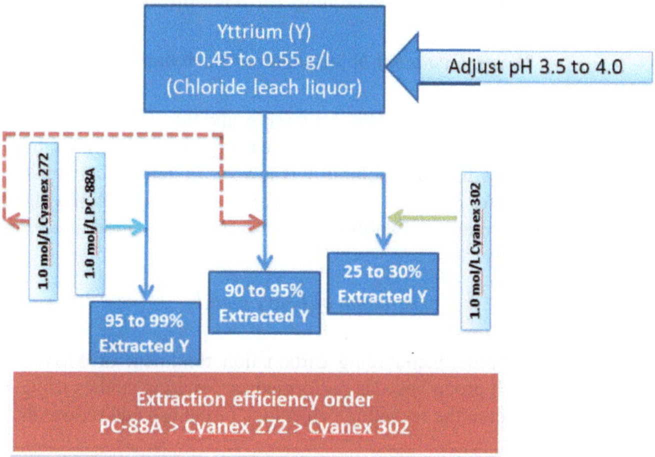

Figure 8. Flow sheet of Yttrium extraction

Convergence Technologies

The new technology of accelerated carbonation is widely used in many industries as an alternative method to natural carbonation which proceeds very slowly in Nature.

Natural carbonation

Natural carbonation occurs by the reaction between atmospheric CO_2 and alkaline materials, which is called "weathering". It is well known that the "weathering" depends on the initial chemical composition, the characteristic of minerals in alkaline materials and the amount of CO_2 uptake. Natural carbonation generally progresses very slowly over a long term.

Accelerated carbonation

In accelerated carbonation, CO2 in a high purity is artificially injected into solid wastes to make the reaction much faster than that by atmospheric CO2, and the reaction can be finalized within a few minutes or hours. This technology is applied to the treatment of solid wastes in which toxic metals are stabilized by carbonated materials, so that the treated solid wastes can be reutilized in other fields.

Accelerated Carbonation Technology for Alkaline Waste (KIGAM)

Waste Treatment by Accelerated Carbonation

The accelerated carbonation of solid wastes containing alkaline minerals such as Ca and Mg before their landfill treatment is effective for decreasing the mobility of heavy metals by adjusting pH to below 9.5 at which their solubility is lowest. In general, an acidic condition may have the risk of causing releases of heavy metals for a long period, accompanying the decrease in buffering capacity by the decrease in alkalinity, whereas the carbonation at an acidic range can increase the buffering capacity.

Carbonation of heavy metals

In the worldwide trend towards encouraging carbonation treatment of MSWI bottom ash as described above, our research group has also performed carbonation studies using MSWI bottom ash and CO_2 in order to increase the recycling percentage of the ash and reduce the concentration of atmospheric CO_2 in Korea.

Wastewater Treatment by Accelerated Carbonation

The formation of solid carbonates between aqueous solutions containing divalent cations and CO_2 is a complicated and important process in Nature and industrial areas. Compared to the aforementioned studies on the carbonation of MSWI bottom ash, however, aqueous carbonation using wastewater is still poorly known and rarely reported in the literature. In this section, we review two representative studies which have reported on the carbonation of cations or anions in wastewater using CO_2.

In Korea, Institute of Geoscience and Mineral Resources (KIGAM), Daejeon, established a successful pilot plant for multi-processing of MSWI ash as shown below in the Figure 9. The capacity of this pilot plant is 200 kg/hour. This pilot plant was the first one in the waste recycling field in Korea and even throughout the Asian countries, and includes multi-processes of particle separation, heavy metal stabilization, chloride removal, and green aggregate/concrete manufacture.

Figure.9. 200 kg/hour MSWI bottom ash multi-process Figure.10. 20 tonns/hour Removal process model of heavy metals chlorides from MSWI bottom ash

The higher capacity (200 kg / hour (Figure.9) plant, 20 tonns / hour plant (Figure.10), (50,000tonns / year (Figure.11)) for real all inorganic waste treatment by accelerated carbonation for real landfill sites for CCS model in South Korea .

Fig.11. CO_2 sequestration pilot plant

This technology can be also considered as an accelerated carbonation in terms of making the reaction shorter using high purity CO_2. One of advantages of this technology compared to other CO_2 storage technologies is that CO_2 is stably stored in final products such as $CaCO_3$ and $MgCO_3$.

Currently, accelerated carbonation processes have been focused on assessing and maximizing the storage of CO_2 by optimizing the operating conditions including pressure, temperature, liquid-to-solid (L/S) ratio, gas humidity, gas flow rate, liquid flow rate, particle size, and solid pretreatment in numerous experimental investigations. The processes via which the carbonation of the minerals or alkaline solid wastes takes place in a single route step are referred to as direct carbonation. Usually, these processes can be divided into two types: gas-solid (dry) carbonation and aqueous (wet) carbonation; these processes are generally operated at an L/S ratio of less than 0.2 and more than 5, respectively.

Quantitative Evaluation

Thermo-gravimetric (TG) analysis of CaCO3

As mentioned, a variety of research on carbonation has been performed worldwide. It is well known that thermo-gravimetric (TG) analysis is one of the most widely used methods for quantitative evaluation of CO2 in carbonated products. The analysis is based on measuring the mass change of a sample caused by heat at temperatures in various ranges. The mass change can be attributed to moisture evaporation or chemical breakdown of compounds into gaseous components.

TG analysis of carbonated products (KIGAM, South Korea)

Based on the previous studies of TG analysis, we also performed quantitative evaluation of CO2 in carbonated MSWI bottom ash. MSWI bottom ash sample containing water of 30-50% was collected in South Korea and was then dried for a day at 100 °C in an oven to reduce water content to less than 0.1%. Dried sample was sieved into three particle sizes using a standard sieve: (1) below 4 mesh, (2) 4–100 mesh, and (3) over 100 mesh. The bottom ash sample in the particle size of over 100 mesh was selected for the carbonation treatment.

For the carbonation treatment, the sample was suspended with a stirrer in deionized water at a ratio of 1:10 (solid: liquid). Using a gas diffuser, the stirred suspension was treated with pure CO2 gas at a rate of 0.5 L/min until pH of the suspension becomes less than 7, at which point the carbonation is completed. After finishing the reaction, the carbonated sample was dried and the mass change of the final product was then measured at various temperatures using a thermo-gravimetric (TG)/derivative TG (DTG) analyzer (TG 209 F3, NETZSCH). The specific analysis conditions were as follows: (1) temperature range: ambient to 1,000 °C, (2) heating and cooling rates: 0.001 and 50 K/min, and (3) cooling time: 20–25 min (from 1,000 to 100 °C).

As aforementioned, the temperature-mass change curves shown in TG analysis indicate thermal and chemical composition changes of a sample. They also show the thermal properties of intermediate processes that occur during sample heating. The DTG curve from the derivation of a TG curve presents both mass change and decomposition rates in its slope. CaCO3 decomposition rate by TG analysis can be considered to determine mass (wt. %) of captured CO2 in a sample, because a final product of the carbonation reaction using CO_2 is $CaCO_3$. We also assumed that mass change of the TG curve at 550–800 °C can be attributed to CO_2 release by $CaCO_3$ decomposition. As shown in Figure. 12, TG curves of raw and carbonated sample present mass losses of −16.27% and −19.74%, respectively, at 550–800 °C. Thus, we confirmed in our experiment that 3.47 wt.% of CO2 was captured in the sample by the carbonation treatment, and concluded that the carbonation could contribute to CO_2 reduction from the atmosphere by sequestrating CO_2 into stable $CaCO_3$[6].

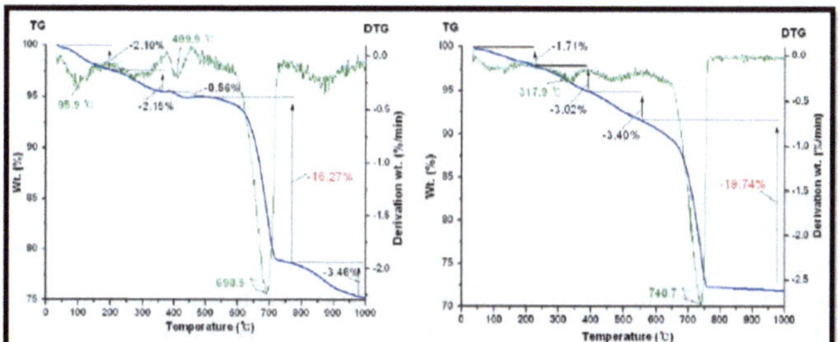

(a) Raw sample: −16.27% mass change at 550−800 °C. (b) Carbonated sample: −19.74% mass change at 550−800 °C.

Figure.12 Quantitative evaluation of CO_2 in municipal solid waste incineration (MSWI) bottom ash before (a) and after accelerated carbonation (b), using TG/DTG analysis at various temperatures (TG = thermo-gravimetric; DTG = derivative thermo-gravimetric) (Unpublished graphs).

CO_2 Solidification of Hazardous Elements from Korean Red Mud

The accelerated carbonation of hazardous elements from wastes such as Korean red mud waste is a controlled accelerated version of the naturally occurring process. In addition, in many cases, binding of toxic metals may occur as the carbonated product rapidly solidifies.

The novel method of the simultaneous calcium hydroxide carbonation with compressed carbon dioxide in order to stabilize the solid material. A number of anions, particularly inorganic oxyanions can be toxic to humans and wildlife at micrograms to milligrams per liter range. We reported in this paper brief data about arsenic oxyanions removal from waste. The calcium hydroxide carbonation with CO_2 was used in order to estimate a removal efficiency of oxyanions (arsenate (V)) from a waste. Preliminary carbonation of arsenate shows 85 ~ 89% removal efficiency during the calcite formation (Figure.13).

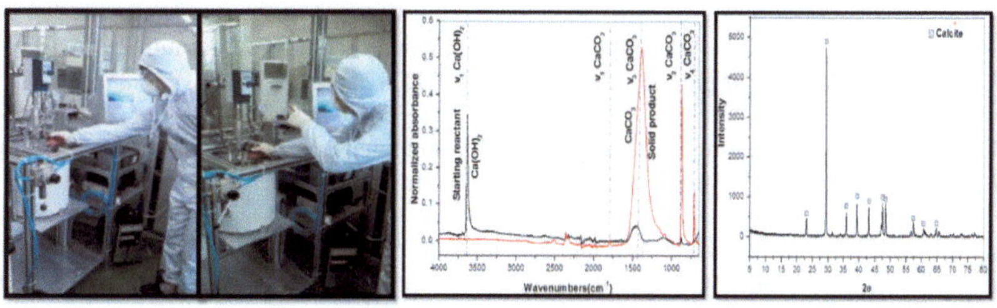

Figure.13. (a). acceleared carbonated reactor (b) FT-IR analysis (c) calcite formation conformation by XRD

CONCLUSIONS

The main purpose of our study was to investigate the possibilities and global trend of rare earths. The objectives of this investigation is recovery of rare earth element from Korean red mud and heavy metals removal from red mud.

The aqueous carbonation of calcium hydroxide with compressed CO_2 could be an economic method to remove various heavy metals oxyanions as well as to allow the mineral sequestration of CO_2. CO_2 capture, utilization, and storage (CCUS) is a promising technology wherein CO_2 is captured and stored in solid form for further utilization instead of being released into the atmosphere in high concentrations. Under this concept, a new convergence technology called accelerated carbonation process has been widely researched and developed.

Acknowledgments

This research was conducted in 2013 with support from the Energy Technology Development Pr oject [2013T100100021] of the Korea Institute of Energy Technology Evaluation and Planning, f inanced by the Ministry of Trade, Industry and Energy.

References

1. Richard Martin, Rare Earth Struggle Intesifies(rare earth metals demand for the cleantech industry, world markets:2011-2017), Article, Sep, 2011.Source: Pike Research).

2. A dam Currie: Rare Earth from Fly ash. A Method Explored, *Rare Earth/Investing News, June 25* (2012).

3.Christopher R. Chitambar, Medical applications and toxicities of gallium compounds, International Journal of Environmental Health, 7(5), 2010, pp.2337-2361.

4.Zhuo Zhao, Yongxiang Yang, Yanping Xiao, Youqi Fan, Recovery of gallium from Bayer liquor: A review, Hydrometallurgy, 125-126, (2012), pp. 115-124.

5.Ryan Castilloux, Yttrium shortages are imminent: Illuminating opportunities for players to stay engaged, Adams Intelligence, Critical metals and minerals research, March 1 (2013).

6. Mihee Lim, Gi-Chun Han, Ji-Whan Ahn and Kwang-Suk You, Environmental remediation and conversion of carbon dioxide (CO_2) into useful green products by accelerated carbonation technology, *Int. J. Environ. Res. Public Health* 7(2010), pp. 203.

RARE EARTH ELEMENT RECOVERY AND RESULTING MODIFICATION OF RESIN STRUCTURE

Sean Dudley[1], Maureen Chorney[1], William Gleason[1], Ed Rosenberg[2], Larry Twidwell[1], Courtney Young[1]
[1]Montana Tech of the University of Montana; 1300 West Park; Butte, MT, 59701
[2]University of Montana; 32 Campus Drive; Missoula, MT, 59812

Keywords: Resin, Rupture, Modification, Rare Earth Elements, Characterisation, Recovery, Concentration

Abstract

Experimental and industrially produced polystyrene and silica resins are tested for recovery capabilities of Rare Earth Elements (REE). Testing regimes being used are typical of resin analysis. Inductively Coupled Plasma-AES results indicate that preferential separation and recovery is possible in the adjusted pH range of2 to 10, solution conditions, and resin type. Testing conditions are resulting in structure modification of the composite resin as indicated by X-Ray Diffraction, Differential Scanning Calorimetry, Scanning Electron Microscopy, Mercury Porisometry, and density analysis. Resin modification indicates both surface and internal structure alteration outside of reported standard behavior of resins. Structural changes have ramification for both Rare Earth Recovery RER and traditional resin operations. Analysis shows that the resin uptake of REE can be manipulated for concentration. Further analysis with SEM work indicates that widespread surface and interior modification is taking place as resins load with REE. This modification is leading to pore rupture and particle breakage. Further investigation is being performed to determine the mode of breakage.

Introduction

Montana Tech of the University of Montana conducts multiple programs for Rare Earth Element (REE) recovery. The objective of this research is to investigate cost effective means to recover and concentrate REE. Ion exchange resins have been widely used in mining and wastewater industries where selective recovery and concentration is needed. Ion exchange resin technology, to date, is not being employed in the REE industry due, in part, to poorly understood interactions between the resin structures and REE chemistry. In order to make upcoming REE operations cost effective, more efficient separation and concentration operations are needed. This project focusses on improved understanding of resins and their application to REE resource recovery.

The use of ion exchange resin technology for separation and concentration is of primary objective of this research project. Multiple industrial and experimental ion exchange resins have been included for testing and analysis. Ion exchange resins were studied under a blind review, including both silica and polystyrene based resins with differing functional groups including amines and phosphates. Phase I testing focuses on batch tests of 17 resins, using a range of light and heavy REE.

Analysis and characterization techniques that were utilized to characterize resin phenomena include X-Ray Diffraction (XRD; Rigaku Ultima IV), Differential Scanning Calorimetry (DSC; TA Instruments SDT Q600), Inductively Coupled-AES (ICP; Thermo Scientific iCAP 6000), Pycnometry, Mercury

Porosimetry (Micrometrics Autopore IV), Scanning Electron Microscopy (SEM; FEI Phenom), and Energy Dispersive Spectrometry (EDAX; Genesis XM 4).

Characterisation

ICP Analysis

ICP-AES is performed on REE and ion exchange batch solutions after 48 hours of stirring to achieve max loading. The average loading of the specific REE onto the ion exchange resin structure (mg REE/gm resin) is calculated from these results.

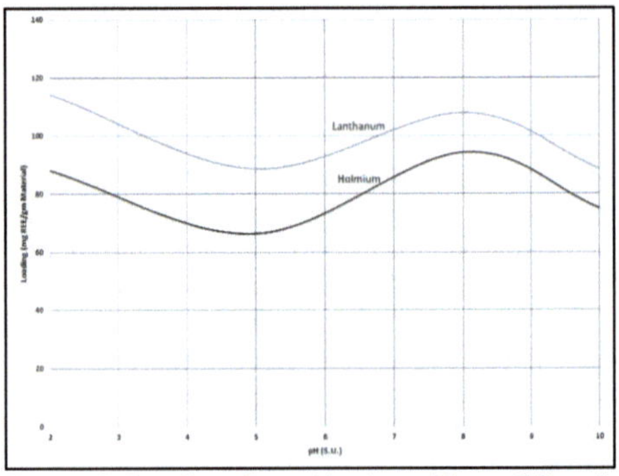

Figure 1: Loading Over Altering pH's on Silica-Based Resin

An industrial silica based ion exchange resin is tested using a light (lanthanum, light grey) REE and a heavy (Holmium, black) REE. Recovery profiles, Figure 1, show variable loading over the pH range. The variability of adsorption characteristics provides an opportunity for selective recovery. Maximised recovery conditions occurred at pH 2, 115 mg/L for Lanthanum and pH 8, 95 mg/L for holmium. Conditions of pH were adjusted using hydrochloric acid and sodium hydroxide.

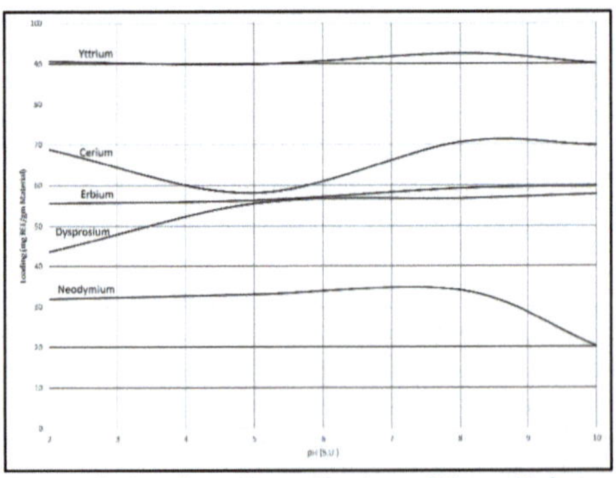

Figure 2: Loading Over Altering pH's on Polystyrene-Based Resin

An industrial polystyrene ion exchange resin is tested over a pH range of 2 to 10. Loading results show variability, seen in Figure 2. Variability presents opportunity of possible selective recovery in future experiments. Yttrium has the greatest recovery at an average of 90 mg/L, but shows no real pH dependence. Neodymium represents a challenge in recovery, averaging 32 mg/L from pH 2-8. Cerium and lanthanum show variability that can be manipulated for preferential recovery. Other REE tested show equivalent recoveries.

DSC Analysis

Calorimetry is used to determine if loading and modification is occurring. Heat flow differentiation indicates loading. Modification is evident in the heat flow shifts, especially that of pH 10. The resin tested is a silica based resin and heat flow changes may be related to phase transitions of the structure. The phase transitions appear to be altered by the occurrence of REE loading.

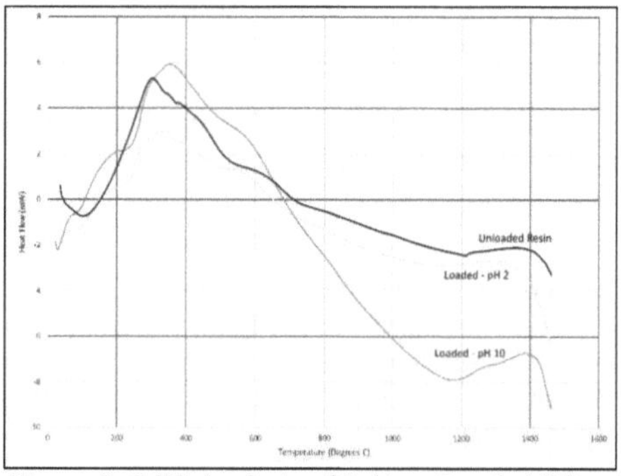

Figure 3: DSC Analysis of Erbium Loaded Silica Ion Exchange Resin

Erbium, Figure 3, shows heat flow profiles for loading conditions of two pH's. The black line (unloaded resin), dark grey line (loaded resin of pH 10) and light grey line (loaded resin of pH 2), show the differences in calorimetry heat flow by pH. The loaded and unloaded ion exchange resins have different calorimetric heat flow responses.

The unloaded ion exchange resin heat flow shows an initial linear endothermic increase with a steep decrease in heat flow. The resins loaded under pH 2 and 10 conditions have an altered heat flow profile. This alteration indicates the change in heat flow profiles due to REE loading. The underlying physical resin structure is being changed such that the profiles are showing characteristic heat flow profile differences.

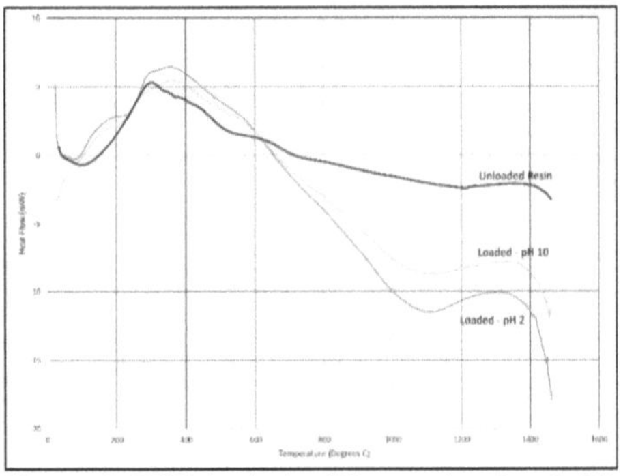

Figure 4: DSC Analysis of Dysprosium Loaded Resin

Dysprosium, Figure 4, like erbium, shows calorimetric modification resulting from dysprosium loading. Differential scanning calorimetry is used as a qualitative indicator of REE loading and modification.

XRD Analysis

XRD is used to identify the presence of rare earth elements on resin structure. The scan of the unloaded ion exchange resin shows an amorphous structure and analysis indicated no readily identifiable phase. Scans of resins loaded with Terbium indicate similar amorphous structures with no discernable sign of a rare earth phase. The lack of a rare earth phase does not preclude rare earth presence; the rare earth is present in a finite dispersed manner and not a defined crystalline phase.

Analysis is carried out comparing shift intensity between unloaded and loaded ion exchange resins. The results indicate an alteration of the amorphous material and is consistent with surface and structural modifications. The percent shift ranges from 20 to negative 35 percent shift. This wide band of intensity shift confirms modification of the structure but XRD analysis is incapable of elucidating more information.

Mercury Porisometry

Mercury porisometry is performed on the highest loaded silica ion exchange resin for each REE to determine a change in porosity of the structure and evaluate loading. From XRD and DSC data, it is expected that there would be a defined trend between percent change in pore area, loading, and SG.

Experiments were designed to determine the relationship between porosity, SG, and loading. Results show a definite link between the three variables. Pore area decreases, signifying the loss of access to pore space from REE loading.

Porisometry will not characterise surface interactions unless sheeting is occurring and subsequently blocking pores. With a silica ion exchange resin, pore area is reduced as loading of REE increases from 100 to 140 mg/gm of REE.

Pycnometry

The mass of silica ion exchange resin will change as loading increases. The volume occupied should not change unless large surface modifications are occurring. Larger specific gravity (SG) changes are thus due to the change in volume of the internal pore structure. Large fluctuations, seen in this research, are considered to be linked to widespread change in volume of the particle. Relatively large changes are expected from polystyrene ion exchange resins because of swelling but the changes seen are not expected of the silica resin tested.

SEM Analysis

Batch Testing and analysis shows that the underlying ion exchange resin substrates are being compromised as rare earths load. Figure 5 shows resin (grey) rupture during metal (white) loading. Such behavior is not reported in literature to date. Figure 5 shows the rupture of a particle loaded withdysprosium under a pH 2 condition.

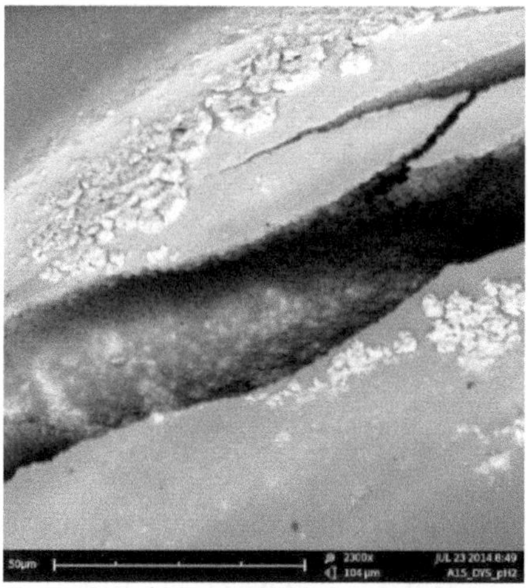

Figure 5: Polystyrene Resin Rupture Showing Interior Loading

EDAX analysis confirms that interior and exterior loading of REE on ruptured particles is present. This rupture occurrence is seen over a wide range of conditions: pH, loading, REE, and type of ion exchange resin. Destructive mechanisms are being investigated, but no published research found to date describes this interaction of REE, pH, and resin type.

Discussion

The suite of analysis that is discussed indicates that modification is occurring to ion exchange resins while being loaded that is not been previously reported in the literature. The original project objective is to maximize conditions for REE recovery and concentration. The ICP data indicates that recovery and separation can be done by manipulating conditions such as pH. These results also indicate that selective uptake is possible. This greater update may lead to optimization for light REE. Further work will include variables such as temperature and oxidative and reductive conditions.

Analysis, including XRD, DSC, SG, Porisometry, SEM and EDAX indicate resin modification as surface and interior pore structure alterations. SEM shows that interior loading is occurring and overloading the pore structures to the point of rupture. SEM shows that surface deposition is resulting in sheet growth that may be a contributing factor to surface tension and contraction. Surface cracking, combined with, interior pore contraction and rupture is leading to particle breakage. Porisomitry and SG data support interior loading observations. DSC and XRD support both interior loading and surface loading considerations.

Conclusions

ICP data indicates that separation and concentration of REE using ion exchange resin technology is possible. XRD, DSC, SG, Porisometry, SEM and EDAX are being used to elucidate chemical interactions between REE and the resin structure.

Particle behavior discovered, shows anomalies occurring over a wide range of conditions. These conditions are basic and fundamental to an operation to recover REE. Results of the other characterization and analysis testing support the breakage mechanisms seen in SEM analysis. It is suspected that widespread alteration of the interior pore structure is leading to steric stress. This stress contorts the interior pore walls leading to localized rupture that starts a chain of ruptures resulting in complete particle breakdown. Future work will focus on identifying the route of particle rupture.

Loading results do indicate the ability to preferentially separate and concentrate REE. Further work will focus on column testing and multi-element loading scenarios. This component will aid in scale-up experimentation.

Acknowledgements

Gratitude is expressed to the funding partners:

- Office of Naval Research; Project: Technical Progress Report on REE Selective Processing by Leaching and Chelating SPCs; Project Number: N000141210592
- Army Research Laboratory; Project: Task 4: Investigation Recovery of Rare Earth Metals Using SPC's; Project Number: W911NF-10-2-0025

References

- Cardarelli, F. (2008). Materials Handbook. Springer.
- Cotton, A. F., & Wilkinson, G. (1976). Basic Inorganic Chemistry. John Wiley & Sons, Inc.
- CRC. (2007-2008). CRC Handbook of Chemistry and Physics (Vol. 88th Edition). (D. R. Lide, Ed.) CRC.
- Emsley, J. (2001). Natures Building: An A-Z Guide to the Elements. Oxford University Press.
- Folger, T. (2011). Rare Earth Elements. National Geographic.
- Goonan, T. G. (2011). Rare Earth Elements - End Use and Recyclability. USGS. USGS.
- Grasso, V. B. (2011). Rare Earth Elements in National Defense: Background, Oversight Issues, and Options for Congress. Congressional Research Service.

- Haxel, G. B., Hedrick, J. B., & Orris, G. J. (2002). Rare Earth Elements - Critical Resources for High Technology. USGS.
- Humphries, M. (2012). Rare Earth Elements: The Global Supply Chain. Congressional Research Service.
- Kailasam, V., Rosenberg, E., & Nielsen, D. (2009). Characterization of Surface-Bound Zr(IV) and Its Application to Removal of As(V) and As(III) from Aqueous Systems Using Phosphonic Acid Modified Nanoporous Silica Polyamine Composites. Ind. Eng. Chem. Res. , 48, 3991-4001.
- MolyCorp (provided by Los Alamos National Labs). (n.d.). About Rare Earths. Retrieved March 25, 2013, from http://www.molycorp.com/resources/the-rare-earth-elements
- Rosenberg, E., & Hart, C. (Unknown). Mine Waste Clean-Up with Novel Hybrid Organic-Inorganic Materials. Missoula: Unknown.
- Takeno, N. (2005). Atlas of Eh-pH Diagrams - Intercomparison of Thermodynamic Databases. Geological Survey of Japan Open File Report No. 419, National Institute of Advanced INdustrial Science and Technology, Research CEnter for Deep Geological Environments.
- U.S. Geological Survey. (2013). Mineral Commodity Summaries - Rare Earths. United States Government.

Rare Metal Technology 2015
Edited by: Neale R. Neelameggham, Shafiq Alam, Harald Oosterhof, Animesh Jha, David Dreisinger, and Shijie Wang
TMS (The Minerals, Metals & Materials Society), 2015

ULTRA HIGH TEMPERATURE RARE EARTH METAL EXTRACTION BY ELECTROLYSIS

Bradley R. Nakanishi, Guillaume Lambotte, and Antoine Allanore

Massachusetts Institute of Technology; 77 Massachusetts Ave., Bldg. 13-5095; Cambridge, MA, 02139-4307, USA

Keywords: molten oxide electrolysis, rare earth oxide electrolyte, recycling, extraction

Abstract

Current industrial methods used for rare earth element (REE) extraction involve: 1) ore enrichment, 2) separation of rare earth oxides (REOs), 3) chlorination or hydrofluorination, and 4) individual electrowinning of REEs from a molten halide electrolyte. The complexity of REE extraction is inherited from their electronic configuration. Recently, molten oxide electrolysis (MOE) has been used to produce reactive metals directly from their oxides, e.g. titanium. As a single-step alternative to processes 3) and 4), or laboratory has investigated rare earth extraction by MOE. A key challenge is to find a molten electrolyte more stable than REOs. One possibility is to use binaries of REOs directly as a solvent. We have, therefore, developed two experimental approaches for studying molten REOs at temperatures exceeding 2200°C. The present work reports the most recent experimental results obtained with La_2O_3-Y_2O_3. Those promising results demonstrate potential for operating with molten REOs and refine the underlying materials challenge for electrodes to enable metal recovery.

Introduction

The rare earth elements (REEs), e.g. dysprosium or praseodymium, possess unique properties that are critical to a diversified energy portfolio [1]. Additionally, REEs are increasingly used in applications ranging from consumer (cellphones) to defense (lasers). Current consumption rates stand at a meager 130 kt annually, but are expected to increase substantially in the next decade with expansion of high growth markets for REEs, i.e. high technology sectors [2]. The similar chemistry of the 17 rare earth oxides (REOs) means that they occur together in the Earth's crust, and the extraction of a single REE requires separation from the 16 others via a complex series of processes [3]. We propose investigating molten oxide electrolysis (MOE) as a simplifying and efficient alternative to current methods of rare earth metal extraction and recovery.

During MOE, metal cations are reduced to metal directly from a molten oxide electrolyte at the cathode using electricity. If an inert anode is used [4], oxygen gas is the lone byproduct of the anodic reaction. A schematic of a MOE cell is shown in Figure 1. This technique has been studied at the laboratory scale for producing reactive metals, e.g. titanium [5].

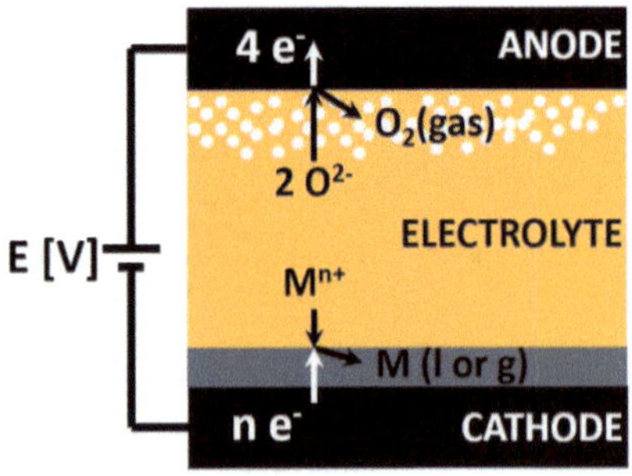

Figure 1: Simplified schematic of an MOE cell illustrating important features

A key challenge for rare earth MOE is determining a suitable oxide supporting electrolyte which is more stable than the rare earth oxides (REOs). Figure 2 is a diagram à la Ellingham-Richardson-Jeffes showing the minimum cell potentials for decomposition and relative stability of the REOs in comparison with a selection of other reactive metal. Certainly, choices of a supporting electrolyte for rare earth MOE are limited to the most stable oxides, obviating silicates or borates. The approach presented here studied the use of an electrolyte composed of the REOs themselves. Given their ultra high melting points [6], operation with a REO-based supporting electrolyte necessitates cell operation at temperature above 2200°C. Operation at such temperatures would enable additional features for separation including exploitation of differences in vapor pressures of the products. Figure 3 is a modified version of Figure 2, which focuses on decomposition potential of only the rare earth oxides in the temperature range of interest to this study.

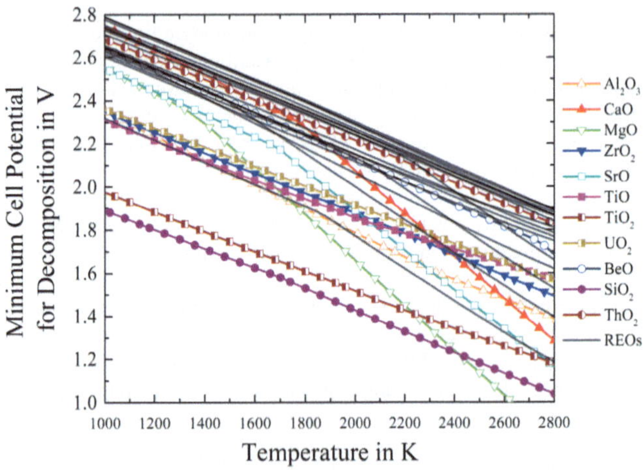

Figure 2: Minimum cell decomposition potential vs. temperature of a selection of the oxides of the reactive metals including the REOs, assuming unit activity for reactants and products [7].

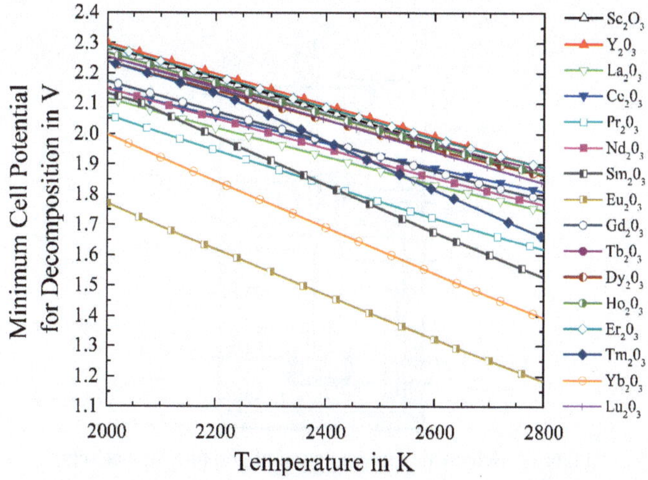

Figure 3: Minimum cell decomposition potential vs. temperature of the REOs in the vicinity of their melting temperatures, assuming unit activity for reactants and products [7].

Experimental Methods

Given the lack of information on the electrolytic nature of molten rare earth oxides in the literature [8] and the challenging nature of high temperature experiments, the development of a setup supporting electrochemical experimentation at temperatures above 2200°C is an important feature of this work. We have developed two laboratory-scale approaches for studying molten rare earth oxides using electrochemistry techniques using i) graphite furnace and ii) floating zone furnace (FZF) methods.

Graphite Furnace Approach

A CG26-6x12-1Z graphite crucible furnace (Mellen Company, Inc.) equipped with 3/8 in. graphite rod heating elements and capable of temperatures up to 2600°C was used (see Figure 4). The graphite furnace is equipped with a pyrometer that is focused on a block of graphite positioned at the center of the hot zone. The electrolyte was prepared by weighing and mixing by hand 99.9% purity powders of La_2O_3 and Y_2O_3 (Alfa Aesar) followed by sintering in a furnace (Lindberg/Blue M) at 1600°C for 48 hours to obtain maximal densification prior to experiment. Chunks of the sintered electrolyte were weighed and placed in a tungsten crucible (Sage Industrial Sales, Inc.) and the assembly was loaded into the graphite furnace with an EDM graphite (GraphiteStore.com) secondary crucible. The furnace was evacuated and purged with ultra high purity argon gas (Airgas) and ramped to 700 °C under vacuum to remove volatiles followed by refilling and ramping to the desired process temperature. In the preliminary stages of this approach, the crucible was used also as the cathode by running a tantalum cathode lead wire to the crucible base from a port in the cap (see Figure 4). Iridium (Furuya Metals Co. Ltd.), tungsten (Midwest Tungsten Service Inc.), and EDM graphite (MWI Inc.) rods have been tested as anodes. Subsequent testing used a three electrode setup in an effort to increase the cathodic current density. A Reference 3000 Potentiostat (Gamry Instruments) was used for electrochemical measurements.

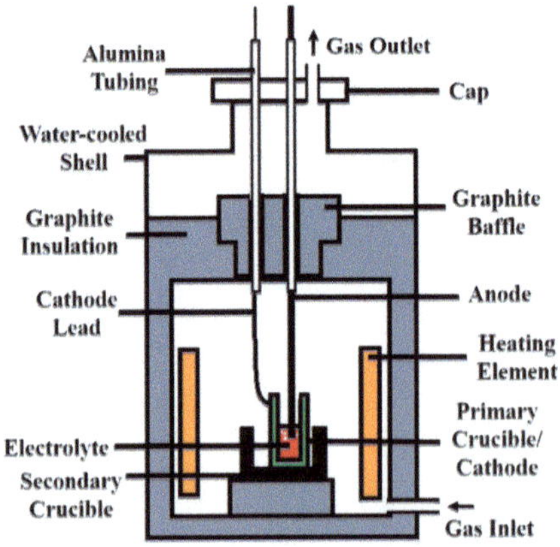

Figure 4: Schematic representation of graphite furnace setup

FZF Approach

A Crystal Systems, Inc. type FZ-T-12000-X-S optical floating zone system equipped with four 25kW xenon lamps capable of heating samples to 3000°C was used (see Figure 5). The floating zone system is equipped with top and bottom rods that allowed for precise, automated control of the sample and electrode positioning within the furnace. Additionally, video feed is provided for direct monitoring of the experiment by a filtered camera. Sample rods of REOs with the dimensions 3-5mm diameter x 75-150mm length were prepared. First, 99.9% purity powders of La_2O_3 and Y_2O_3 (Alfa Aesar) were weighed and mixed to the desired composition by hand. Next, the mixture was ground to a finer powder and mixed by mortar and pestle. Then, the powder was hydrostatically pressed in a taught latex tube and removed in preparation of a green body ready for sintering. Sintered sample rods of greater than 90 percent densification were prepared by heating in a furnace (Lindberg/Blue M) at 1600°C for 48 hours. One end of the rod was wrapped in nickel wire for suspending in the FZF.

The sintered REO rods were loaded into the FZF. A custom quartz tube (Technical Glass Products, Inc.) sealed with rubber O-rings and fitted with four ports allowed for electrode connections (the three ports near the bottom), a thermocouple probe or light guide for optical pyrometer (one port near the top), and vacuum-purging with ultra high purity argon (Airgas). The electrodes, comprised of 99.95% purity tungsten (Rembar Company) and iridium (Furuya Metals Co. Ltd.), were threaded through a multi-bore alumina tube fixed to the bottom positioning rod in the FZF. Electrical connections were made by welding nickel wire to the base of the wires and threading the wires from the multi-bore tube through fittings (Swagelok) on the bottom three ports. A Reference 3000 Potentiostat (Gamry Instruments) was used for electrochemical measurements. Following measurements, samples were quenched by powering down the lamps.

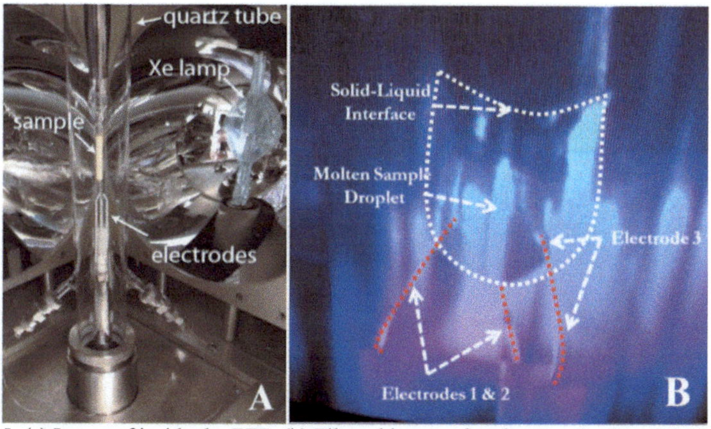

Figure 5: (a) Image of inside the FZF; (b) Filtered image of molten La_2O_3 (60 at%)-Y_2O_3 (40 at%) droplet with iridium electrodes inserted during electrochemical measurements. The white and red dotted lines are present for clarification purposes.

Results and Discussion

With the graphite furnace, significant progress has been made with materials compatibility testing for crucible and electrode materials. Refinement of the electrochemical laboratory cell with this approach is still in progress. Tungsten has shown great promise as a crucible and cathode material. Figure 6 shows a polarized optical micrograph of the cross section of a premelt. Anode material tests demonstrate challenges with operating with carbon (graphite electrodes).

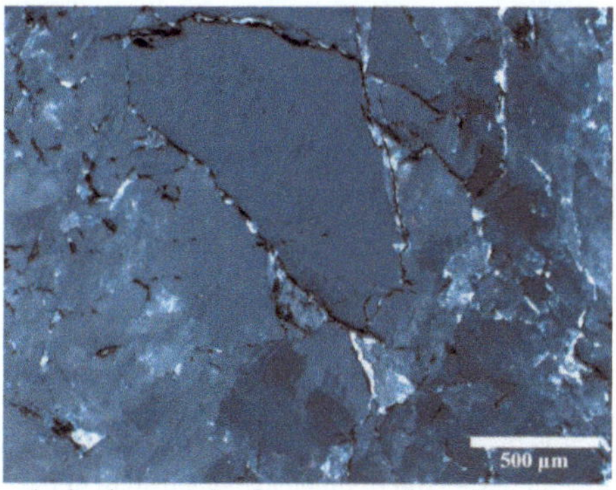

Figure 6: Polarized optical micrograph of a cross-section of La_2O_3 (60 at%)-Y_2O_3 (40 at%) premelted in a tungsten crucible in the graphite furnace at 2250°C.

For the first time, electrochemical measurements have been performed in molten rare earth oxides. The La_2O_3 (60 at%)-Y_2O_3 (40 at%) electrolyte used demonstrated relatively high electrical conductivity, with a bulk resistance between the wires located 2 mm apart of 1.34 ohms (see Figure 7). Preliminary bulk electrolysis experiments with three iridium wires and observations of the electrolyte post-experiment show the formation of metallic droplets in the vicinity of the cathode (see Figure 8), indicative of the production for metal during electrolysis. This results, which further analysis is ongoing, points to the key challenge of alloying/melting of the cathode material during electrolysis. Further developments are on the way to find more stable materials to work as a solid cathode in such conditions.

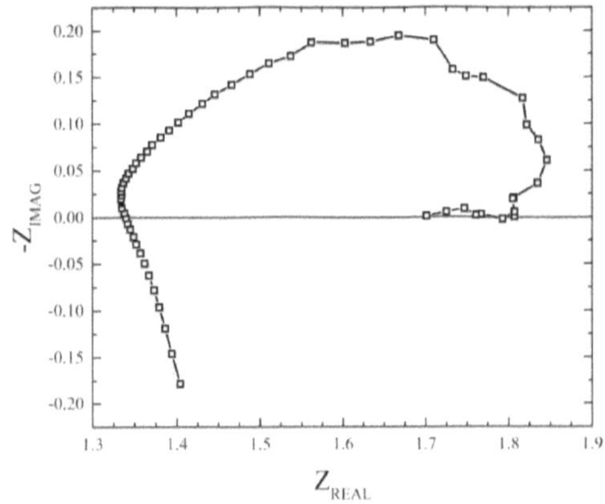

Figure 7: Results of potentiostatic electrochemical impedance spectroscopy (EIS) shown in a Nyquist plot from FZF experiment three iridium wires and La_2O_3 (60 at%)-Y_2O_3 (40 at%) electrolyte. Electrodes were 2 mm apart in molten droplet during data collection.

Figure 8: Polarized optical micrograph of quenched electrolyte in the vicinity of the cathode following FZF experiment with three iridium electrodes. The light regions are metallic products of the cathodic reaction while the light grey areas are the electrolyte. The dark regions are epoxy used in preparing sample cross sectioning.

Conclusion

- ❖ REOs have been identified as potential candidate for a rare earth MOE supporting electrolyte
- ❖ Graphite furnace and floating zone furnace methods have been developed in an effort to probe molten REOs with pioneering in situ electrochemical measurements at ultra high temperatures
- ❖ Three iridium electrodes measurements in the FZF have been performed, offering promising results toward the development of t electrochemistry in molten REO supporting electrolyte at a temperature in excess of 2200°C.

Acknowledgements

This research was made possible by the support of the Office of Naval Research (ONR) under contract N00014-11-1-0657.

References

[1] R. Jaffe, "Energy Critical Elements: Securing Materials for Emerging Technologies," Washington D.C., 2011.
[2] "Critical Materials Strategy," 2011. U.S. Department of Energy.
[3] C. K. Gupta and N. Krishnamurthy, *Extractive Metallurgy of Rare Earths*. New York: CRC Press, 2004, p. 504.
[4] A. Allanore, L. Yin, and D. R. Sadoway, "A new anode material for oxygen evolution in molten oxide electrolysis.," *Nature*, vol. 497, no. 7449, pp. 353–356, May 2013.
[5] N. A. Fried and D. R. Sadoway, "Titanium Extraction by Molten Oxide Electrolysis," *TMS Conf. (Charlotte, NC)*, 2004.
[6] M. Zinkevich, "Thermodynamics of rare earth sesquioxides," *Prog. Mater. Sci.*, vol. 52, no. 4, pp. 597–647, May 2007.
[7] C. W. Bale, "FactSage 6.2." Thermfact and GTT-Technologies, Montréal, 2013.
[8] E. E. Shpil'rain, D. N. Kagan, L. S. Barkhatov, and L. I. Zhmakin, "Electrical conductivity of yttrium and scandium oxides," *Rev. Int. Hautes Temper. Refract.*, vol. 16, pp. 233–236, 1979.

Rare Metal Technology 2015

Vanadium-Molybdenum-Tungsten

Rare Metal Technology 2015
Edited by: Neale R. Neelameggham, Shafiq Alam, Harald Oosterhof, Animesh Jha, David Dreisinger, and Shijie Wang
TMS (The Minerals, Metals & Materials Society), 2015

A NOVEL TECHNOLOGY OF VANADIUM EXTRACTION FROM STONE COAL

Mingyu Wang[1], Xuewen Wang[1], Bowen Li[2]

[1]School of Metallurgy and Environment, Central South University, Changsha, Hunan 410083, China
[2]Department of Materials Science and Engineering, Michigan Technological University, Houghton, MI 49931, USA

Keywords: Stone coal; Vanadium; Comprehensive utilization

Abstract

A novel technology characterized by higher extraction of vanadium and recovery of valuable elements was developed to recover vanadium from stone coal. The brief flow of the novel technology mainly includes low temperature sulphation roasting, water leaching, recovering potassium, aluminum, copper and selenium, solvent extraction enrichment vanadium, precipitation of red vanadium, removal sodium from red vanadium and calcinations. The total recovery of vanadium reached 76.3%, which was higher than the current applied technology. The novel technology was environmentally-friendly and could extract vanadium from the stone coal, in which the content of V_2O_5 was less than 0.8%.

Introduction

Vanadium is an important by-product that is used almost exclusively in ferrous and non-ferrous alloys due to its physical properties as adding to tensile strength, hardness, and fatigue resistance. China, as a major player in the global supply of vanadium bearing products has two important vanadium sources, one is vanadium-titanium magnetite ore, and the other is vanadium bearing stone coal. With the rapid growth of demand of vanadium products, the study on the extraction of V_2O_5 from stone coal attracts more and more attention.

From 1970s on, some companies began to extract vanadium from stone coal by the classical NaCl roasting technology. However, the released exhaust gas containing Cl_2 and HCl during roasting pollutes the environment seriously. In the recent years, many technologies, such as oxygen pressure acid leaching [1], roasting-alkaline leaching [2,3], calcium salt roasting-sulfuric acid leaching [4] and sulfuric acid with villiaumite leaching [5], have been used to extract vanadium from stone coal. The vanadium leaching percentage of these methods is high, but the operations have high costs or high water pollution for treating stone coal. A productive process of V_2O_5 from stone coal was proposed in 1996, which included grinding, two step counter-current sulfuric acid leaching, solvent extraction, precipitation and pyrolysis [6]. This technology avoids waste gas pollution and the product V_2O_5 met the standard specification, however, it only can be used in partial stone coal, in which the vanadium mainly existed in amorphous phase form. This process was only successfully applied to commercial production in the south region of Shanxi province. In these stone coals, the vanadium existed in amorphous phase form accounts for about 66%, and the rest of vanadium existed as vanadium bearing minerals. All the above technologies were only used for stone coal, in which the content of V_2O_5 was more than 0.8%. In China, stone coal containing more than 0.8% V_2O_5 accounts for about 10% of the total stone coal reserve. Furthermore, all the above technologies were only used for extracting vanadium from stone coal. In fact, stone coal is complex ore, in which many other elements also can be extracted.

Combining the advantages of the two processes of direct leaching and pressure leaching with sulfuric acid, a novel technology, named as low temperature sulphation roasting-water leaching was developed, which characterized by higher recovery of vanadium, low production costs, no pollution to environment as well as comprehensive recovery of copper, selenium, potassium and aluminum.

Materials and methods

Materials

The raw stone coal used in this study was taken from Xiangxi district, Hunan province in China and the main composition of stone coal is shown in Table 1.

Table 1 Main chemical composition of stone coal, wt%

V_2O_5	SiO_2	Al_2O_3	CaO	Fe_2O_3	MgO	K_2O	C	S	Cu	Se
0.65	73.22	6.51	0.26	2.57	0.32	0.86	12.81	1.82	0.058	0.027

Methods

After being crushed and ground, the particle size of stone coal is less than 0.18 mm. Prior to roasting process, 85 Kg ground stone coal was mixed with sulfuric acid solution in a desired ratio. After mixing, there is no flow dynamic solution in sample. The mixture was put in a stainless steel crucible, which can be heated by oil bath. After the mixture was roasted at a desired temperature and time, the calcine was taken out for leaching by water at room temperature to extract vanadium. After filtering, potassium and aluminum were recovered from solution by cooling to obtain the crystal of potassium aluminum sulfate.

Iron shavings was subsequently employed to reduce Fe^{3+} and recovery copper and selenium in the solution, then the pH of the solution was adjusted to about 2.0 by adding potassium carbonate before enriching vanadium. After enriching vanadium with P204 from the solution, H_2SO_4 solution was used as the stripping agent to strip vanadium from the loaded organic phase.

The oxidative of vanadium from the stripping solution was performed by using $NaClO_3$ as the oxidation agent at 60 °C for 1 h, and then 4 mol/l NaOH was added to adjust the pH to about 2.0 for the precipitating of red vanadium. After removing sodium, high quality V_2O_5 was acquired by calcining the red vanadium at 530°C for 3h. Figure 1 is the flow sheet of vanadium extraction from stone coal.

Results and discussion

Roasting

During sulphation roasting, the main chemical reactions of vanadium can be expressed by the following equations:

$$V_2O_3 + 3H_2SO_4 = V_2(SO_4)_3 + 3H_2O \quad (1)$$
$$VO_2 + H_2SO_4 = VOSO_4 + H_2O \quad (2)$$

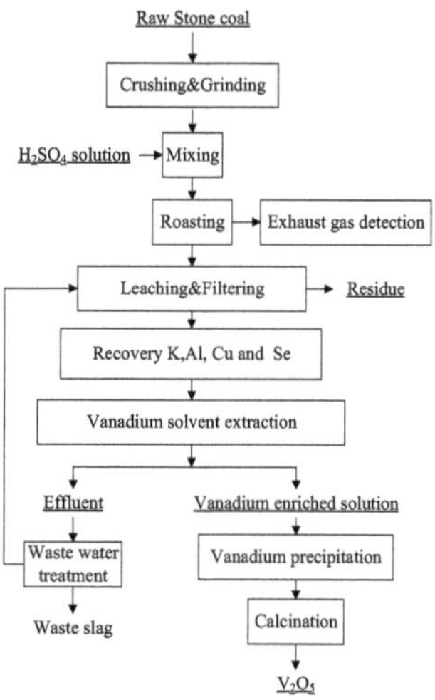

Figure 1. Flow sheet of vanadium extraction from stone coal

One can see from the equations, the vanadium in stone coal will produce vanadium sulphate, which is easy to be dissolved in water. The purpose of the roasting was to improve the reaction temperature of Eq.(1) and Eq.(2), so as to improve its reaction rate. In this process, roasting temperature was very important. Too high roasting temperature will lead to the decomposition of sulfuric acid, as the decomposition temperature of sulfuric acid is only 330°C [7]. Combined with the vanadium leaching percentage, the proper roasting condition was determined: roasting temperature 140°C, roasting time 3h, and the addition of sulfuric acid was 20wt%.

For the extraction of vanadium, environmental is very important. The reason of many former process cannot be applied, is environmental pollution relatively large, particularly air pollution. The composition of roasting exhaust gas during low temperature sulphation roasting was illustrated in Table 2 (The composition was measured by Changsha Institute of Environmental Science).

Table 2 Composition of roasting exhaust gas, mg/Nm^3

SO_2	HF	HCl	Sulfuric acid mist	As	Hg	Pb
4.21	4.06	3.46	2,041	6.63×10^{-3}	7.80×10^{-5}	0.020

As can be seen from Table 2, in the roasting exhaust gas, the contents of harmful components were very low except sulfuric acid mist. Therefore, after water spray, the exhaust gas can be discharged.

Leaching

After sulphation roasting, the calcine was leached in a reaction kettle with water. In this process, $VOSO_4$, $V_2(SO_4)_3$ and others soluble sulfate were dissolved by water. The effects of liquid to solid ratio, leaching temperature and leaching time on vanadium leaching percentage, were studied. The suitable leaching condition was obtained by a series of experiments. The proper leaching condition was: room temperature leaching for 2 h with liquid to solid ratio of 1.6 ml/g. Under this condition, the leaching percentage of vanadium reached 78.3%. The concentrations of the main metal components of the leach solution are shown in Table 3, and pH of leaching solution is 1.1.

Table 3 Composition of the leach solution of stone coal, g/L

V_2O_5	TFe	Al	P	Ti	Ca	Mg	K	Na	Cu	Se	Ni	Zn	Mo
2.81	9.61	13.16	2.43	0.11	0.54	0.31	2.51	0.08	0.24	0.21	0.15	0.32	0.04

Usually, only the stone coal in which the content of V_2O_5 was more than 0.8% can be used. It can be seen from Table 1, in this study the content of V_2O_5 is only 0.65%, which is far below 0.8%. That is to say, the novel technology can extraction vanadium from the stone coal, in which the content of V_2O_5 is less than 0.8% and also can get a high leaching percentage.

Recovery potassium, aluminum, copper and selenium

During leaching, K, Al and Fe in the calcine were dissolved along with vanadium into water. In order to avoid the accumulation of Al and K in water circulating system, they were recovered by cooling crystallizing process for precipitating potassium aluminum sulfate, and the optimum cooling temperature was chosen as 10 °C, based on the whole process requirements.

$$K^+ + Al^{3+} + 2SO_4^{2-} + 12H_2O = KAl(SO_4)_2 \cdot 12H_2O\downarrow \quad (3)$$

In order to recover Cu and Se, iron shavings were employed as the reductant. The suitable pretreatment condition was obtained by a series of experiments and the proper condition was: 0.5% iron shavings addition. In reduction process, Cu and Se also were precipitated.

$$Fe + Fe^{3+} = 2Fe^{2+} \quad (4)$$
$$Fe + Cu^{2+} = Fe^{2+} + Cu \quad (5)$$
$$2Fe + 6H^+ + SeO_3^{2-} = Se + 2Fe^{2+} + 3H_2O \quad (6)$$

560 g dry precipitate can be obtained per 1 m^3 leach solution being reduced. After drying, the XRF analysis (X-Ray Fluorescence) of the precipitate is shown in Table 4. As can be seen, the main component of the precipitate is copper. The content of oxygen is high, which may be caused by drying process. The grains of the precipitate are fine, which is easy oxidized.

Table 4 XRF analysis of the precipitate, wt%

O	Mg	Al	Si	P	S	Cl	K	Ca	Ti	V
18.8	0.38	3.29	3.90	1.57	1.87	0.18	0.22	1.30	0.29	0.41
Cr	Mn	Fe	Ni	Cu	As	Se	Zr	Mo	Ag	Pb
0.75	1.88	2.65	0.016	49.27	4.51	2.6	0.05	0.19	0.03	0.49

Enrichment of vanadium

The concentration of vanadium in leach solution (2.81 g/L) was too low to precipitate red vanadium. Therefore, it was necessary to enrich the vanadium. According to the literature [8] and our former work, 10 vol% P204 (D2EHPA), 5 vol% TBP (tributyl phosphate) and 85 vol% sulfonated kerosene were chosen as solvent extraction organic phase. During the extraction, P204 (D2EHPA) reaction with vanadium can be expressed as follows:

$$VO^{2+} + 2\,(HR_2PO_4)_{(o)} \rightarrow VO \cdot (R_2PO_4)_{2(o)} + 2H^{2+} \quad (7)$$

Where R is C_8H_{17}. The optimum operating condition for the extraction was: pH 2.0 with the addition of potassium carbonate, O/A ratio 1.5:1, ambient temperature and extraction time 7 min. The extraction of vanadium was 98.92% by eight counter-current extraction stages using the optimum conditions.

1.5 mol/L H_2SO_4 solutions was used as stripping agent, and the optimum operating condition for stripping was: O/A ratio 2:1, stripping time 5min, flow ratio 10:1, ambient temperature. The stripping yield of vanadium was above 99.14% by five counter current stripping stages. The reaction of stripping between the P204 (D2EHPA) and the vanadium ions is shown as follows:

$$VO \cdot (R_2PO_4)_{2(o)} + H_2SO_4 \rightarrow 2\ (HR_2PO_4)_{(o)} + VOSO_4 \quad (8)$$

Production of V_2O_5

After stripping, the enriched vanadium solution was oxidized by 200g/L $NaClO_3$ solution at 50-60 °C for 1 h. After that, red vanadium was precipitated with the addition of 4 mol/l NaOH solutions at 90°C for 3h and the end pH was 1-2. In order to removal the effect of sodium, the red vanadium was rinsed with the mixed solution of Ammonium chloride and ammonia [9]. After roasting the rinsed red vanadium at 530 °C for 3 h, the purity of V_2O_5 was 99.2%, meeting the standard specification.

Conclusions

It is proved from the experiments that it is an effective technology to recover vanadium from stone by sulphation roasting, water leaching, solvent extraction enrichment vanadium, precipitation of vanadium and calcinations. After sulphation roasting at 140 °C for 3 h with 20wt% sulfuric acid addition, vanadium in which can be leached with water at room temperature and the vanadium leaching percentage reached 78.3%. The technology can extract vanadium from the stone coal, in which the content of V_2O_5 was less than 0.8%. The total recovery of vanadium reached 76.3%. In addition, K, Al, Cu and Se in stone coal can also be recovered.

References

1. Z.G. Deng et al.," Extracting Vanadium from Stone-Coal by Oxygen Pressure Acid Leaching and Solvent Extraction," *Transactions of Nonferrous Metals Society of China*, 20(2010), s118-s122.
2. D.S. He et al.,"An Environmentally-Friendly Technology of Vanadium Extraction from Stone Coal," *Minerals Engineering*, 20(2007), 1184-1186.
3. M.Y. Wang, X.L. Li, and X.W. Wang," Study on Leaching Vanadium from Stone Coal by NaOH," *Journal of Northeastern University (Natural Science)*, 30(2009),81-83.
4. X.Y. Zou et al.," Research on the Roasting Process with Calcium Compounds for Silica Based Vanadium Ore," *The Chinese Journal of Process Engineering*,1(2)(2001),189-192.
5. X.Y. Zhou et al., "Leaching of Vanadium from Carbonaceous Shale," *Hydrometallurgy*, 99(2009),97–99.
6. Z.L. Lu, "Investigation and Industrial Practice on Extraction of V_2O_5 from Stone Coal Containing Vanadium by Acid Process," *Hydrometallurgy of China*, 21 (4)(2002),175-183.
7. Anon., *Inorganic chemistry* (Beijing, Higher Education Press, (2001) 468.

8. M.T. Li et al., "Extraction of Vanadium from Black Shale using Pressure Acid Leaching," *Hydrometallurgy*, 98(2009), 308-313.
9. X.W. Wang and M.Y. Wang," A Technology for Producing V_2O_5 Product from the Solution Containing Vanadium" (China patent: 201110262556.4,2011).

Rare Metal Technology 2015
Edited by: Neale R. Neelameggham, Shafiq Alam, Harald Oosterhof, Animesh Jha, David Dreisinger, and Shijie Wang
TMS (The Minerals, Metals & Materials Society), 2015

MECHANICAL ACTIVATION OF PROCESSING OF EGYPTIAN WOLFRAMITE

Aly M. Abdel-Rehim and Mohamed Y. Bakr
Alexandria University, 69 Sultan Hussein St., Shallalat, Alexandria, Egypt

Keywords: Wolframite, Ball-milling, Alkaline Processing, Kinetics.

Abstract

The present work is related with mechanical activation of processing of Egyptian wolframite by simultaneous ball-milling and pressure alkaline leaching, to improve the recovery of tungsten from wolframite. The results show a total recovery of tungsten from wolframite (97.8- 98.7%) by simultaneous ball–milling and alkaline leaching at temperature 175-200°C.

It is concluded that processing of wolframite by simultaneous ball- milling and leaching is mechanically activated and accelerated as a result of the grinding action of steel balls that caused the immediate rupture by continuous removing the product layer of the reaction (sodium tungstate and iron and manganese hydroxides) from reacting particle surfaces. This leads to the exposing of fresh and activated wolframite particle surfaces to alkaline leach solution and consequently the acceleration of the reaction of wolframite leaching .The results of the kinetic study Of alkaline processing of wolframite and its mechanical activation show that the system is under kinetic control.

1. Introduction

Wolframite is one of the most important commercial source of tungsten, its alloys and compounds. In Egypt, wolframite occurs in considerable amounts in Igla mineralization, Eastern Desert. It commonly occurs as disseminated prismatic to tabular crystals in quartz veins and less common in the granitic rocks. Cassiterite is the most common associated ore mineral with wolframite, forming intergrowth texture with it. The most common gangue minerals associated with wolframite are quartz and muscovite (Said, 1990). Several methods based on sintering process are: 1) with sodium carbonate at 850-950°C; 2) with sodium hydroxide at 400-500°C; 3) with sodium sulphate and coal at 660°C; 4) with sodium bisulphate at 800°C; 5) with calcium carbonate or calcium oxide at 850-1000°C; 6) with calcium oxide and flux (calcium chloride) at 700-775°C. In all of these methods, sodium tungstate is the final product , from which tungstic acid or other tungsten compounds are produced (Abdel–Rehim 2001; Bolotinkova 1973; Burwell, 1966; Gomes, Daniel and Andrea, 1979; Klatshko et al., 1979; Meerson and Zelikman 1965; Reznitshenko, 1967; Srinivas et al., 2000; Zelikman et al., 1964). Chlorination of wolframite is of interest for its application to low grade ores, tails and complex tungsten–tin ores and concentrates. Different chlorinating agents are used such as chlorine, hydrogen chloride, carbon tetrachloride and sulphur dichloride.

Tungsten chlorides are easily hydrolyzed in water with the formation of tungstic acid. Chorination of wolframite is carried out by different ways: 1) chlorination of its mixture with coke and volatilization of tungsten. chloride and oxichlorides; 2) its preliminary roasting at 700°C, followed by its chlorination with sulphur dichloride and chlorine (Barsukova and Zverev, 1977; Gokhale, and Bhat, 1967; Meerson and Zelikman, 1965).

Acid leaching of wolframite include its leaching with hydrochloric or sulphuric or nitric acids. Hydrochloric acid is mainly used for leaching of tungsten ores, especially for scheelite. Alkaline leaching of wolframite is successfully carried out with sodium hydroxide at 175-190°C or with sodium carbonate solutions at 180-310oC with the formation of sodium tungstate and iron and manganese hydroxides or their hydrated oxides.

Commercially, both sintering of wolframite with sodium carbonate or leaching with sodium hydroxide solution are commonly used for production of tungsten trioxide in industry (Abdel-Rehim, 1973; Amer, 2000; Balaz, 2000, 2008; Bolotinkova, 1973; Doyle and Duyvesteyn, 1993; Habashi, 1999; Kalashnik et al., 1973; Klatshko et al., 1979; Lassner and Schubert, 1999; Meerson and Zelikman, 1965; Osseo-Asare, 1992; Paul and Dale, 1982; Quatrine et al., 1982; Rysegz and Mazapek, 1979; Sakharkar et al., 1992; Vladimir et al., 1978; Zhao et al., 1995)

Sintering methods of wolframite are carried out at high temperatures and have many disadvantages, the most important of which are preliminary high degree of grinding of wolframite concentrate, difficulty of equipment design, rapid corrosion of furnaces and their coatings, high temperature and consequently high expense of electric energy. Alkaline leaching of wolframite is most common process used as it has many advantages in comparison with sintering methods. The main advantages are low temperature of leaching, simultaneous removal of iron and manganese during leaching and the regeneration of alkali.

The sodium tungstate solutions produced from processing of wolframite are directed to purification processes for the removal of silicon, phosphorous, arsenic and molybdenum. The precipitation of tungsten compound from sodium tungstate solution is carried out either by the immediate precipitation of tungstic acid by acid treatment of solutions, or precipitation in the form of calcium tungstate, followed by its decomposition with acids, or crystallization in the form of hydrated sodium tungstate or paratungstate, from which tungstic acid is produced by treatment of their solutions with hydrochloric acid. These three methods of separation of tungsten from solution are used in industry but the first and second ones are widely used (Balaz, 2000 ; Bolotinkova, 1973; Klatshko et al., 1979; Lassner and Schubert, 1999; Meerson and Zelikman, 1965; Rysegz and Mazapek, 1979: Vladimir et al., 1978).

The present work illustrates a study of new technique of mechanical activation of processing of Egyptian wolframite by its simultaneous ball- milling and pressure alkaline leaching, to improve the recovery of tungsten from wolframite. In this method, grinding and leaching of wolframite are combined together in one process at lower temperature and suitable time

2. Experimental Techniques

This research was carried out using wolframite separated from Igla mineralization, Eastern Desert. The concentrate has black colour with metallic luster

2.1. Characterization of Wolframite Sample

Wolframite crystals are thick tabular or short columnar, dark grey, brownish black, or iron black colour, black streak, greasy to submetallic luster. The specific gravity determined for wolframite crystals ranges from 7.1 to 7.5. In polished sections, wolframite is commonly greyish white in colour with faint brownish tint, subhedral form, well developed two sets of cleavage in some crystals, low to medium reflectivity, deep brownish internal reflections, faint polarization colours from deep grey to reddish grey, with weak anisotropism and inclined extinction. The chemical composition of wolframite concentrate is given in Table 1.

Table 1: Chemical composition of wolframite Concentrate.

Chemical component	Content %
WO3	62.86
Fe2O3	19.35
MnO2	14.41
SiO2	1.10
Al2O3	0.20
SnO2	0.62
CuO	0.21
As2O3	0.25
S	0.72
P2O5	0.10

It is observed that the tungsten trioxide content of the concentrate is 62.86% and the remainder 3.14% included associated impurities. The X-ray diffraction data of Igla wolframite are consistent with the corresponding ASTM data. The X-ray powder diffraction pattern of the concentrate shows only the characteristic wolframite peaks, which are sharp and intense, suggesting good crystallinity. No peaks of any mineral impurity were detected by X-ray diffraction study.

2.2. Apparatus and Procedure

Experiments of ball-milling and alkaline leaching of wolframite were carried out in stainless steel ball-mill autoclaves of cylindrical shape with a 400 g charge of balls was used, which was equivalent to 45% of the total autoclave volume. The weight ratio of steel balls of diameters 5, 10 and 15 mm was 0.5 : 1.5 : 2. The number of revolutions of the ball-mill autoclaves was taken as 70% of the critical velocity (186 rpm) and equals 130 rpm.

2.3. Alkaline Processing of Wolframite

A 10g wolframite concentrate at 96.2% -2 + 0.125 mm was charged into the autoclave without preliminary grinding with sodium hydroxide solution of definite amount and concentration. After a predefined stirring period at a given temperature, the autoclave was cooled to room temperature. The pulp obtained after leaching was discharged from the autoclave into a beaker and diluted with hot water to a ratio S : L = 1:4 to improve the solubility of sodium tungstate. To avoid hydrolysis of sodium tungstate, a certain alkalinity of solution was fixed at sodium hydroxide content 50 g/l NaOH. The pulp was filtered under vacuum and the iron-manganese cake was washed with dilute sodium hydroxide solution (50 g/l NaOH). Filtrate and washed solution were added together and analyzed for the tungsten content and its recovery from wolframite was determined.

2.4. Measurement of Vapor Pressure

Before starting the experiments of wolframite processing, the vapor pressure over sodium hydroxide solutions of different concentrations was measured at different temperatures, in a specially designed autoclave connected to a manometer.

3- Determination of Thermodynamic Constants

Before the study of the mechanical activation of processing of wolframite, an attempt was made for approximate calculation of thermodynamic constants of the reaction of wolframite leaching. The thremodynamic data given in Table 2 were used in calculations.

Table 2: Thermodynamic data used

Compound	ΔGo, kcal mol-	References
WO3 (c)	182.470	Weast et al., 1997
FeWO4 (c)	24 9 .800	Ivanov, 1972
MnWO4 c)	278.7 0	Ivanov, 1972
Na2WO4(aq)	345.180	Ivanov, 1972, Robie , 1 968
NaOH (aq)	100.184	Weast et al ., 1997
Fe(OH)2 (c)	115.570	Weast et al., 1997
Mn(OH)2(am)	145. 00	Weast et al., 1997

The reaction of leaching of wolframite (as iron tungstate) with sodium hydroxide may be represented as:

$$FeWO_4 + 2\ NaOH \rightarrow Na_2WO_4 + Fe(OH)_2 \quad (1)$$

$$\Delta G^{O}_{reaction} = \Delta G^{O}_{Na_2\,WO_4} + \Delta G^{O}_{Fe\,(OH)_2} - \Delta G^{O}_{FeWO_4} - 2\Delta G^{O}_{NaOH} = -10.582\ \text{kcal/mol.} \quad (2)$$

The equilibrium constant (K) of the alkaline processing of wolframite (as iron tungstate) may be calculated from the equation relating to 25oC.

$$\Delta Go =- RT\ \ln K \quad (3)$$

$$\text{Log K} = \frac{10582}{4.5756 \text{ x } 398} = 7.761$$

$$K = 5.76.10^7$$

The reaction of leaching of manganese tungstate with sodium hydroxide may be represented as:

$$MnWO_4 + 2NaOH \rightarrow Na_2\,WO_4 + Mn(OH)_2 \quad (4)$$

$$\Delta G^{O}_{reaction} = \Delta G^{O}_{Na_2\,WO_4} + \Delta G^{O}_{Mn(OH)_2} - \Delta G^{O}_{MnWO_4} - 2\Delta G^{O}_{NaOH} = -11.912\ \text{kcal/mol.} \quad (5)$$

The equilibrium constant (K) of alkaline leaching of wolframite (as manganese tungstate) may be calculated as the following:

$$\text{Log K} = \frac{11912}{4.5756 \text{ x } 398} = 8.736 \quad (6)$$

$$K = 5.45.10^8$$

The equilibrium constants (K) are large and the reaction of leaching of wolframite with sodium hydroxide may be considered in practice to be irreversible.

4. Results and Discussion

The mechanical activation of processing of Egyptian wolframite by its alkaline leaching in ball-mill autoclaves is based on the formation of sodium tungstate and iron and manganese hydroxides and may be represented by the chemical reactions (Eqs. (1) and (4)). Besides iron and manganese hydroxides, their hydrated oxides and sodium ferrite are formed (Meerson and Zelikman, 1965).The separation of tungsten from iron and manganese is based on the dilution of the pulp of leaching with water, where sodium tungstate being soluble is separated in solution, while iron and manganese compounds are insoluble and remain in the residue.

4.1.Vapor Pressure Data

The vapor pressure data measured over sodium hydroxide solutions of different concentrations at different temperatures are shown in Fig.1. The vapor pressure over sodium hydroxide solutions generally increases as the temperature rises. A sharp increase of vapor pressure for all concentrations was observed at temperature higher than 190°C. On the other hand, as the concentration of sodium hydroxide solution increases, the vapor pressure decreases.

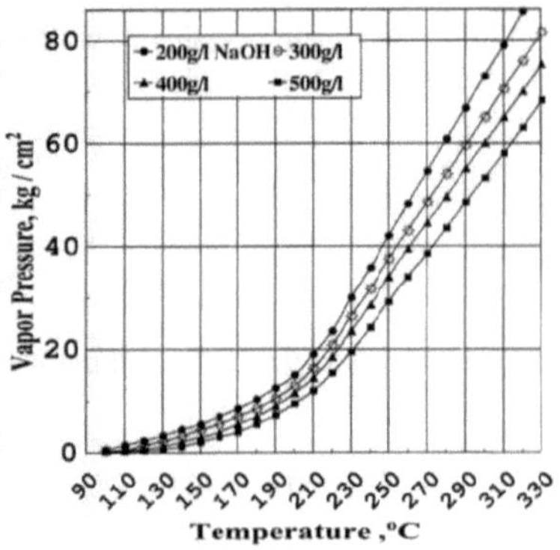

Fig. 1. Relation between the vapor pressure over sodium hydroxide solutions of different concentrations (200, 300, 400 and 500 g /l NaOH) and temperature.

2. Alkaline Processing of Wolframite

4.2.1. Effect of Alkali Concentration. Fig. 2 shows the results of alkaline processing of wolframite at different concentrations of sodium hydroxide solution at (100 – 500 g/l NaOH) at 125 and 175°C within 4 hours (using 150% theoretical requirement of sodium hydroxide). The recovery of tungsten from wolframite sharply increases with increasing the concentration of sodium hydroxide solutions from 100 to 400 g/l NaOH. Then, this increase is gradual up to concentration 500 g/l NaOH, at which high leaching of wolframite is attained. On the other hand, as the concentration of sodium hydroxide solutions increases, the vapor pressure decreases (as shown in Fig.1). The increase of recovery of tungsten from wolframite with concentration follows the law of mass action.

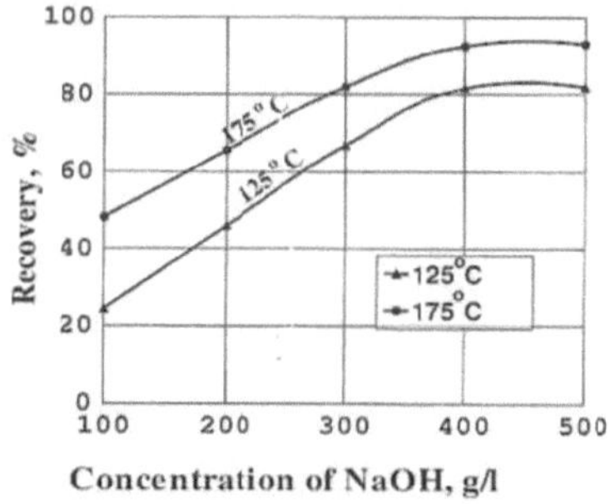

Fig. 2. Relation between the recovery of tungsten from wolframite and concentration of sodium hydroxide solutions at 125 and 175°C within 4 hours using 150% theoretical requirement of sodium hydroxide.

Using alkali concentration 500 g/l NaOH, the recovery of tungsten increases from 81.8% to 92.6% as the temperature increases from 125 to 175°C, due to the increase of vapor pressure over the pulp of wolframite eaching from 2.4 to 8.1 kg/cm^2.

4.2.2. Effect of Amount of Sodium Hydroxide. The effect of amount of sodium hydroxide on alkaline processing of wolframite was studied at 125 and 175°C within 4 hours using alkali concentration 500 g/l NaOH and amounts of sodium hydroxide, ranging 100 to 200% theoretical requirement. Fig. 3 shows that the recovery of tungsten sharply increases with increasing stoichiometric amount of sodium hydroxide up to 150%, beyond which the effect is negligible. At 175°C, the recovery of tungsten is higher than that at 125°C due to the increase of vapor pressure over the pulp of wolframite leaching (see Fig.

1). The suitable amount of sodium hydroxide, satisfying high recovery of tungsten, may be considered as 150% theoretical requirement.

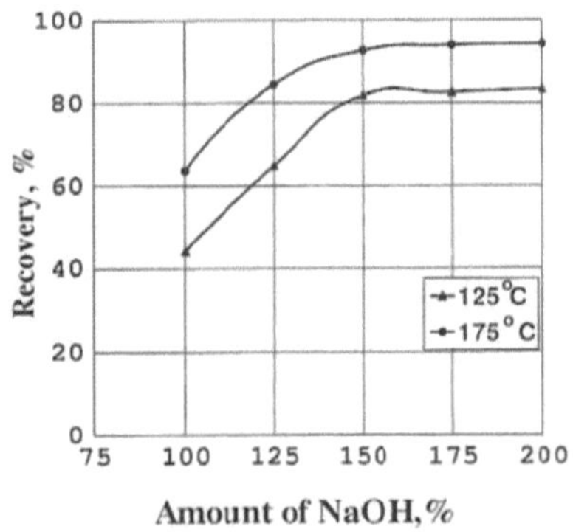

Fig. 3. Relation between the recovery of tungsten from wolframite and amount of sodium hydroxide at 125 and 175°C within 4 hours using 500 g/l NaOH solution strength.

4.2.3. Effect of Temperature and Time. The effect of temperature and time on alkaline processing of wolframite was investigated at different temperatures from 100 up to 225°C at different times, using 150% theoretical requirement of sodium hydroxide, at concentration 500 g/l NaOH. From the results obtained (Fig.4 and 5), the following conclusions may be drawn:

(1) The recovery of tungsten from wolframite generally increases as the temperature and time increase, except at 225°C and long time (6h), it decreases slowly. The increase of tungsten recovery is at different rates, depending on the temperature and retention time. The higher the temperature the larger the increase of tungsten recovery.

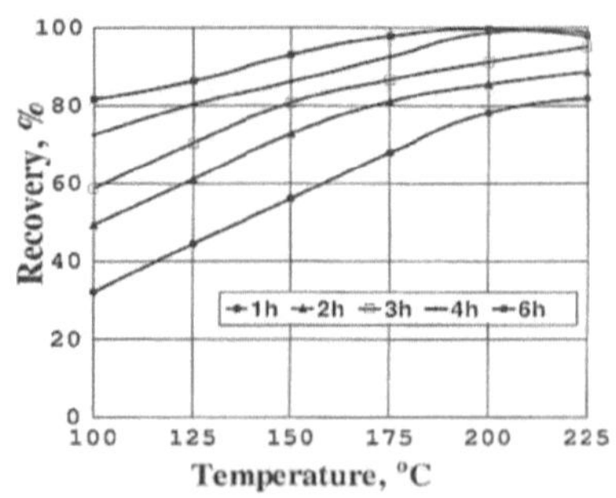

Fig. 4. Relation between the recovery of tungsten from wolframite and temperature at different times using 150% theoretical requirement of NaOH and 500 g/l NaOH solution strength.

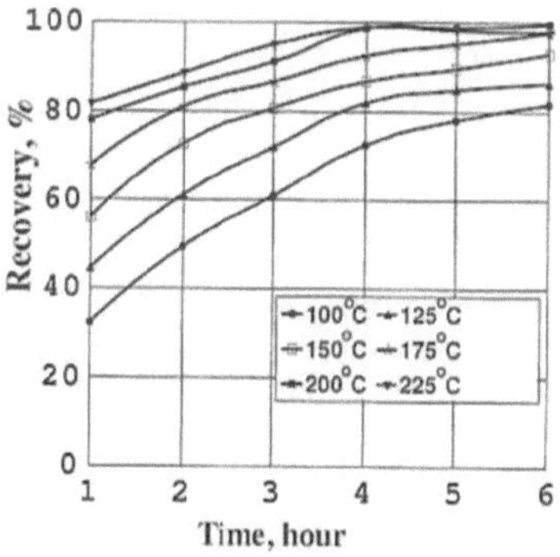

Fig. 5. Relation between the recovery of tungsten from wolframite and time at different temperatures using 150% theoretical requirement of NaOH and 500 g/l NaOH solution strength.

(2) The increase of tungsten recovery from wolframite with temperature is due to the increase of vapor pressure over the pulp (at 150, 175, 200 and 225°C, the vapor pressure is 3.2, 5.6, 10 and 17.8 kg/cm^2, respectively) as shown in Fig.1.

(3) At 100 and 125°C, the recovery of tungsten from wolframite sharply increases during the first 4 h, then gradually increases. t these temperatures, high recovery of tungsten is attained only at long time (6 h) (81.77% and 86.4% at 100 and 125°C, respectively).

(4) At 150-225°C, the recovery of tungsten from wolframite is between 73.5% and 88.7% within 2 h but it reaches higher values 86.6- 98.8 % at retention time 4h. After 4 h, there is a gradual and small increase of recovery of tungsten. Complete leaching of wolframite is attained at 175-200°C within 6 and 4 h, respectively. At 225°C and at longer time than 4h, there is a gradual and small decrease of tungsten recovery, which may be due to the reversible reaction starting between the products of leaching, namely, sodium tungstate and insoluble iron and manganese hydroxides and hydrated oxides. This leads to the formation of the insoluble iron and manganese tungstates or their hydrated forms.

Form these data, it is concluded that complete leaching of wolframite (97.8- 98.7%) is attained at 175 and 200°C within 6 and 4 h , respectively. These conditions can be improved by carrying out the wolframite processing by simultaneous ball-milling and pressure leaching under oxidizing conditions at much lower temperature and shorter time.

5. Kinetics of Alkaline Processing of Wolframite

The kinetics of the alkaline processing of wolframite with sodium hydroxide by simultaneous ball-milling and leaching was studied over temperature range 100-225°C during different times, according to the experimental data obtained (Fig.5). The kinetics of alkaline leaching of wolframite is governed by the structure of wolframite and the condition of the surface coating of the products of leaching, namely sodium tungstate iron and manganese hydroxides. The rate of the reaction of leaching of wolframite was studied for two models controlled by chemical reaction or diffusion through the product layer depending on the temperature of leaching.

5. 1. Rate Controlled by Chemical Reaction

The alkaline leaching of wolframite is mechanically activated by its processing in ball-moll autoclaves. At 150-225° C, the surface coating of sodium tungstate and iron and manganese hydroxides are easily and continuously removed as formed, by the grinding action of steel balls. This leads to the exposing of fresh wolframite particle surfaces to the alkali leach solutions. In this condition, there will be no resistance to the flow of alkali solution reaching the interface and the rate of the reaction is not affected by the surface coatings. The process of leaching is therefore controlled by the chemical reaction at the interface and the kinetics of the reaction may be described by the following equation (Habashi, 1980) :

$$k\,t = [\,1 - (\,1 - \alpha\,)^{1/3}\,] \tag{7}$$

where k = rate of the reaction and α = fraction of wolframite reacted by time of leaching t.

The experimental data in Fig. 5 were plotted as a function of $[\,1 - (\,1 - \alpha\,)^{1/3}\,]$ vs. time t (Eq. 7) for temperature range 150-225° C, as shown in Fig. 6. It is observed that the data display a linear relation with respect to the chemical model. The slopes of such straight lines determine the rate of the reaction (k). The rate of wolframite leaching increases as temperature increases and reaches its maximum $3.5\,.\,10^5\ s^{-1}$ at 225°C and lower value $1.4\,.\,10^5\ s^{-1}$ at 150°C. At the optimum temperature of leaching 200oC, representing complete extraction of tungsten, the rate of the reaction reaches a quite well value $2.7\,.\,10^5\ s^{-1}$. From Arrhenius plot (Fig. 7), the calculated apperent activation energy (E) of the reaction of wolframite leaching is 21.622 kcal mol^{-1} (90.488 kJ mol^{-1}).

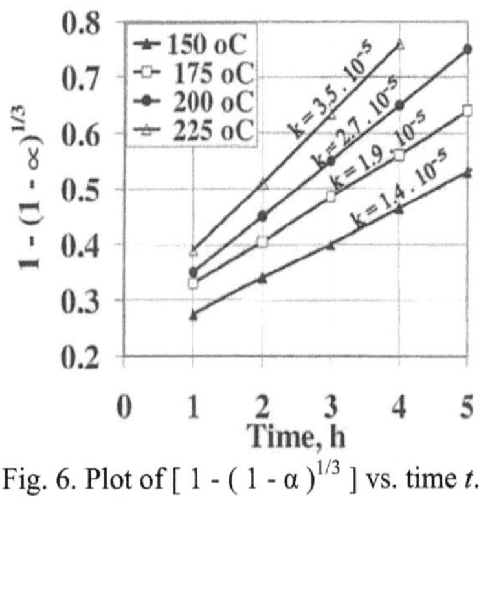

Fig. 6. Plot of $[\,1 - (\,1 - \alpha\,)^{1/3}\,]$ vs. time *t*.

This value is consistent with the values of activation energy reported for chemically controlled reactions. This confirms that alkaline processing of wolframite at 150°C and higher temperatures is controlled by the chemical reaction at the surface of wolframite particles.

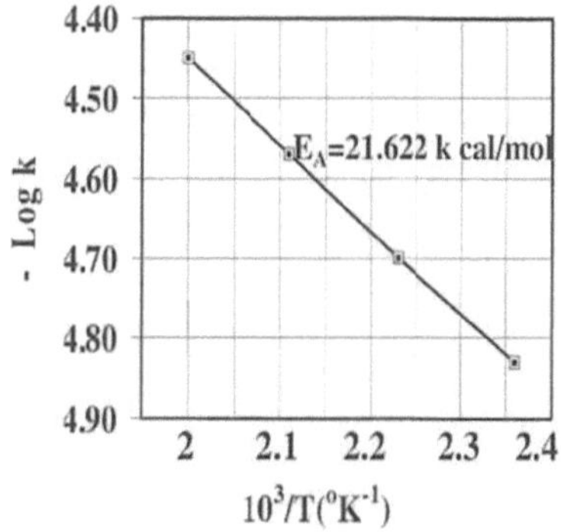

Fig. 7. Arrhenius plot of wolframite leaching as a function of log K vs. 10^3 /T (K^{-1}).

5. 2. Rate Controlled by Diffusion Through the Product Layer

At 100 and 125°C, the alkaline processing of wolframite leads to the formation of dense thick suspension of fine particles of wolframite and the reaction products. This phenomenom was observed in spite of performing the leaching of wolframite in ball-mill autocalves. The fine wolframite particles are surrounded by compact saturated layer of sodium tungstate and hydroxides of iron and manganese. For this suspension of very fine particles and products, the velocity of rotation of autoclave and balls are beyond the critical value such that diffusion through the product layer is rate controlling. The alkali solution reaches the mineral with difficulty. In order that leaching of wolframite proceeds, the alkali must diffuse through the product layer to reach the surface of the mineral. Therefore, the process is controlled by the rate of diffusion of alkali through this layer. In this case, the rate of the reaction may be determined by application of Jander's equation and Crank-Ginstling-Brounshtein's equation (Eq. 8 and 9, respectively (Habashi, 1980) as the following :

$$k_t = [\,1 - (\,1 - \alpha\,)^{1/3}\,]2 \qquad (8)$$

$$k_t = 1 - 2/3\,\alpha - (\,1 - \alpha\,)^{2/3} \qquad (9)$$

where α = fraction of wolframite reacted by time t and k = rate constant of the reaction.

If leaching of wolframite is controlled by diffusion through the product layer, the experimental data in Fig. 5 must give a linear relation, when the right-hand side of Eq. 8 and 9 are plotted vs. time t. As shown in Figs. 8 and 9, they give straight lines at 100 and 125oC, indicating a diffusion controlled model, where the slopes of the straight lines give the rate constant of the reaction (k). The calculated values of the reaction rates at 100 and 125°C for Fig. 8 ($6.5 \,.\, 10^6$ and $8.1 \,.\, 10^6 \, s^{-1}$) using Jander's equation are in good agreement with those for Fig. 9 ($6.6.10^6$ and $7.9 .10^6 \, s^{-1}$, respectively) using Crank-Ginstling-Brounshtein's equation.

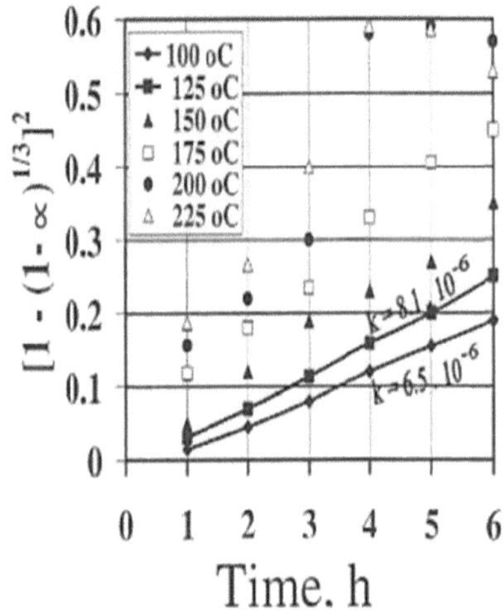

Fig. 8. Plot of $[1 - (1 - \alpha)^{1/3}] 2$ vs. time *t*.

It is concluded that at 100 and 125oC,the alkaline processing of wolframite is a diffusion controlled process, depending on the rate of flow of alkali solution and its velocity of diffusion through the product layer to the surface of wolframite particles. The diffusion resistance through the coherent layer controls the overall kinetic process.

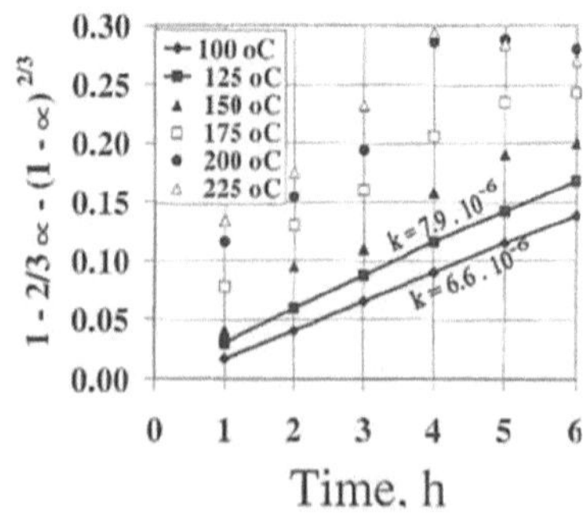

Fig. 9 Plot of $[1 - 2/3\,\alpha - (1 - \alpha)^{2/3}]$ vs. time *t*.

6. Conclusions

The study of mechanical activation of processing of Egyptian wolframite by its simultaneous ball-milling and pressure alkaline leaching revealed the following conclusions:

(1) In general, the recovery of tungsten from wolframite increases at different rates, depending on the temperature and time. The higher the temperature and the longer the retention time, the large increase of tungsten recovery will be. The increase of tungsten recovery with temperature is due to the increase of vapor pressure over the pulp of wolframite leaching.

(2) At 100 and 125°C, the recovery of tungsten is low, but at long time (6 h), it reaches high values (81.7% at 100°C and 86.4% at 125°C). This is due to the low vapor pressure over the pulp of leaching.

(3) At 150-225°C, the recovery of tungsten gradually increases and it reaches high values 88.5- 98.8% within 4h. The high recovery of tungsten at such temperatures is due to the increase of vapor pressure over the pulp of leaching. At 225°C and longer time than 4 h, a gradual and small decrease of recovery is observed. This decrease of recovery may be due to the reversible reaction starting between the products of wolframite leaching with the formation of insoluble iron and manganese tungstates or their hydrated forms .

(4) The calculated values of the standard free energy (ΔG^o) and equilibrium constants (K) of the reaction of alkaline processing of wolframite as iron or manganese tungstate at 298.15 K are – 10.582 and – 11.912 kcal mol^{-1} and $5.76 \cdot 10^7$ and $5.45 \cdot 10^8$, respectively. These values reflect the irreversibility of the reaction.

(5) The kinetic study of the mechanical activation of processing of wolframite by simultaneous ball-milling and leaching is governed by the structure of wolframite and the condition of the product surface coating. The experimental and kinetic data reveal that the rate of the reaction of wolframite leaching is controlled by diffusion through the product layer at 100 and 125°C. At 150°C and higher temperatures, it is chemically rate-controlled as confirmed by the value of the calculated apparent activation energy (21.622 kcal mol^{-1}) at temperature range 150-225°C and other observations. The rate of leaching is therefore a function of temperature, time and solid-liquid contact. The results of the kinetic study of alkaline processing of wolframite and its mechanical activation show that the system is under kinetic control.

(6) It was found that complete leaching of wolframite (97.8- 98.7%) is attained at 175 and 200°C within 6 and 4h, respectively. These conditions may be improved by carrying out the wolframite processing by the ball-milling and alkaline leaching under oxidizing conditions at much lower temperature and shorter time

(7) Processing of wolframite by simultaneous ball-milling and pressure leaching is mechanically activated and accelerated in ball-milling autoclaves as a result of the grinding action of steal balls, that caused the immediate rupture and continuous removal of sodium tungstate and iron–manganese hydroxides from the reacting particle surfaces. This leads to the exposing of fresh wolframite particle surfaces to alkali leach solution and consequently the acceleration of the reaction of wolframite processing .

References

Abdel-Rehim, A.M., 2006. Mechanical activation of monazite processing by simultaneous ball-milling and leaching In: Kongoli, F. and Reddy, R.G., (Eds.), Advanced Processing of Metals and Materials, vol. 6, Aqueous and Electrochemical Processing , TMS Publ., USA, pp. 65-174.

Abdel-Rehim, A.M., 2001. Thermal analysis and X-ray diffraction of synthesis of scheelite from wolframite. J. Thermal Analysis and Calorimetry, 65(2), 1283-1296.

Abdel-Rehim, A.M., 2005. A new technique for extracting zirconium from Egyptian zircon concentrate. Int. J. Miner. Process, 76, 234-243.

Abdel-Rehim, A.M., 1973. Autoclave leaching of some economic minerale. Acta Geol. Acad . Sci. Hung., Acad. Sci. Publ, Budapest, 17,307-318.

Amer, A.M., 2000, Investigation of hydrometallurgical processing of mechanically activated low grade wolframite concentrate. Hydro-metallutgy, 58 (3), 251-259.

Balaz, P., 2000. Extractive Metallurgy of Activated Minerals. Elsevier, Amsterdam, p. 270.

Balaz, P., 2008. Mechanochemistry in Nanoscience and Minerals Engineering, Springer, Berlin Heidelberg.

Barsukova, Z.S. and Zverev, L.V., 1977. Chlorination of wolframite in alkali and alkaline earth metals chlorides. Izvestia Vyshikh Utsheb. Zavedeny, Nonferrous Metals, 7 (Russian).

Bolotinkova, L.E., 1973. Technological Production of Rare Metals. Izdatelstvo Metallurgia Publ., Moscow, p. 470.

Burwell, B., 1966. Process for Recovery of Tungsten from Scheelite and Wolframite. US Patent N. 3256058, June 1966, p. 7.

Doyle, F.M. and Duyvesteyn, S., 1993. Aqueous processing of minerals, metals and materials. JOM, TMS Publ., 45(4), 46-54.

Gokhale, Y.W. and Bhat, T.R., 1967. Studies on the chlorination of wolframite and low grade tungsten ores. Ind. J. Techn, 5(9), 280-282.

Gomes, M.J., Daniel A.O. and Andrea, E.R., 1997 . Extraction of Tungsten from Ores. US Patent N. 4167555, Iss. Sept. 11, 17 , p.14.

Habashi, F., 1999. A Textbook of Hydrometallurgy, 2nd Edition. Metallurgie Extractive Quebec, Sainte Foy, Canada, Distributed by Laval University Bookstone.

Habashi, F., 1980. Principles of Extractive Metallurgy, vol. 1, Gordon and Breach, New York, pp. 223-251.

Ivanov, G.F. 1972.Geochemical conditions of formation of wolframaite deposits. Izdatelstva Nauka, Publ. Science, Moscow, pp. 89 -116.

Kalashnik, A.N., Resniakova, A.A. and Mirken, L.A., 1973. Processing of tungsten-molybdenum bearing products of industry. Nonferrous Metals, Moscow, 13, 57-65.

Klatshko, L.I., Levtonov, I.P.and Umansky, A.M., 1997. New Technology of Tungsten and Molybdenum. Metallurgyia Publ., Moscow. P, 18 .

Lassner, E. and Schubert, W.D., 1. Tungsten, Properties, Chemistry, Technology of the Element, Alloys and Chemical Compounds. New York, Plenum Publ.

Meerson, G.A. and Zelikman, A.N., 1965. Metallurgy of Rare Metals. Metallurgizdat, Moscow, pp. 49 -85.

Osseo-Asare, K., 1992. Solution chemistry of tungsten leaching system. J. Electronic Materials. Springer,21(1), 555-564.

Paul, B.Q., Dale. K.H., 19982. Combined Autoclave Soda Digestion of Wolframite and Scheelite.US Patent N. 43200 2, Issue date March 16, 1982.

Quatrini, R.V., Martin, C. and Martin, B.E., 1982. Tungsten Recovery from Tungsten Ore Concentrate by Caustic Soda Digestion . WIPO/patent WO/1 82/0038 , Int. Applic. N. PCT/US 1 82/060405.

Reznitshenko, N.A. 1967. Metallurgy of Tungsten and Molybdenum and Niobium. Nauka. (Science), Moscow.

Rysegz, l. and Mazapek, C., 1997. Hydrometallurgical methods of processing of ores containing Mo, W and Re. Rudi I Metali Niezel (Polish), 24(5), 245-248.

Robie, R.A., Waldbaum,D.R., 1968. Thermodynamic Properties of minerals and related substances at 298.15 K(25oC)-and 1atm. U.S. Geol. Surv. Bull. (125), 16-19

Said, R. 1990. The Geology of Egypt. Balkema, Rotterdam.

Sakharkar, M.V, Sharma, J., Ray, R.K., Biswas, A.K., 1992. Characterization of synthesized wolframite and their leach residues. Hydrometallurgy, 44(1-2), 65-69 .

Srinivas, K., Srineeva, T., Natarajan, R. and Padmanabhan, N.P., 2000. Studies on the recovery of tungsten from composite wolframite- scheelite concentrate. Hydrometallurgy, 58(1), 43-50.

Vladimir, Z., Zdenka, Z. and Daniel, A.B., 1978. Process for Recovery High Purity Tungsten Composition from Wolframite. Patent N. 40, 2400 Issue date May 30, 1978, Int. classif. C010-4100.

Weast, Robert C., et al., 1997. CRC Handbook of Chemistry and Physics, 5th ed. CRC Press, Inc., pp. D67-D78.

Zelikman, A.N., Samsonov, G.V. and Krein, O.E., 1964. Metallurgy of Rare Metals. Izdatelstvo Metallurgia Publ., Moscow., pp. 25- 39.

Zhao, Z.W., Li, Y.J., Li, H.Q. and Sun, P.M., 1995. Digestion of scheelite- wolframite mixed concentrate with caustic soda. J. Central South Univ. Technology, 26 (3), 349-357.

LEACHING OF VANADIUM FROM THE ROASTED VANADIUM SLAG WITH HIGH CALCIUM CONTENT BY DIRECT ROASTING AND SODA LEACHING

Xiao-Man Yan[1], Bing Xie[1], Lu Jiang[1], Hai-Peng Guo[1], Hong-Yi Li[1]

[1]College of Materials Science and Engineering, Chongqing University, Chongqing 400044, China

Keywords:calcified vanadium slag,directroasting,soda leaching

Abstract

Traditional, calcium compounds cannot be used in vanadium extraction processing due to the limitation of sodium roasting technology, which results in unsatisfied dephosphorization for vanadium-containing hot metal in converter and thus increasing the dephosphorization burden for subsequent steelmaking. The calcium roasting technology is an alternative for vanadium extraction and allows adding lime into vanadium slag in converter. In the present study, vanadium was leached by soda from roasted vanadium slag with high calcium content. Effects of roasting temperature and time, leaching temperature, Na_2CO_3 concentration andliquid to solid ratio on vanadium leaching rate were studied.

The results show that 94% of vanadium was extracted by roasting the vanadium slag at 950℃ for 2.5h and then leaching in 160g/L Na_2CO_3 liquid at 95℃ for 1h with liquid to solid ratio of 10:1 ml/g. Impurities of Si and P reacted with calcium and formed insoluble Ca_2SiO_4 and $Ca_3(PO_4)_2$ respectively.

Introduction

Vanadium is a strategic element that is widely used in many fields, such as steel industry, chemical industry and aerospace industry [1-2]. Vanadium is relatively abundant in the earth's crust and distributed in nature in a scattered form. It often occurs in combination with various minerals, vanadium titanomagnetite ore is one of the main sources of vanadium[3]. For recovery of vanadium from vanadium titanium magnetite, the blast furnace reduction and converter oxidizing technology is used. During converter oxidizing process, the vanadium slag and semi-steel were obtained by blowing oxygen into vanadium containing hot metal. The semi-steel was poured into another converter for steelmaking, while the vanadium slag was used for extracting vanadium by subsequent roasting-leaching process.Sodiumroasting-water leaching is the traditional technology, but it has some limitations. For example, it could produce contaminated gas such as HCl andCl_2.

This requires low CaO content in vanadium slag, when CaO content increases 1%, V_2O_5 will lose 4.7%-9%[4],and result in unsatisfied dephosphorization for vanadium-containing hot metal in converter and thus increasing the dephosphorization burden for subsequent steelmaking.In recent years, the development of calcium roasting technology makes it possible to add lime into converter during vanadium extracting process; the vanadium slag will contain CaO.

In this paper, a novel processing of direct roasting and soda leaching for extracting vanadium from vanadium slag with high calcium content is presented. The factors influencing vanadium recovery in the roasting and leaching process were investigated; the optimal conditions for roasting and leaching were obtained.The results show 94.13% of vanadium was extracted by roasting the vanadium slag at 950℃for 2.5h and then leaching in 160g/L Na_2CO_3 liquid at 95℃ for 1h with liquid to solid ratio of 10:1 ml/g,while the leaching rate of phosphorus is only 10.41%.

Experimental

Materials

It is noted that the best vanadium and phosphorus extracting will be obtained when the basicity of slag reached up to 1.5[5]. Therefore the basicity of the sample employed in this study was fixed at 1.5. All of the ingredients used in the sample were analytical pure grade. The mixed sample was put into corundum crucible surrounded by graphite crucible. When temperature of the MoSi2 furnace was 1173K, the graphite crucible was placed into theMoSi$_2$ furnace and heated to 1723K. After the sample melting, cooled in furnace for 30 minutes and let temperature of the MoSi2 furnace fall to 1373k slowly and then removed the graphite crucible which was allowed to cool at air temperature. After cool down, the vanadium slag with high calcium content was obtained and milled and screened by 200-mesh sieve.

Table I. Chemical composition of vanadium slag (mass%)

FeO	SiO_2	Al_2O_3	CaO	V_2O_3	TiO_2	MgO	MnO	P_2O_5
29.01	15.86	1.67	16.47	14.08	10.63	3.25	6.50	2.53

Experimental procedure

The vanadium slag without any additive was direct roasted in muffle furnace at a certain temperature and time under air atmosphere by opening the door of muffle furnace. To prevent the sample from sintering, the vanadium slag should be stirred at 15-minute intervals. When the roasting process was finished, the roasted material was taken out and cooled at room temperature and milled and screened by 200-mesh sieve.
The roasted slag was leached by the Na_2CO_3 solution at certain temperature, S/L ration and time. The sample and Na_2CO_3 solution were placed into the three-necked flask by mixing. The reactor had 3 necks, one for a condenser to keep the concentration of Na_2CO_3 invariant and other two for cork. The reaction mixture was agitated with a magnetic stirrer at 400r/min and heated on a hot plate indirectly through a water bath. After leaching, the leach slurry was filtered immediately via vacuum filtration and washed several times with distilled water.
Ammonium ferrous sulfate titration was used to analyze the concentration of vanadium in the solution. The phosphorus in the leach solution was determined by ICP-AES.

Results and discussion

Results of vanadium slag roasting experiments
Roasting

Vanadium slag is mainly composed of iron spinel phase, olivine, metallic iron, free silica, etc. Roasting is high-temperature oxidation behavior of vanadium slag and in the processing the mainreactions are as follows[6]:

$$FeO.V_2O_3+FeO+1/2O_2=Fe_2O_3{\cdot}V_2O_3 \quad (1)$$
$$Fe_2O_3{\cdot}V_2O_3+1/2O_2= Fe_2O_3.V_2O_4 \quad (2)$$
$$Fe_2O_3{\cdot}V_2O_3+1/2O_2= Fe_2O_3{\cdot}V_2O_5 \quad (3)$$
$$Fe_2O_3{\cdot}V_2O_5=Fe_2O_3+V_2O_5 \quad (4)$$
$$V_2O_5+CaO=CaO{\cdot}V_2O_5 \quad (5)$$

In addition, the forming of calcium vanadate during roasting processing depends on roasting temperature and the content of calcium salt[7].
The mass and mole ratio for CaO/V_2O_5 in experimental materials are 0.96 and 3.13 respectively. According to the aforesaid reactions, calcium content is sufficient relative to vanadium content. So any additives need not to be added and direct roasting is employed. Fig.1 shows the XRD patterns of samples before roasting and roasted at 950℃. It can be seen that vanadium mainly exists in vanadium spinel, and olivine phase and pyroxene phase also existed in vanadium slag before roasting. After roasting, Fe_2O_3 and Fe_2TiO_5 and $Ca(VO_3)_2$ was found.

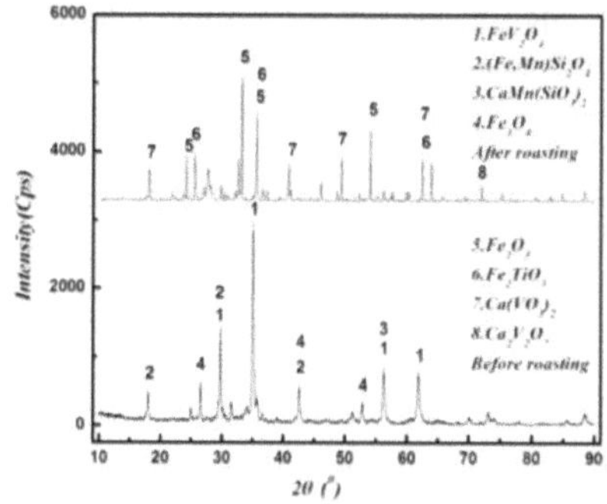

Fig 1. XRD Pattern of vanadium slag and the roasted vanadium slag

Effect of roasting temperature

The effect of temperature on the roasting of the slag sample was investigated at the temperature range of 300℃-1000℃ with fixed roasting time of 2.5h. Then the roasted slags were leached with 160g/L Na_2CO_3 in liquid/solid ratio of10:1 (mL/g) at 90℃for 1 hr. The results are shown in Fig.2. As shown in this figure, vanadium extraction increased with the roasting temperature increasing from 300℃ to 950℃, vanadium extraction reached up 88.61% at 950℃. However, with a further increase in roasting temperature, the leaching rate of vanadium decreased. This phenomenon was caused by slightly solubility vanadate being formed at higher temperature and morecomplex silicates also being produced at higher temperature and wrapped vanadium for the roasting of vanadium slag with high content of CaO.

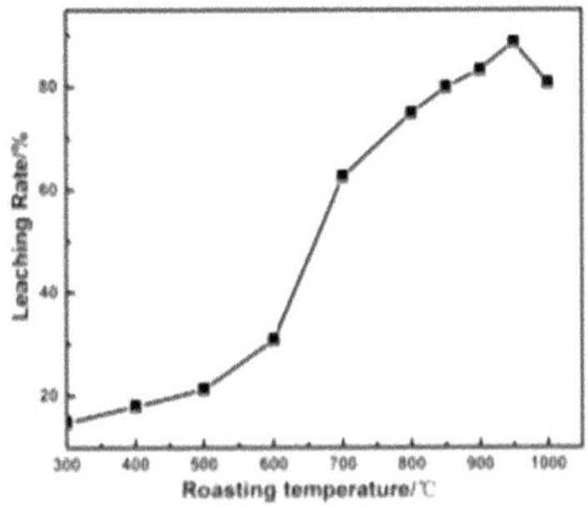

Fig 2. Effect of roasting temperature on leaching rate of vanadium

Effect of roasting time

Roasting time is another important factor in the roasting process. In order to study the effect of roasting time, a series of experiments was conducted. The roasting time for vanadium slag was varied from 0.5h to 3h at 950℃ andleaching condition was the same as described in studying of roasting temperature.The results of subsequent leaching (Fig.3) clearly show that the optimized roasting time was 2.5h. Further Increasing the roasting time would cause the sample being sintered [9], resulted in decreasing of vanadium leaching rate.

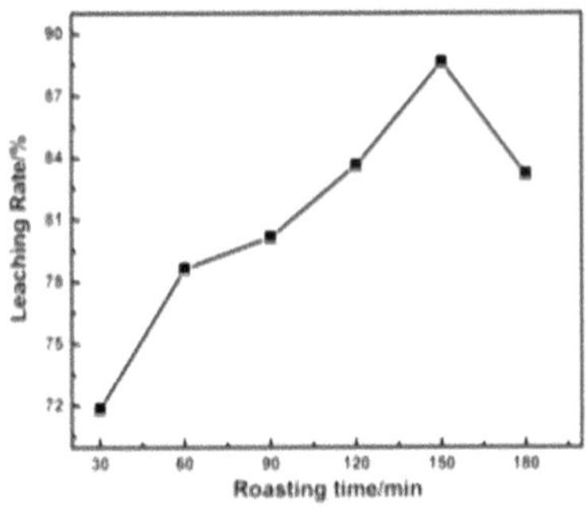

Fig 3. Effect of roasting time on leaching rate of vanadium

Results of sodium carbonate leaching

Calcium vanadatewhich formed in roasting process could be leached by acid or alkaline for vanadium extraction. Comparing with acid leaching, alkaline leaching can decrease dissolving of impurities, which is beneficial to subsequent vanadium precipitation. In alkaline leaching, exchange reaction between CO_3^{2-} andVO_3^- makes $Ca(VO_3)_2$turn to CaCO3 with smaller solubility product and let vanadium enter into the filtrate[10]. In general case,Na_2CO_3, $NaHCO_3$ andNH_4HCO_3 solutions are used for leaching roasted vanadium slag. The roasted slag was leached by the mentioned three kinds of solution at 95℃ for 1h with liquid to solid ratio of 10:1mL/g under the same experimental conditions, and the filtrates was gotten by filtering. The

experimental results shown that the leaching rate of vanadium and phosphorus are 94.13%, 10.41%;90.49%, 13.98%; 88.64%, 15.30% when Na_2CO_3, $NaHCO_3$ and$NH_4HCO_3$are used as leaching solution respectively, which indicate that Na_2CO_3 leaching is a more effective leaching process compared with other two kinds of solutions.
The main reaction happened in leaching reaction is as follow:

$$Ca(VO_3)_2+Na_2CO_3=CaCO_3+2NaVO_3 \quad (6)$$

Then the effect of Na_2CO_3 concentration, leaching temperature and liquid to solid ratio on leaching rate of vanadium were studied as described in table II.

Table II.The experiment scheme of Leaching roasted vanadium slag by sodium carbonation

Influence factors	Na_2CO_3 concentration /g/L	Leaching temperature/℃	Liquid to solid ratio/mL/g
Na_2CO_3 concentration /g/L	40,80,120,160,200	80	10: 1
Leaching temperature/℃	160	60,70,80,90,95,100	10: 1
Liquid to solid ratio/mL/g	160	80	4:1,6:1,8:1,10:1,12:1

Effect of Na_2CO_3 concentration

Fig.3 illustrates the effect of Na_2CO_3 concentration on vanadium extraction at 90℃ for 1h with a liquid to solid ratio of 10:1mL/g. As shown in this figure, the leaching rate of vanadium reached a peak of 88.61% when Na_2CO_3 concentration was 160g/L. However, further increasing Na_2CO_3 concentration leads to lower vanadium extraction. In this situation, the unreacted roasted samples were covered by a layer of solid products, which prevented the further leaching of vanadium at higher Na_2CO_3concentration, resulting to the decreasing of vanadium leaching[7].

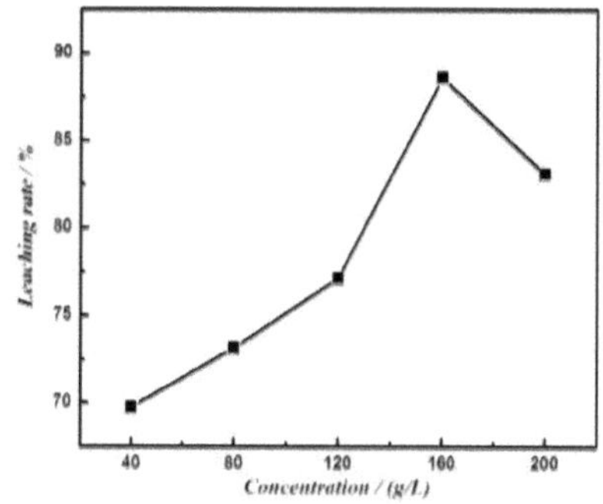

Fig 4. Effect of Na_2CO_3 concentration on leaching rate of vanadium

Effect of liquid to solid ratio

Fig 5. shows the relationship between liquid to solid ratio and the leaching rate of vanadium.The roasted slag was leached by 160g/L Na_2CO_3 concentration at 90℃ for 1h at the liquid to solid ratio from 4:1mL/g to 12:1mL/g. The leaching rate of vanadium increased with increasing the liquid to solid ratio up to 10:1ml/g.

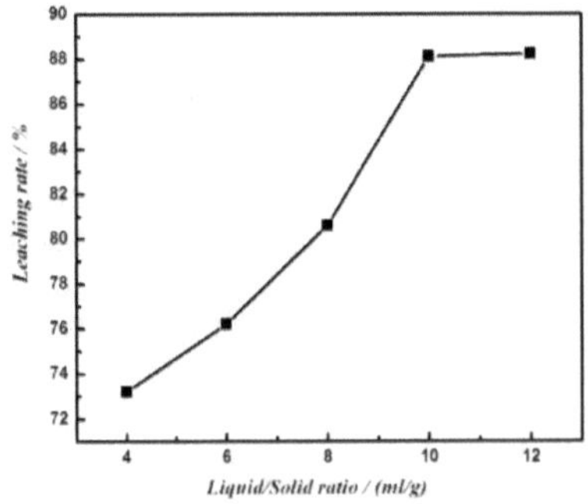

Fig 5. Effect of liquid to solid ratio on leaching rate of vanadium

Effect of leaching temperature

Fig.6 shows the experimental results of the roasted slag leached with 160g/L Na_2CO_3 concentration for 1h at the temperature range of 60℃ to 100℃ with liquid to solid ratio 10:1mL/g. As shown in Fig.6, the leaching rate of vanadium increased from 68.42% to 94.13% with the leaching temperature increase from 60℃to 95℃, and further increasing leaching temperature, the leaching rate of vanadium almost kept the same.

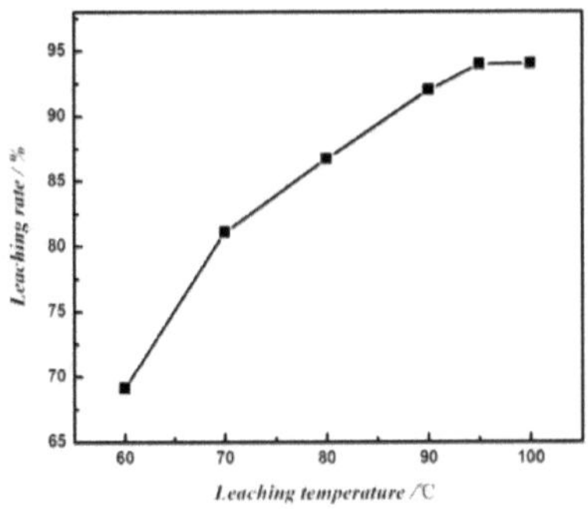

Fig 6. Effect of leaching temperature on leaching rate of vanadium

XRD pattern of Leaching residue and the roasted slag

Fig.7 shows XRD pattern of leached residue and the roasted slag, $CaCO_3$ was found and $NaVO_3$cannot be foundin the residue. The leaching rate of phosphorus was 10.41% through ICP-AES detection. Calcium phosphate had been found in the residue, it is probably that under the experimental conditions, the reaction between Calcium phosphate in the vanadium slag and sodium carbonate was relative weak so thatphosphoruswas difficult to be transferred into the leachate.

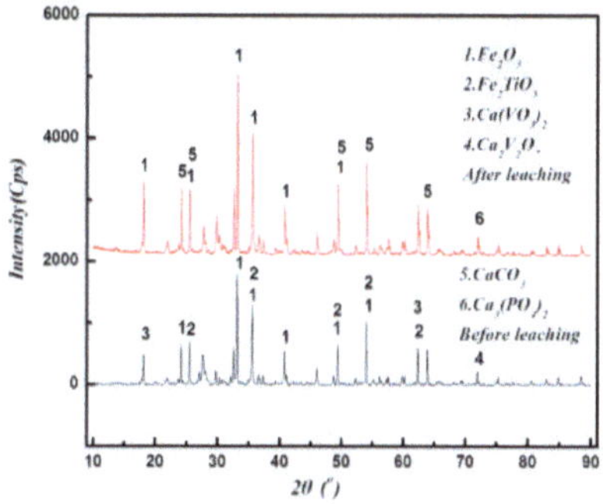

Fig 7. XRD pattern of Leaching residue and roasted vanadium slag

Conclusions

In order to effectively extract vanadium and prevent phosphorus leaching during leaching processing of vanadium slag with high CaO, the effect of roasting and leaching process parameters on vanadium extraction were studied. The experimental results illustrated that direct roasting and soda leaching was an effective method of extracting vanadium and preventing phosphorus from vanadium slag with high content of CaO.

(1) Roasting temperature and time are 950℃, 2.5hrs respectively, the Na_2CO_3 concentration is 160g/L, leaching temperature and time are 95℃ and 1h respectively, and the liquid to solid ratio is 10:1mL/g.

(2) Under the optimum conditions the leaching rate of vanadium could be reached up to 94.13% and the leaching rate of phosphorus was only 10.41%.

References

[1] Moskalyk, R.R., and Alfantazi,A.M, "Processing of vanadium: a review, "Minerals Engineering, 16(2003), 793-805.

[2] Taylor P R et al., "Extractive metallurgy of vanadium containing titaniferrous magnetite ores: a review," Minerals & Metallurgical Processing, 23(2006), 80-86.

[3] M.S. Liao, Foreign vanadium metallurgy. (Beijing: Metallurgical Industry Press, 1985), 270-297.

[4]M. L. Wang, Oxidation of vanadium. (Beijing: Metallurgical Industry Press, 1982), 121-130.

[5] Tao Zhao,"Experimental study of phosphorus removal and vanadium extraction from vanadium-containing hot metal,"The 18th annual national conference on steelmaking, (2014), 270-274.

[6] S.L.Yang, Vanadium Titanium. (Beijing: Metallurgical Industry Press, 2009),28-31.

[7] S.X.Li, Mechanism research on oxidation roasting and leaching process of high calcium low-grade vanadium slag.(Chongqing: Chongqing university)

[8] S.D.He, "An Environmentally-friendly Technology of Vanadium Extraction from Stone Coal,"Miner. Eng, 20(2007), 1184–1187.

[9] X.J. Li, "The effect of roasting parameters in calcium roasting on the leaching rate of vanadium, " Journal of process engineering, 12(2012), 56-57.

[10] J. C, Vanadium and vanadium metallurgy. (Pan zhihua:Comprehensive utilization of Pan zhihua resource leading group office, 1983)

Rare Metal Technology 2015
Edited by: Neale R. Neelameggham, Shafiq Alam, Harald Oosterhof, Animesh Jha, David Dreisinger, and Shijie Wang
TMS (The Minerals, Metals & Materials Society), 2015

SOLVENT EXTRACTION OF VANADIUM FROM CONVERTER SLAG LEACH SOLUTION BY P204 REAGENT

ZHANG Ting 'an, ZHANG ying, LV Guozhi, LIU Yan, ZHANG Guoquan, LIU zhuolin

Key Laboratory of Ecological Utilization of Multi-metal Intergrown Ores of Ministry of Education, Northeastern University, Shenyang 110819, China

Keywords: Vanadium slag; without roasting; Pressure leaching; Extraction separation; Distribution factor

ABSTRACT

In view of the high consumption and contamination of traditional vanadium extraction technology, a new process of leaching the converter vanadium slag without roasting with titanium dioxide waste was proposed. Extraction mechanism of P204 in the leaching solution had been researched and single factor experiments show that after a prereduction of Fe, and under conditions: ambient temperature, leaching solution original pH=2.5, phase ratio (A/O)=1:3, shaking time 8min, the extraction rate of V is more than 98.62% meanwhile the separation rate between V and Fe is up to 135.3 in the 20%P204-10%TBP-Sulfonated kerosene extraction system.

1. Introduction

Vanadium is an important strategic metal as it is widely used in alloy preparation, petrochemical, aerospace and other fields. Vanadium is always found in combination with various minerals such as vanadium-titanium magnetite and black shale. More than 88% [1-2] of the vanadium feedstock in China is sourced from the converter slag produced from vanadium-titanium magnetite which are commonly at Panzhihua in Sichuan province and at Chengdu in Hebei [3]. At the present time, the main treatment for the converter slag include a roasting process whereby in one case the slag is mixed with sodium salts then roasted and in the other case the slag is roasted with calcium slats. But they always suffered from low vanadium recovery or produce waste water and off gas needing cleaning treatment [4]. An alternative process based on waste acid leaching of the vanadium containing slag without roasting has been proposed [5].

Since vanadium is extracted from converter slag including different elements as well as the chemical composition of titanium dioxide waste is multiplex and the acid leaching process is not selective so many purities such as Fe (III), Fe (II), Mg (II), Al (III) and Mn (II) .Thus the purification step for the leaching solution has a variety of components are always a challenging part of the production process. This paper described that solvent extraction is an effective method for extracting vanadium (IV) from a multi-element mixed acidic leaching solvents. (2-ethylhexly) phosphoric acid (P204) is chosen as the extractant for vanadium in the low pH leaching solution due to its high selectivity and operability. The objective of the present investigation concentrates on mechanism of vanadium selective extraction and effects of several issues (such as aqueous phase pH value, extractant dosage, A / O phase ratio) on the elements extraction.

2. Experimental

2.1 Materials

In the present work, a titanium-ferrous BOF slag from Panzhihua iron and steel co., LTD in Sichuan province was used in the experiment. The chemical composition of the slag had been analyzed as shown in Table1.

Table 1. Chemical composition of converter vanadium slag (mass, %)

Fe_2O_3	SiO_2	V_2O_5	MnO	TiO_2	Cr_2O_3	CaO	Al_2O_3	MgO	P_2O_5	SO_3	K_2O	Nb_2O_5
41.79	14.76	13.22	11.70	9.03	3.43	3.16	1.79	0.67	0.25	0.09	0.06	0.05

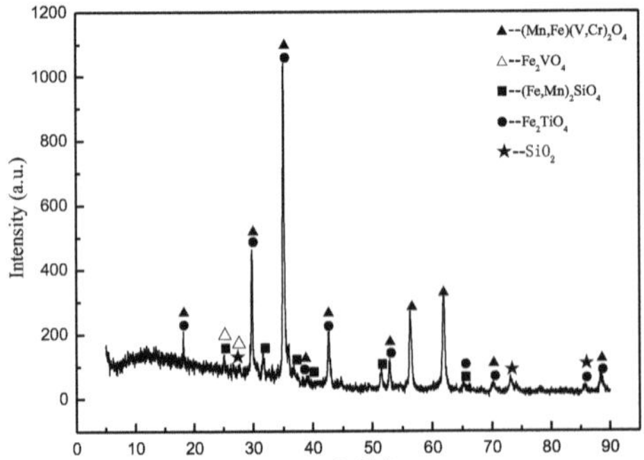

Fig 1. XRD analysis of vanadium slag

The XRD of converter slag(Fig. 1) show that the main phase are Vanadium-titanium-iron phase and Silicate phase with vanadium mainly in the iron olivine .The acid for leaching was simulated waste acid from the titanium dioxide production process of which the composition as shown in Table 2.

Table 2. Chemical composition of Titanium Dioxide Waste (Concentration, *g/L*)

chemical content	H_2SO_4	Fe^{2+}	Mg^{2+}	Al^{3+}	Mn^{2+}
g/L	200	30.5	2.5	1.57	2.7

Previous studies [6]have indicated that the optimum parameters during leaching were: 130 ℃, leaching time 90min,vanadium slag particle-0.075 ~ +0.055 mm (-200 ~

+260mesh), stirring rate 500rpm, initial acid concentration 200g/ L and a liquid to solid ratio(L / S) of 10: 1, under these appropriate conditions, the composition of leaching solution is shown in Table 3

Table 3. Content of valuable elements in the leaching solution（g/L）

chemical content	V	Fe	Mn	Al	Mg	Cr	Ti
g/L	7.06	69.86	8.75	3.83	2.94	1.96	0.19

Under the above leaching condition, a series of the pH-potential diagram[7] indicate that the solution after leaching contain vanadium in form of VO^{2+} and other elements as Fe(III), Fe(II), Mg(II),Al(III) and Mn(II) etc. Solvent extraction technique was used to treat the leaching solution, and in the organic phase we used (2-ethylhexly) phosphoric acid (P204) as extraction agent, tributyl phosphate (TBP) as accelerant and sulfonated kerosene as thinner.

2.2 Method

Solvent extraction experiments were carried out in 125ml separating funnels by mechanically shaking the organic and aqueous phases in a SHA-C type water bath oscillator at room temperature (298±1 K) for 6 min. It was found that the vanadium extraction reached an equilibrium after 6 min but the iron extraction slowly increased with longer shaking times, possibly due to slow oxidation of Fe(II) to Fe(III). Therefore, an equilibration time of 6 min was chosen as optimum. Plasma emission spectrometer (ICP) was used for measuring the content of valuable elements in the aqueous phase to facilitate the calculation of element extraction rate.

2.2.1 P204 extraction mechanism Wang Y X, Zeng P [7] had researched P204 extract V (IV) process, considered the extract reaction under acidic conditions is as below:

$$VO^{2+}_{(a)} + 2\,H_2A_{2(o)} = VO\,(HA_2)_{2(o)} + 2\,H^+_{(a)} \quad (1)$$

$$Fe^{3+}_{(a)} + 3\,H_2A_{2(o)} = [FeA_3 \cdot 3HA]_{(o)} + 3\,H^+_{(a)} \quad (2)$$

In the above formula, H_2A_2 represents P204, (O) represents organic phase, (a) represents aqueous phase. And the structure of P204 is as below:

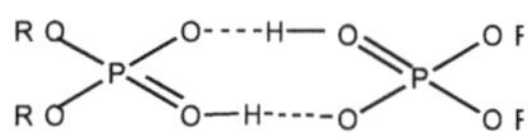

（R== C4H9CH (C2H5) CH2）

Fig 2. Structure of P204

The vanadium extracted substance and Fe (III) extracted substance have the structure as below:

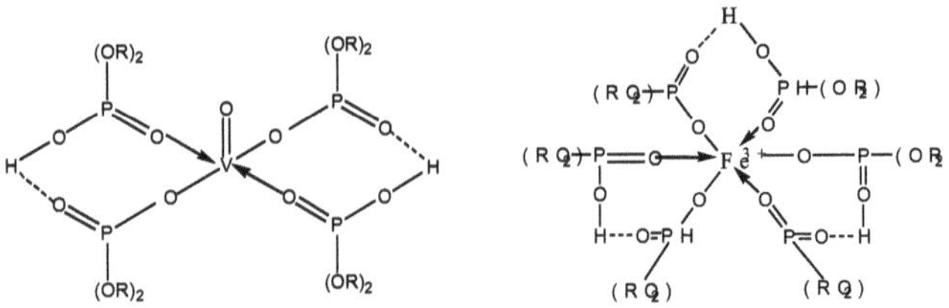

Fig 3. Structure of extracted substance

We can see that P204 is a dimer from its structure due to the two hydrogen bonds. In the extraction process, metal ions to be extracted will replace H in the O-H group and form new covalent bonds, while coordinating with lone pair of electrons O in the P=O group and form a stable coordinate bond [8]. Fig.3 shows that vanadium extracted substance was consisted of two eight-member rings and Fe(III)extracted substance was consisted of three eight-member rings.

2.2.2 Reduction pretreatment Extractant (P204) employed in this study can coextract vanadium and trivalent iron ion, which can lead to a low service efficiency of the extractant and a product with high impurity content (Fe). Hence, it is necessary to reduced Fe(III) to Fe(II) while V(V) was reduced to V (IV)before extraction as P204 hardly extraction Fe(II)extraction and its extraction ability for V (IV) is much greater than the extraction ability for V(V) [9]. Sodium sulfate was employed as the reductant, the main reaction as follows:

$$2VO_2^{+}+SO_3^{2-}+2H^{+}=2\ VO^{2+}+H_2O+SO_4^{2-} \tag{3}$$
$$2Fe^{3+}+SO_3^{2-}+H_2O=SO_4^{2-}+2H^{+}+2Fe^{2+} \tag{4}$$

Previous experiments have shown that Fe (III) and V (V) in the leaching solution will be fully reduced when the amount of reducing agent is 0.2Fe (/V mol, hereinafter kept the same). In addition, the organic phase capacity of titanium, chromium, manganese and other impurity elements such as phosphorus is weak in the extraction system, which will benefit a higher purity vanadium extraction.

3. Results and discussion

3.1Effects of pH value

Under the experiment condition: ambient temperature,20%P204-10%TBP-Sulfonated kerosene extraction system, phase ratio (A/O)=1:1, contact time 6 min, the effect of aqueous phase pH on the extraction of vanadium and other elements in the leaching solution was investigated.

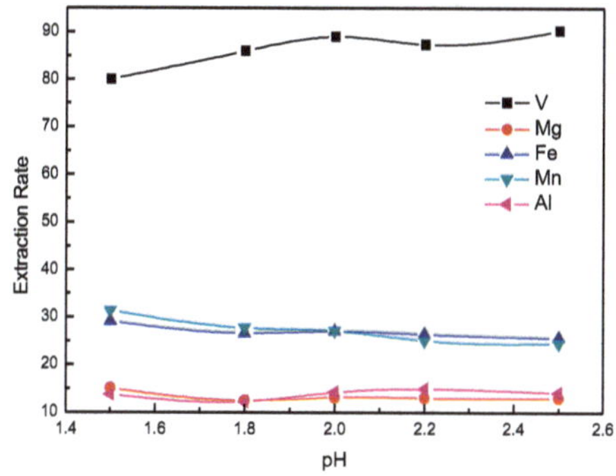

Fig 4. Effect of initial pH value of liquid phase on the extraction rates of V, Mg, Fe, Mn and Al

It is showed in Fig.4 that the vanadium extraction efficiency is greatly affected by pH value. Vanadium extraction efficiency increase significantly from 80.06% to 90.28%, when the pH value of the leaching solution ranged from 1.5 to 2.5, which could be explained by reaction (1), the release of H^+ would be enhanced in faintly acid so the higher the initial pH value is, the easier it would be for the vanadium to be extracted. A further increase of pH barely improve the vanadium extraction efficiency since vanadium extraction equilibrium is established when the pH value is 2.5 and vanadium loss may exist attributed to the adsorption effect of the impurity elements hydrolysate in the aqueous phase if the pH is higher [10] .The extraction rates of other elements fluctuate slightly and then steady when pH ranges from 1.5 to 2.5. Therefore, initial pH value should be adjusted to exactly 2.5 for the following experiments.

3.2 Effects of P204 dosage

Under the experiment condition: ambient temperature, pH =2.5, phase ratio (A/O) =1:1, contact time 6 min, the effects of P204 dosage (P204/Organic volume ratio) on the extraction of vanadium and other elements was investigated.

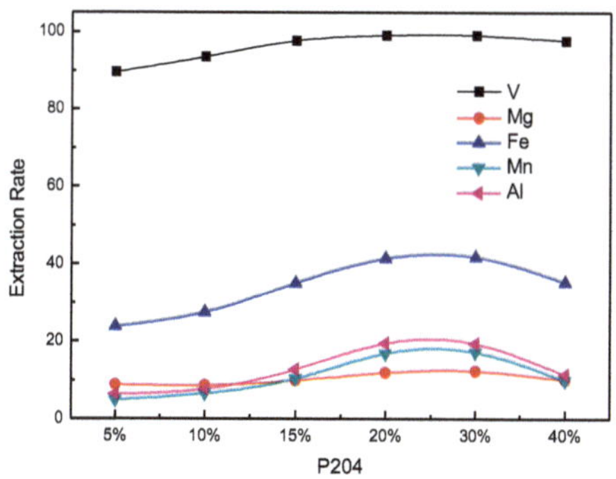

Fig 5. Effect of different amount of P204 on the extraction rates

of V, Mg, Fe, Mn and Al

Fig.5 shows that vanadium extraction rate initially increase then remain stable with the increase of P204 dosage. The vanadium extraction rate increased from 89.60% to 99.10% continuously when the P204 dosage range from 5% to 20%.No significant change for vanadium extraction rate is observed when the P204 dosage continue to increase which could attributed to the fact that nearly quantitative extraction was achieved and the viscosity of the extractants increased. The extractants with higher viscosity were more difficult to mix with aqueous solutions and it spent more time to reach equilibrium[11].And more impurity elements is extracted to the organic phase when the P204 dosage is more than 20%,which will not conducive to the stripping process. Thus in this work, 20% P204 dosage was considered optimal and chosen for the subsequent steps.

3.3 Effects of A / O phase ratio

Under the experiment condition: ambient temperature, pH =2.5, 15%P204-10%TBP-Sulfonated kerosene extraction system, contact time 6 min, we have also investigated the effect of A/O phase ratio on the extraction efficiency of vanadium and other elements in the leaching solution.

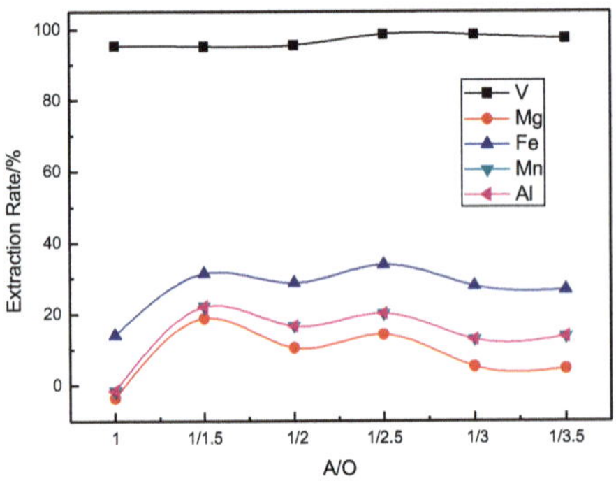

Fig 6. Effect of different (A/O) ratio on the extraction rates of V, Mg, Fe, Mn and Al

As was shown in Fig.6, vanadium extraction rate is 95.26% at the point A/O =1:1, and rise to the highest point 98.62% when A/O =1:3, indicate that a higher A/O phase ratio contribute to the vanadium extraction. Then it is hardly affected by the A/O phase ratio when it is further increased. The extraction rate of other elements rise with the increase of A/O ratio then decreased slightly after A/O =1:3. This may be explained by the unstable value of the elements' distribution ratio with different A/O ratio. So A/O=1:3 was considered optimal due to a low A/O ratio would lead to a huge consumption of the organic phase.

Under the optimal technological condition: ambient temperature, leaching solution original pH=2.5, phase ratio (A/O)=1:3, contact time 6min, in the 20%P204-10%TBP-Sulfonated kerosene extraction system the separation rates between V and other elements could be calculated from the formula(5)and(6)[12],as are shown in Table 4.

$$D_A = \frac{C_{[A](O)}}{C_{[A][W]}} \qquad (5)$$

$$\beta_{(A/B)} = \frac{D_A}{D_B} \qquad (6)$$

Among the above formulas, $C_{A(O)}$refers to the concentration of A in the organic phase, $C_{A[W]}$ refers to the concentration of A in the aqueous phase, D_A refers to the distribution ratio of A.$\beta_{(A/B)}$ refers to the separation coefficient between A and B.

Table 4. Separation rates between V and other elements

$\beta_{(V/Fe)}$	$\beta_{(V/Mn)}$	$\beta_{(V/Al)}$	$\beta_{(V/Mg)}$	$\beta_{(V/Fe、Mn、Al、Mg、)}$
135.30	1151.5	329	255.89	65.76

The result showed that the separation rates between V and other main elements in the leaching solution are all achieve a great value, an optimal separation effect of V have been attained in the extraction system under the optimal technological condition.

4. Conclusions

In the presented work, 20%P204-10%TBP- Sulfonated kerosene was employed as the extraction organic phase. The extraction mechanism of P204 in the leaching solution had been researched and excellent vanadium extraction efficiency can be reached while high purity of organic phase containing vanadium was obtained. About 98.62% of vanadium could be extracted after the organic phase was contacted with the aqueous phase (initial pH=2.5) for 6 min at room temperature when A/O=1:2.5 in the 20%P204—10%TBP synergistic extraction system.

Acknowledgements

This work was financially supported by the Chinese National Programs for High Technology Research and Development (Nos. 2010AA03A405 and 2012AA062303), the National Natural Science Foundation of China (Nos. U1202274, 51004033, 51204040, and 50974035), the National Science and Technology Support Program (No. 2012BAE01B02), the Fundamental Research Funds for the Central Universities (No. N100302005), and the Doctoral Fund Project (No. 20120042110011).

References

[1] Non-ferrous Metals Extraction Technique Manual Editorial Board. Rare High Melting Point Metal, Metallurgical Industry Press, Beijing, 1999, 276-350.

[2] Chen J L. Hydrometallurgy Manual, Vanadium, Chromium Hydrometallurgy, Metallurgical Industry Press, Beijing, 2005, 935-999.

[3] Yang S Z. Vanadium Metallurgy, Metallurgical Industry Press, Beijing, 2010, 22-42.
[4] Moskalyk R R, Alfantazi A M. Processing of vanadium: a review [J]. Mineral Engineering, 2003, 16(9): 793-805.
[5]Zhang T A, MU W Z, Dou Z H, Lv G Z,Liu Y,Zhao Q Y,He J C. A wet extraction method of vanadium used vanadium titanium slag [P].CN201010514573.8.2011,02,09. [6] Zhang G Q, Zhang T A, Zhang Y, Lv G Z, Liu Y, Liu Z L, Pressure Leaching of Converter Vanadium Slag with Waste Titanium Dioxide[J]， Rare Metals, DOI 10.1007/s12598-014-0225-3.
[7] ZHANG Ting-an, MU Wang-zhong, DOU Zhi-he, LÜ Guo-zhi, LIU Yan, Potential—pH diagrams for V-Fe-H2O system during oxygen pressure acid leaching of vanadium-bearing converter slags[J], The Chinese Journal of Nonferrous Metals,2011,21(11).
[8]Wang Y X, Zeng P, Mechanism Research on V (IV) Extraction by P204 [J], Rare Metals, 1990,14(3) : 172-175.
[9]Cheng R X. The Microscopic Mechanism of Gallium Group Elements Extraction by P204, Rare Metal, 1984, 03:66-71.
[10]Li X B, Wei C, Fan G, Deng Z G, Li M T, Li C X, Vanadium Pentoxide Extraction from Stone Coal Acid Leaching Solution used Ammonium Salt[J], Mining and Metallurgy,2010,03:49-53
[11] Guoping Hu, Desheng Chen, Lina Wang, Jing Chong Liu, Extraction of vanadium from chloride solution with high concentration of iron by solvent extraction using D2EHPA[J], Separation and Purification Technology,2014.01(31).
[12]Xu G X, Wang W Q, Wu J G, Gao H C, Shi D, Extraction Chemistry, Shang Hai: Shanghai Scientific and Technical Publishers, 1984,35-36.

Rare Metal Technology 2015
Edited by: Neale R. Neelameggham, Shafiq Alam, Harald Oosterhof, Shijie Wang, and David Dreisinger
TMS (The Minerals, Metals & Materials Society), 2015

EFFECT OF SOLUTION COMPOSITIONS ON OPTIMUM REDOX POTENTIAL IN BIOLEACHING OF CHALCOPYRITE BY MODERATELY THERMOPHILIC BACTERIA

Hongbo Zhao[1,2*], Jun Wang[1,2], Wenqing Qin[1,2], Guanzhou Qiu[1,2]

[1]*School of Minerals Processing & Bioengineering, Central South University, Changsha 410083, Hunan, PR China*

[2]*Key Lab of Bio-hydrometallurgy of Ministry of Education, Changsha 410083, Hunan, PR China*

Hongbo Zhao: alexandercsu@126.com Jun Wang：wjwq2000@126.com;

Wenqing Qin: csuqinwenqing@126.com; Guanzhou Qiu: qiuguanzhoucsu@126.com

Currently, high grade copper resources have been extensively exploited, leaving behind low grade copper ores which are difficult and costly to be processed with traditional technologies. Bio-hydrometallurgy is considered as a simple, efficient, low-cost and eco-friendly technology, which is widely considered as a promising technology in processing low grade ores. Chalcopyrite ($CuFeS_2$) is the most abundant copper-bearing minerals in the world, but it is refractory to bio-hydrometallurgy. Moderately thermophilic bacteria, including *Acidithiobacillus caldus* (*A. caldus*) and *Leptospirillum ferriphilum* (*L. ferriphilum*), are considered to have a good prospect for industrial application due to their advantages in accelerating the kinetics of chalcopyrite dissolution. Therefore, how to enhance the dissolution process of chalcopyrite in bioleaching by moderately thermophilic bacteria is an important topic. Many researchers concluded that low redox potential is beneficial to accelerating dissolution, while high redox potential cause rapid passivation. Some researchers proposed that chalcopyrite can be reduced to chalcocite (Cu_2S) at a low redox potential and the formed chalcocite (Cu_2S) can be rapidly oxidized. The process can be interpreted by chemical reactions as Eq. (1)-Eq. (4).

$$CuFeS_2 + 3Cu^{2+} + 3Fe^{2+} \rightarrow 2Cu_2S + 4Fe^{3+} \quad (1)$$

$$CuFeS_2 + 3Cu^{2+} + 4e^- \rightarrow 2Cu_2S + Fe^{2+} \quad (2)$$

$$Cu_2S + 4Fe^{3+} \rightarrow 2Cu^{2+} + 4Fe^{2+} + S^0 \quad (3)$$

$$2Cu_2S \rightarrow 4Cu^{2+} + 2S^0 + 8e^- \quad (4)$$

Therefore, there are two valuable questions: What is the optimum redox potential, and what is the effect of solution compositions on the optimum redox potential during bioleaching

* Corresponding author: Dr. Hongbo Zhao, Email: alexandercsu@126.com, Tel: 86-15116335985

of chalcopyrite by moderately thermophilic bacteria.

We mainly used electrochemical measurements, leaching experiments, XRD analysis and SEM analysis to investigate the above two questions.

The corresponding critical redox potential for the reaction of Eq. (1)-Eq. (4) can be calculated by thermodynamic calculation. It can be concluded that chalcopyrite can be reduced to chalcocite (Cu_2S) which can be further dissolved in the optimum range of redox potential (E_L~E_H).

CV tests were carried out at sweep rate of 20 mV/s, two distinct peaks were detected. Peak C was mainly considered as the reduction of chalcopyrite to chalcocite (Cu_2S). Therefore, the peak potential can represent the critical potential of E_H.

Anodic polarization tests and rest potential tests were conducted to verify the accuracy of E_H. For tests in the presence of sole Cu^{2+}, the presence of Cu^{2+} accelerated the corresponding reaction. The peak potential was proportional to the logarithms of concentration of Cu^{2+} and the coefficient was about 59.12. The rest potential was proportional to the logarithms of concentration of Cu^{2+} and the coefficient was about 35.07.Therefore, the average value of coefficients was 47.1, which was in agreement with the result of thermodynamic calculation (47.3).

For tests in the presence of Fe^{2+} and Cu^{2+}, the peak potential was proportional to the logarithms of concentration of Cu^{2+} and the coefficient was about 80.23. The rest potential was also proportional to the logarithms of concentration of Cu^{2+} and the coefficient was about 20.23. Therefore, the average coefficient was about 50.23, which was approximately equal to the value of thermodynamic calculation (47.3).

Sole Fe^{2+} was not able to accelerate the reduction of chalcopyrite. In the presence of Cu^{2+}, a negative logarithmic relationship existed between the high critical redox potential (E_H) and the concentration of Fe^{2+}. It was in accordance with the results of thermodynamic calculation.

We further use chemical leaching experiments to verify the accuracy of proposed optimum redox potential range. The slope of the redox potential to the logarithm of concentration ratio was almost the same with the theoretical value calculated by Nernst equation (60 mV/decade), proving that the redox potential of leaching solution was mainly determined by the concentration ratio of Fe^{3+} to Fe^{2+}. Results of chemical leaching showed that E_{L2}~E_H was the optimum range of redox potential for chemical leaching of chalcopyrite.

We further use bioleaching experiments to verify the accuracy of proposed optimum redox potential range. As Cu^{2+} and Fe^{2+} can promote the reduction of chalcopyrite according to the electrochemical measurements, so we conducted the bioleaching experiments in the presence and absence of Cu^{2+} and Fe^{2+}. Cu^{2+} and Fe^{2+} can quickly cause the chemical equilibrium of reduction reactions of chalcopyrite, and large amount of Cu^{2+} and Fe^{2+} was able to maintain the redox potential of leaching solution at a relatively high and stable value.

Bioleaching of chalcopyrite by *A. caldus* showed that the redox potentials of bioleaching with no addition of Cu^{2+} or Fe^{2+} were too close to the low critical redox potentials (E_{L3}), thus resulting in a steady but not high leaching rate. The addition of Cu^{2+} and Fe^{2+} can significantly accelerate the dissolution of chalcopyrite at the initial stage of bioleaching, but the further dissolution was inhibited. XRD and SEM analysis showed that large amount of jarosite and elemental sulfur formed at the initial stage of bioleaching in the presence of 0.1 M Cu^{2+} and Fe^{2+}. In addition, the formed jarosite and elemental sulfur almost totally covered the surface of chalcopyrite in extremely compact and tight form, thus inhibiting the further dissolution.

Similarly, for bioleaching of chalcopyrite by *L. ferriphilum*, the redox potentials of bioleaching with no addition of Cu^{2+} or Fe^{2+} was beyond the optimum redox potential range from the 14th day, thus resulting in stagnation of dissolution The addition of Cu^{2+} and Fe^{2+} can significantly accelerate the dissolution of chalcopyrite at the initial stage of bioleaching, but the further dissolution was inhibited. XRD and SEM analysis showed that large amount of jarosite formed at the initial stage of bioleaching in the presence of 0.1 M Cu^{2+} and Fe^{2+}. In addition, the formed jarosite almost totally covered the surface of chalcopyrite in extremely compact and tight form, thus inhibiting the further dissolution. Therefore, Cu^{2+} and Fe^{2+} should be added periodically to prevent the rapid formation of jarosite and accelerate the dissolution of chalcopyrite.

The main conclusions were as follows:

The optimum redox potential of both chemical leaching and bioleaching of chalcopyrite can be predicted by a proposed mathematic model as follows. The model indicated that the redox potential can be controlled at an optimum range of redox potential (E_L~E_H) to achieve high copper extraction.

$$E_H = 482 + 47.3\lg(aCu^{2+}) - 15.77\lg(aFe^{2+}) / mV$$

$$E_L = 401 + 29.5\lg(aCu^{2+}) / mV$$

The addition of Cu^{2+} and Fe^{2+} significantly promoted the dissolution of chalcopyrite at the initial stage of bioleaching. However, the further dissolution was hindered due to jarosite. Therefore, Cu^{2+} and Fe^{2+} can added periodically to prevent the rapid formation of jarosite and promote the dissolution of chalcopyrite.

Rare Metal Technology 2015

Poster Session

Rare Metal Technology 2015
Edited by: Neale R. Neelameggham, Shafiq Alam, Harald Oosterhof, Animesh Jha, David Dreisinger, and Shijie Wang
TMS (The Minerals, Metals & Materials Society), 2015

Research on Quality Improvement of Titanium Sponge By Process Optimization

Li Liang [1,2], Li Kaihua [1,2], Miao Qingdong [1,2], Wang Cong [1,2]
[1]Pangang Group Research Institute Co., Ltd., Panzhihua 617000, China
[2]State Key Laboratory of Vanadium and Titanium Resources Comprehensive Utilization, Panzhihua 617000, China

Keywords: Titanium Sponge; Process optimization; product grade;

Abstract

The current process of production of titanium sponge involves the use of $TiCl_4$ as a feed material and a huge amount of heat is generated during the magnesiothermic reduction process. For this reason, it is possible to agglomerate into a titanium lump with the partial high temperature. And, it is difficult to obtain soft titanium sponge with low content of impurities. In order to solve the current problems in the titanium sponge production process, process optimization experiments study of $TiCl_4$ feeding speed, vacuum distillation temperature, system pressure and time control had been conducted for obtain high quality titanium sponge. Experimental results show that lower feeding speed can improve the output capacity of titanium sponge and softer products can be obtained. The product grade rate of titanium sponge improved significantly.

Introduction

Titanium has advantages of high strength, low specific gravity, corrosion resistance, non-magnetic, weldability and other characteristics[1]. It is widely used in aerospace, transportation, petrochemical and other fields. At the present time, the only industrial method to produce titanium sponge is magnesium reduction process (Kroll process). It mainly consists of two stages: reduction and vacuum distillation. Liquid magnesium is used to reduce titanium tetrachloride into titanium sponge in the reduction process. The reduction byproduct of $MgCl_2$ is generated too. To obtain pure titanium sponge, vacuum distillation is conducted which separates excess magnesium and magnesium chloride which was generated in reduction process [2,3].

Pangang Group Titanium Industry Co., LTD employs 7.5t distillation furnace of "I" style to produce titanium sponge[4]. This system has advantages of strong production ability and high recovery percentage of titanium. It achieved success in the first feeding trial production process. However, there also exists some--difficulties, such as temperature, time and system pressure control in the production process. These lead to distillation channel blockage, slow evaporation, high chlorine content in titanium sponge, hardness and other problems. This research conducted some optimizing experiments of $TiCl_4$ feeding speed, distillation time and system pressure, and some other important control parameters. The effects to titanium sponge quality were analyzed in order to improve product quality of titanium sponge by the reduction and distillation processes optimization.

Thermodynamics analysis of titanium sponge production

Titanium sponge production mainly contain reduction and distillation processes[5]. It is a typical heterogeneous reaction process in the reduction process. Table 1 shows the main chemical reaction equations of magnesium reduction of $TiCl_4$. Reduction products mainly contain Ti sponge, $MgCl_2$ and a few low-valent titanium chlorides. Some redundant of Mg and magnesium chloride residue in reduction furnace. To obtain pure Titanium sponge, they must be separated from titanium sponge lump.

Table 1 Chemical reaction equations of magnesium reduction of $TiCl_4$

Reaction equation	Reaction number
$1/2TiCl_4(g) + Mg(l) = MgCl_2(l) + 1/2Ti(s)$	(1)
$1/2TiCl_4(g) + Mg(g) = MgCl_2(l) + 1/2Ti(s)$	(2)
$1/2TiCl_4(l) + Mg(g) = MgCl_2(l) + 1/2Ti(s)$	(3)
$1/2TiCl_4(l) + Mg(l) = MgCl_2(l) + 1/2Ti(s)$	(4)

Fig.1 shows the ΔH and ΔG of reaction (1) which is listed in Table 1 .

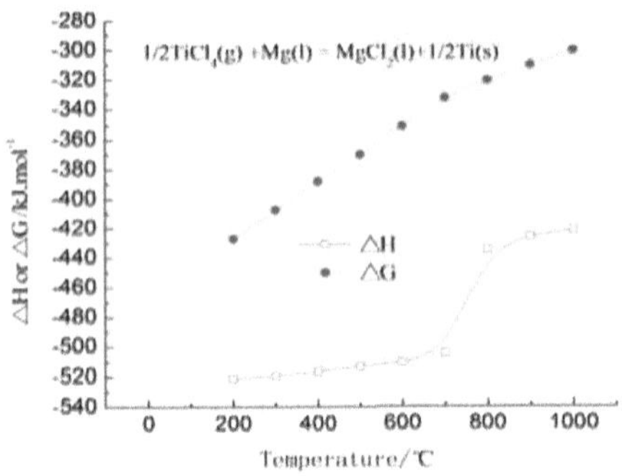

Fig.1 ΔH and ΔG of reaction (1) at different temperature

From Fig.1, the value of ΔH and ΔG of reaction (1) rise with the temperature increased. From the computed results of ΔH at 200℃~1000℃, it can be seen that reaction of (1) is a strong exothermic reaction. Then, it must remove excessive reaction heat to prevent overheating blast in the reduction process.

Ti and Mg will react chemically with O_2, N_2 under high temperature and normal pressure in air. Vacuum distillation always has been used to separate magnesium chloride and excess magnesium in the industry. A comparison of saturated pressure values at the same temperature can be used as a rough guide to determine which composition should exhibit preferential evaporation. The saturated vapor pressures of $MgCl_2$, Mg and Ti can be calculated with the Van Laar Equation [6]. And Fig.2 shows the relationship between lg*p* and temperature.

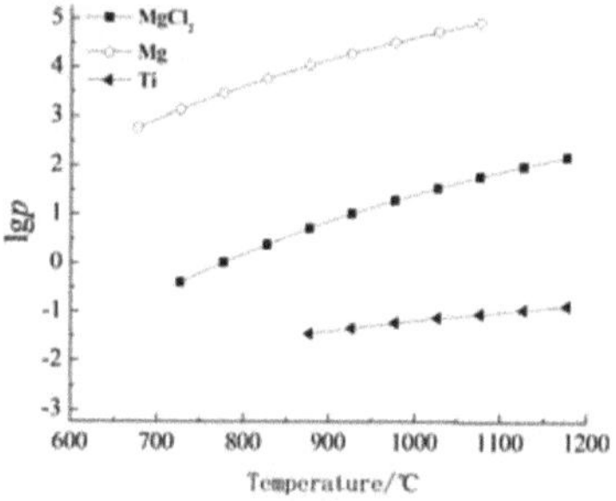

Fig.2 lg*p*-*T* diagram of $MgCl_2$, Mg and Ti

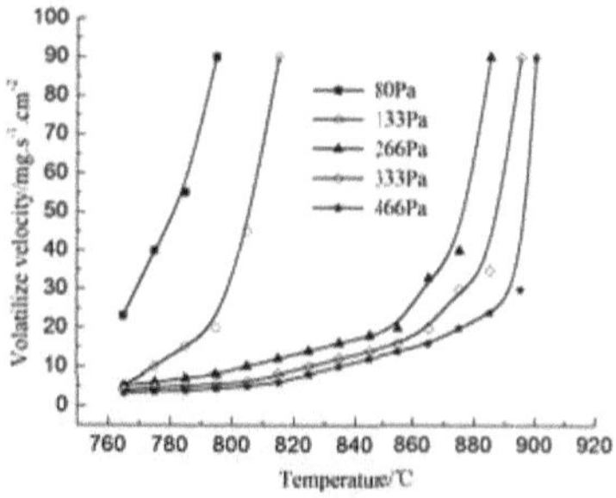

Fig.3 Volatilize velocity of $MgCl_2$

It can be seen from Fig.2 that volatile of Mg and $MgCl_2$ is bigger than titanium at the same temperature. And we can expect a better realization the separation of titanium with Mg, $MgCl_2$ and other reduction products when distillation temperature is controlled between 900℃~1200℃.

In addition, the boiling point of Mg and $MgCl_2$ will reduce under vacuum, relative volatility will enlarge than normal pressures. For instance, the boiling point of Mg and $MgCl_2$ will downgrade to 516℃ and 677℃respectively when the vacuum degree reach to 10.1Pa. Therefore, vacuum distillation can reduce distillation temperature and improve the volatile speed of Mg and $MgCl_2$. Figure 3 presents the volatile speed of $MgCl_2$ at different pressure. As shown in Fig.3, it is a significantly rise of $MgCl_2$ volatilize velocity along with system pressure reduce from 466Pa to 80Pa. The distillation temperature can reduce 110℃ when volatilize velocity of $MgCl_2$ reaches the maximum value. And titanium sponge with less chlorine and magnesium impurity can be obtained when the pressure is lower.

Influence of $TiCl_4$ feeding speed

The feeding speed of $TiCl_4$ will decide the heat release rate of reduction reaction process. It will cause titanium sponge sintering when $TiCl_4$ feed rate is big. Reduction experiments were carried out in "I" type furnace of 7.5 tons per furnace to obtain the influence of different $TiCl_4$ feeding speed on titanium sponge production. Fig.4 shows the different $TiCl_4$ feeding speed at the same reduction stage. The maximum $TiCl_4$ feeding speed was reduced from $320kg.h^{-1}$ to $290kg.h^{-1}$. Fig.5 shows the different output capacity of titanium sponge between high and low $TiCl_4$ feeding speed. Through comparison between the experiment conditions and the results of output capacity of titanium sponge, it is shown that the output capacity of titanium sponge can increase significantly when the maximum $TiCl_4$ feeding speed was reduced to $290kg.h^{-1}$. According to the analysis results thermodynamic, it's a multiphase exothermic chemical reaction when titanium tetrachloride was reduced by magnesium. Reduce $TiCl_4$ feeding speed can thoroughly promote the reduction reaction and output capacity of titanium sponge.

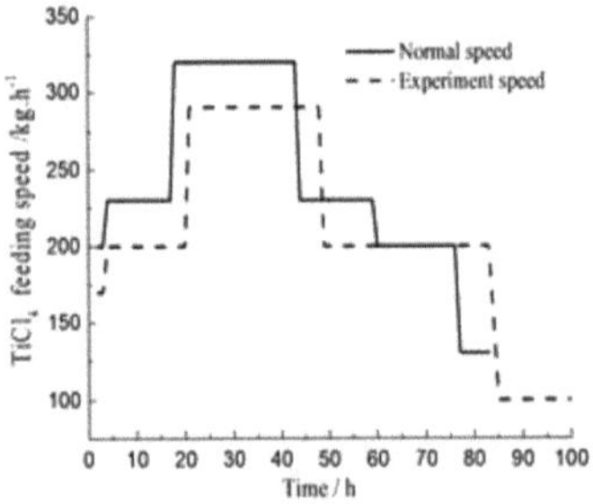

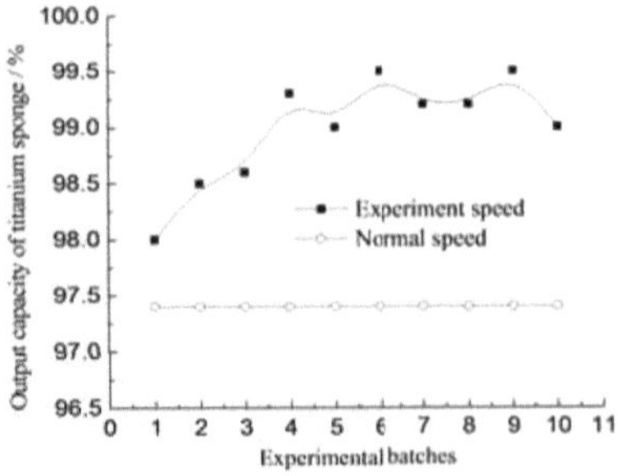

Fig.4 $TiCl_4$ feeding speed at reduction stage Fig.5 Output capacity of titanium sponge

Table 2 list the chemical composition of titanium sponge when reduced feeding speed was reduced. It indicate that the mass fraction of impurities in titanium sponge were lower than 0_A grade titanium sponge which was claimed by Chinese titanium sponge standard of GB/T 2524-2010.

Table 2 Chemical composition of titanium sponge of reduced feeding speed

	Chemical composition (mass fraction)%										
	Ti	Impurities, ≤									
	≥	Fe	Si	Cl	C	N	O	Mn	Mg	H	grade
Chinese standard (GB/T 2524-2010)	99.8	0.03	0.01	0.06	0.01	0.01	0.05	0.01	0.01	0.003	0_A
Experiment product(average)	99.8	0.009	0.005	0.059	0.01	0.006	0.057	0.001	0.002	0.002	0_A

Influence of system pressure and distillation time

To obtain pure titanium sponge, reduction products are separated by vacuum distillation. Evaporation rate of the impurities are determined directly by system pressure when distillation temperature is constant. Mg is easier to separate with titanium sponge in vacuum distillation process. The degree of $MgCl_2$ evaporation will more impact on product quality of titanium sponge. In order to investigate the influence of the system pressure to $MgCl_2$ evaporation, vacuum distillation experiments were carried on in different pressure conditions between 0.8Pa and 9.72Pa. Chlorine content in distilled titanium sponge were detected and analyzed. The results of experiments are shown in Fig. 6.

According to Fig.6, the influence of the system pressure on content of chlorine in titanium sponge is outstanding. Less chlorine titanium sponge product can be obtained under lower distillation pressure. The content of chlorine in titanium sponge increased from 0.01% to 0.091% when distillation system pressure increased from 0.8Pa to

9.72Pa. It increased quickly when the system pressure is more than 4Pa.Therefore, vacuum distillation process should be operated less than 4Pa to obtain lower chlorine content of titanium sponge products. Meanwhile, we can judge the quality of titanium sponge products preliminary through observe the end point of the distillation system pressure.

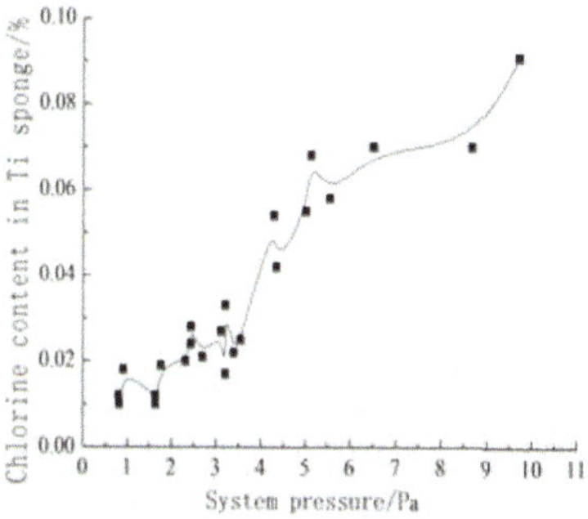

Fig.6 Relationship between content of chlorine with system pressure

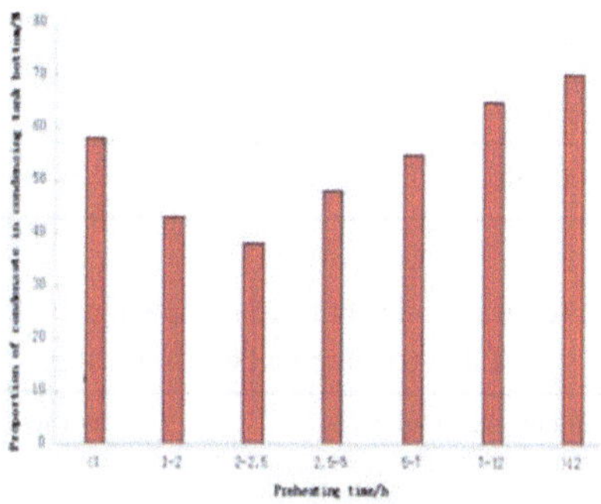

Fig.7 Influence of preheating time on distribution of condensation

Different distillation preheating time experiments were conducted to investigate the influence to vacuum distillation at 770℃to 800℃. The results are shown in Fig. 7. It can be seen that the proportion of condensate in condensing tank bottom increases when preheating time was too long or too short. The steam is less and condenses quickly around the bottom of condensation tank when preheating time was too short. In order to ensure the distillation channel unobstructed, the early stage of the preheating time should control between 2 h to 2.5 h.

Table 4 shows the influence of spray water time on distribution of condensation.

Table4 Influence of spray water time on distribution of condensation

Spray time of spray water /h	Proportion of condensate at bottom/%	Proportion of condensate at middle part/%	Proportion of condensate at upside/%
≤27	48-65	20-31	15-21
27-30	17-39	38-51	23-32
30-33	32-48	32-40	20-28
33-36	36-50	31-39	19-25
≥36	43-55	28-35	17-22

It is shown that the location of the condensate in condensation recovery tank change with the extension of time of spray water at the beginning of the distillation. Proportion of condensate at bottom can reach 43% - 55% when spray water sprayed more than 36 hours. However, it was less than 50% when spray water time is only 27 hours to 33 hours, and proportion of condensate at middle part can reach 32%-51% at the same time. Therefore, spray water continuous spray time should control between

27 to 33 hours at the earlier stage of vacuum distillation.

Optimized experiments of vacuum distillation

Table 5 shows the optimized distillation control parameters that on the basis of condition experiments.

Table 5 Optimum parameters of exploration experiments

Initial temperature/℃	Preheating temperature/℃	preheating time /h	Spray water temperature/℃	Spray time of spray water /h	System pressure / Pa
700-730	770-800	2-2.5	33-40	30-36h	<4Pa

Products grade compare is shown in Table 6.

Table 6 Compare before and after vacuum distillation process optimization

National standard grade	Titanium sponge grade	Varies proportion after optimized /%
——	PTi-90	4.04
0_A	PTi-95	12.12
0	PTi-100	-4.64
1	PTi-110	-8.774
2	PTi-120	0.804
3～5	PTi-130～150	2.84
Poor quality titanium sponge	Poor quality titanium sponge	-6.4

Table 6 shows that optimization of distillation process is helpful to increase the rate of titanium sponge product's grade. The proportion of 0 grade and 0_A grade titanium sponges increased by 16.54%. The poor quality titanium has reduced by 6.4%. At the same time, the distillation process optimization can effectively reduce phenomenon of the distillation channel blockage.

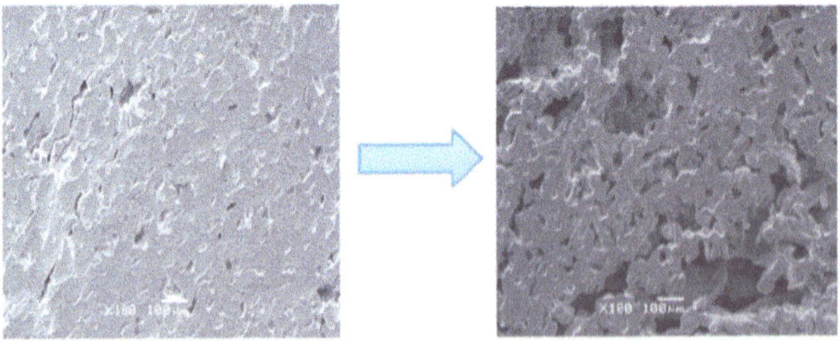

Fig. 8 SEM images of titanium sponge

Figure 8 presents the SEM images of titanium sponge before and after process optimization. It can be seen that the right image is more loose the left one. Softer titanium sponge can be obtained through reduction and vacuum distillation process optimization.

Conclusion

Thermodynamic analysis indicates the value of △H and △G of magnesium thermal reduction $TiCl_4$ reaction rise with the temperature increased. Volatilize velocity of $MgCl_2$ rise significantly along with system pressure reduce from 466Pa to 80Pa.The process optimization experimental results show that softer titanium sponge can be obtained through reduction and vacuum distillation process optimization. The output capacity of titanium sponge can increase significantly when the maximum $TiCl_4$ feeding speed was reduced to 290kg.h^{-1}, and mass fraction of impurities in titanium sponge were lower than 0_A grade titanium sponge. Preheating time at vacuum distillation stage should control between 2 h to 2.5 h to ensure the distillation channel unobstructed.

References

[1]Jing Gao， Nai-qi Qu. 2002 Discussion on Titanium Sponge Production Process. *Iron Steel Vanadium Titanium*, 23 (3), 44-48. (in Chinese)

[2]Guozhu Deng, 2010. Ttanium Metallurgy. Beijing: *Metallurgical Industry Press*, 26-33. (in Chinese)

[3]CH.R.V.S. NAGESH, CH. SRIDHAR RAO, 2004. Mechanism of Titanium Sponge Formation in the Kroll Reduction Reactor. Metallurgy and materials transactions B, 34(2), 66-74.

[4]Liang Li, Kai-hua Li, Xiao-zhe Chen. 2013. Reduction product Separation by Vacuum Distillation in the Process of Titanium Sponge Preparation. *Vacuum*, 50(3), 34-38. (in Chinese)

[5]Wei Mo, Guo-zhu Deng, Fang-chen Luo, 2006. *Titanium Metallurgy*. Beijing: Metallurgical Industry Press, 299-300. (in Chinese)

[6] Zheng S.S, Chen W.H, Cai J, Li J.T, Chen C, Luo.X.T. “Mass transfer of phosphorus in silicon melts under vacuum induction refining”, *Metallurgy and materials transactions B*,2010,41B,1268-1273.

Rare Metal Technology 2015
Edited by: Neale R. Neelameggham, Shafiq Alam, Harald Oosterhof, Animesh Jha, David Dreisinger, and Shijie Wang
TMS (The Minerals, Metals & Materials Society), 2015

RECOVERY OF RARE EARTH ELEMENTS FROM NDFEB MAGNET SCRAPS BY PYROMETALLURGICAL PROCESSES

Yuyang Bian[1], Shuqiang Guo[1], Kai Tang[2], Lan Jiang[1], Changyuan Lu[1], Xionggang Lu[1], Weizhong Ding[1]

[1]Shanghai Key Laboratory of Modern Metallurgy and Materials Processing, Shanghai University, Shanghai 200072, China
[2]SINTEF Materials and Chemistry, 7465 Trondheim, Norway

Keywords: Permanent magnet, Rare earth, Recycling

Abstract

For recovery of rare earth elements from the NdFeB based permanent magnet scraps, a high temperature process was used. Permanent magnets scraps were firstly oxidized in air. The optimal oxidation conditions under air are at temperature of 700°C holding for 30min. After the oxidation process, the oxides were reduced by graphite crucible at 1500°C under Ar atmosphere. By this way, the rare earth elements were recovered in the form of oxides, and Fe was separated to the metal phase. Boron distributed in both the rare earth oxides (REO) containing slag and the metal phase. In order to reduce the boron contents in the rare earth oxides, the carbon saturated iron bath and NdFeB added carbon saturated iron bath were used to refine the slag. By this process, the rare earth oxides with a comparatively high purity of 98.2% were gotten.

Introduction

Rare earth are widely used in green technologies, such as electrical vehicles, novel catalysts, high efficient lighting systems and wind turbines.[1] However, the regional distribution of rare earth ores in the world is uneven. Also caused by the environmental and labors cost of rare earth ore mining, China controls more than 90% supply of rare earth world widely. In 2010, China published the policy to protect the domestic rare earth sources. Because of the increasing demand and supply shortage, additionally triggered by the restriction on the export, the prices of the rare earths have a sharp increase.[2-6]

In order to explore new rare earth resources, plenty of researches have been done on recovery of the rare earth elements, especially from tailings, permanent magnets, waste fluorescent lamps and rechargeable batteries.[3, 7-9] The content of rare earth elements in permanent magnets is more than 30%. During the production of the permanent magnet, including smelting, smashing and grinding, 30%-40% of NdFeB alloys was wasted as useless materials.[10] Because of the oxidation of the NdFeB materials in high service temperature, the life of the magnets is shorten. It is necessary to recover the rare earth elements from the permanent magnet scraps, sludge and the end-of-life magnets.[11-14]

Several methods have been proposed on the recovery of the rare earth elements from the NdFeB magnets. The most traditional methods are based on the hydrometallurgy process, in which the chloric acids, sulfuric acids and oxalic acids will thus be used and introduce some inevitable pollutions to environment. Compared to the hydrometallurgy process, the pyrometallurgical methods are environment friendly. Based on the strong affinity of rare earth elements to oxygen, Nakamoto et al. extracted the rare earth elements as oxides from neodymium magnetic sludge and found that the boron oxides concentration in the slag phase is decreasing with the holding time prolonging in a graphite crucible at 1550°C.[15] The extraction of Nd from waste NdFeB alloys by the glass slag method was also proposed by Saito et al., in which way, boron oxide was used as the extraction agent.[16] Uda used $FeCl_2$ to recover the rare earths from magnet sludge by transforming the rare earth elements into corresponding rare earth chlorides and further separated the rare earth chlorides by vacuum distillation.[17] The metallic alloy methods were also proposed. Takeda et al. used magnesium as extraction medium to extract the rare earth elements by controlling temperature difference inside the reaction vessel. The rare earth elements were separated from the magnet as Mg-Nd alloys, and the rare earth metal was finally gotten by vacuum distillation.[18] The use of silver was also proposed, and the rare earth oxide were gotten as the final production of the process.[19] Some other methods, such as electrolysis method[20] and carbonylation method[21], also provide alternative ways to recover the rare earth elements from the permanent magnet.

In this paper, a pyrometallurgical method was used to recover the rare earth elements from the NdFeB based magnet. The NdFeB materials was firstly oxidized and then reduced in high temperatures, in which process the Fe_2O_3 and part of B_2O_3 were reduced to metal phase by graphite crucible and the rare earth oxides remained in the oxides phase. After separating the oxides and the metal phase, the rare earth oxides are recovered. Novelly, the iron metal and the NdFeB added iron metal bath were used to refine the rare earth containing oxide phase.

Experimental

2.1 Materials

The raw NdFeB materials used in our research are commercial magnet ingots (diameter 9mm × height 12mm) without magnetizing. The elements concentration of the magnets are listed in Table I. The iron used in the experiments for the metal bath is with the purity of 99.99%. The purity of the graphite crucibles are more than 99.99%.

Table I. Composition of the bulk NdFeB magnet (mass %)

Fe	Nd	Pr	La	B	Al
61.60	30.73	4.39	1.58	0.96	0.83

2.2 Experimental process

The process of recovery of the rare earth oxides from the NdFeB materials is displayed in Figure 1(a). Details of the experimental setup in the reduction and refinement processes are shown in Figure 1(b). The electric furnace was heated by $MoSi_2$ heating elements. During the experimental process, Ar (200ml/min) was used as protective gas. The illustrations of the refinement of the rare earth oxides by carbon saturated iron bath or NdFeB added carbon saturated iron bath are given in Figure 1(c).

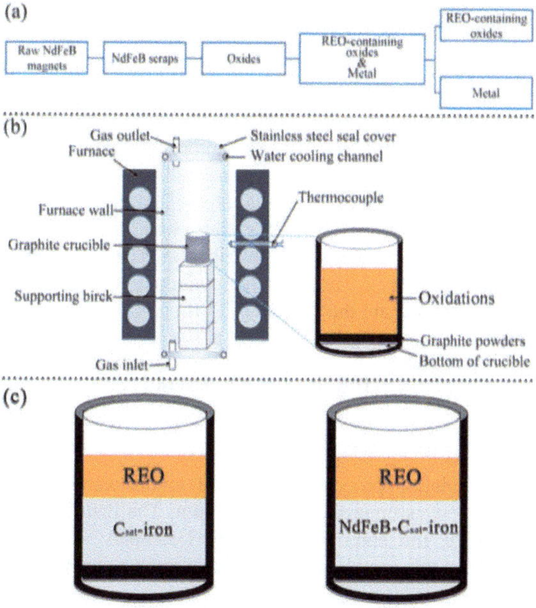

Figure 1. The flowchart of the process of recovering rare earth oxides from permanent magnets (a), the schematic diagram of apparatus used in the reduction and refinement processes (b), and the illustration of the refinement by metal baths.

2.2.1 The oxidation process

Before the reduction process, the rare earth elements should be transformed to corresponding oxides. The oxidation behaviors of the NdFeB materials were investigated by TG-DSC at different temperatures (400, 500, 600, 700, 800 and 900°C).

2.2.2 The reduction process

After the oxidation process, the oxides (REO-Fe_2O_3-B_2O_3) were gotten, and further used in the reduction process. At the reduction temperature of 1500°C, Fe_2O_3 was reduced to iron. The rare earth oxides were still remained in the oxides phase. After this process, the REO containing slag and the iron based metal phase were produced. By separating the slag from the metal, the rare earth oxides was finally gotten.

2.2.3 Refinement of the rare earth oxides

In present experimental temperature, the alumina and boron oxides can hardly be reduced to metal phase efficiently. The carbon saturated iron metal bath and the NdFeB added carbon saturated iron metal bath were used to refine the REO containing slag that gotten in the reduction process.

2.3 Characterization

The particle size of NdFeB powders was measured by particle size analyzer (Mastersizer 2000). The NdFeB samples were tested by differential scanning calorimetry (DSC) and thermogravimetry (TG) with a standby of 20min to stable the facility at 40°C. Then the sample was heated to corresponding temperature with heating rate of 20°C/min in Ar atmosphere, and holding at the experimental temperature for 1 h in Air. The enthalpy curves were normalized to 1 mg, and sapphire was used as reference material to achieve the calibration. Backscattered-electron microscopy (SU-1510, HITACHI) with energy X-ray spectroscopy (Oxford INCA EDS system) was used in characterizing the morphology of the REO-containing slag phase and the elements distribution. The concentration of Nd, Pr, La, Al, Fe and B in both the oxides and metal phase was measured by inductively coupled plasma atomic emission spectrometer (PE400, Perkin Elmer), and the content of Carbon in the metal phase was measured by CS analyzer (CS 600CR, LECO).

Results and discussion

3.1 The oxidation process

In order to accelerate the oxidation process, the NdFeB ingots were pulverized into fine particles mechanically. The particle size was controlled under 150μm. The distribution of the particle size was shown in Figure 2. Obviously, the homogeneous NdFeB samples were gotten. The mean volume diameter is at about 64μm. Then this sample was taken for oxidation experiments using DSC-TG.

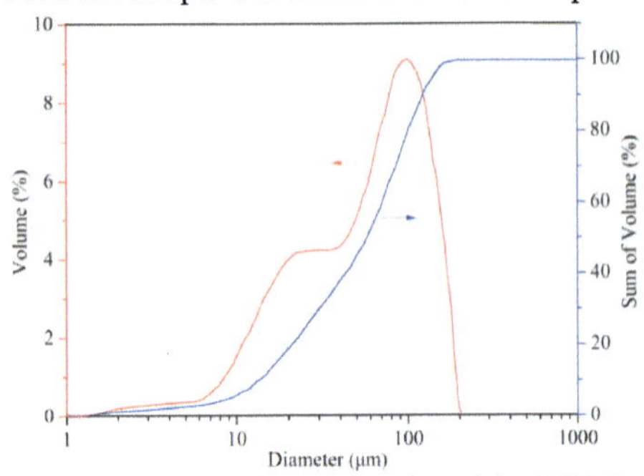

Figure 2. The distribution of particles size of the NdFeB powders.

Figure 3 shows the TG curves of the NdFeB samples under different experimental conditions. Even under the Ar (with purity of 99.999%) atmosphere, the mass increase of the samples is apparently due to the oxidation of NdFeB samples by the residual oxidizing gas (O_2 and H_2O) in Ar. After the oxidation temperature reached, the synthetic air was introduced into the chamber of the TG-DSC facility, and the samples were hold at the oxidation conditions for 1h. Under the air atmosphere, the samples were oxidized intensively with an explosive growth of the sample mass within a few seconds. Then the mass change of the sample slows down in the following holding time. The oxidation process was not finished at temperatures of 400, 500 and 600°C showing by the continuous growing curves in Figure 3. However, the curves of TG at the temperatures of 700, 800 and 900°C present stable tendency after a certain time, which means the oxidation process is basically finished. The oxidation process finished in 35min for experiments both at temperature of 700 and 800°C but in 20min for the experiments at 900°C. The total mass change of the sample at 700, 800 and 900°C are all at about 31% which match well with the calculated value from the composition from Table 1. It confirmed that the NdFeB material can be fully oxidized under 700, 800 and 900°C in a short time arranging from 20 to 35min.

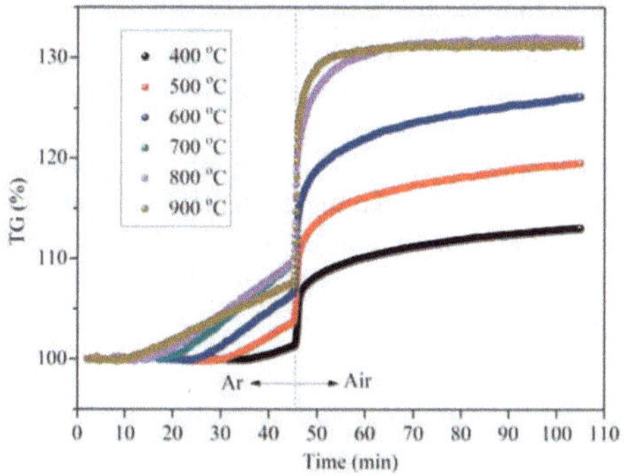

Figure 3. The TG curves of the NdFeB samples heated at different temperatures.

The oxidation process is an intensively exothermic process, as shown in Figure 4(a). The heating program is shown in Figure 4(b). After reaching the isothermal stage, the oxidation process was start under air. The exothermic peaks was shown in each of the DSC curves. Although the ΔH of the oxidation will decrease with increasing of temperature, the main exothermic peak becomes sharper as shown in Figure 4(a). It indicated that the oxidation rate increases with the increasing of experimental temperature. The arrows in Figure 4(a) show small exothermic peaks appears in each of the DSC curves except the one at 400°C. Edgley et al. had reported that the oxidation of NdFeB magnet was a transgranularly oxidation process.[22] Higher reaction temperature improves the reaction rate and accelerate the decomposition of the main $Nd_2Fe_{14}B$ phase. A new surface of the samples will be further oxidized relevant to the second oxidation exothermic peaks.

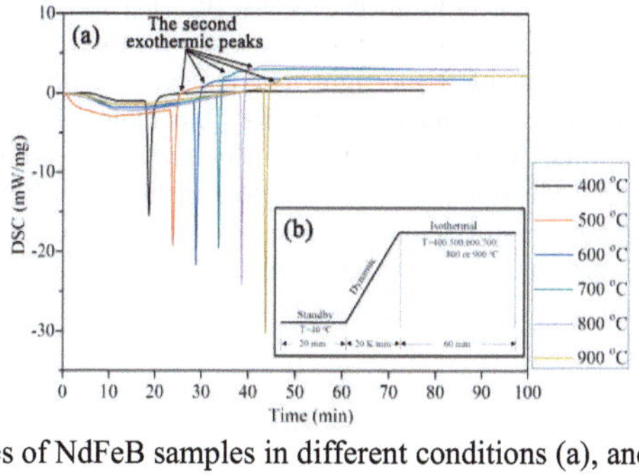

Figure 4. The DSC curves of NdFeB samples in different conditions (a), and the heating program (b)

To ensure the rare earth elements were fully oxidized, the temperature of 700°C and the holding time of 30min are considered as the optimal oxidation conditions. After the oxidation process, higher temperature will lead to further sintering process or formation of $NdFeO_3$.[23]

3.2 The separation of REO and metal phase

After the oxidation process, REO, iron oxides, alumina and boron oxides were produced. Figure 5 shows the standard Gibbs energy for each elements in the raw magnet also with Carbon. The thermodynamic calculations were performed using HSC chemistry 5.1 software for windows.[24] The rare earth elements Nd, Pr and La have very similar thermodynamic properties, so only the oxygen potential of Nd is shown. Thermodynamically, Fe_2O_3 can be reduced to iron easily by carbon around 700°C. B_2O_3 will be reduced by carbon at temperatures over 1650°C. The other oxides, such as alumina and rare earth oxides, are hardly reduced by carbon under the experimental temperatures. Based on the difference of the thermodynamic properties, Fe_2O_3 can be reduced into metal phase and the rare earth elements were remained in oxide phase by controlling the temperature of reduction process.

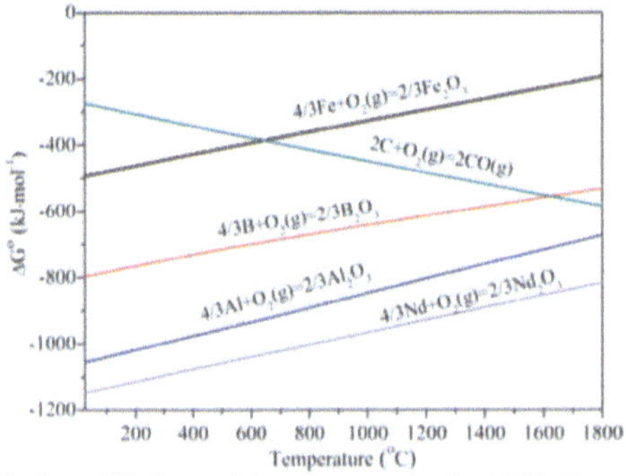

Figure 5. The chemical equilibrium of the elements in the NdFeB materials with oxygen.

The products of the reduction process is shown in Figure 6(a). It displays the slag phase and the metal phase separated clearly. The initial content of the oxides in the crucible was labeled by the white dash line. Different from the traditional metallurgical process, the slag has not flowed up to the top of the metal phase. The density of the REO-containing oxides is similar to the metal phase, which resulting in the slag and the metal phase separated left and right. The temperature of reduction process was program controlled as shown in Figure 6(b). Holding at 1200°C for 2h is mainly for the reduction of iron oxide in solid phase. And the holding time of 4h at 1500°C is for the reduction of boron oxide and separation of the slag and metal. During the reduction process, CO released due to the reaction of Fe_2O_3 and graphite crucible. The emission of gas from solid phase is easy. However, if the reduction happened at 1500°C, the oxides will melt and the release of the gas become hard. And the splash of the melts will happen.

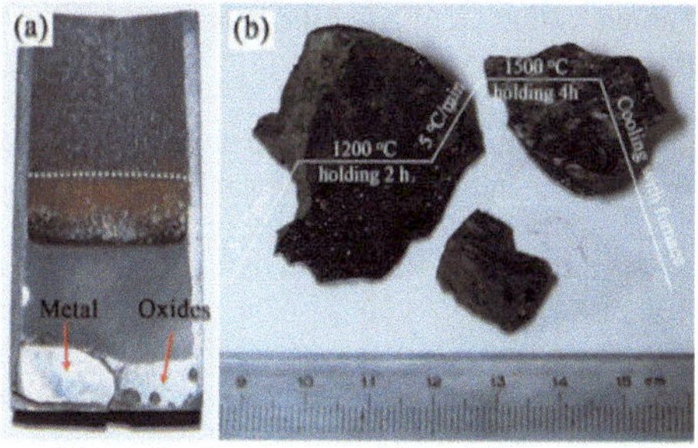

Figure 6. The productions of the reduction process (a), and the rare earth oxides separated from the metal phase also with the reduction heating program (b).

After reduction process, the REO containing oxides phase and the iron based metal phase are separated. By this process, the rare earth oxides were produced, as shown in Figure 6(b). From the BSEM image of the oxide phase, shown in Figure 7, different phases are observed. Fe in the slag phase is the iron droplets that remained in the slag. Aluminum distribution in the oxide phase is because of the $REAlO_3$ peroskite phase.[25]

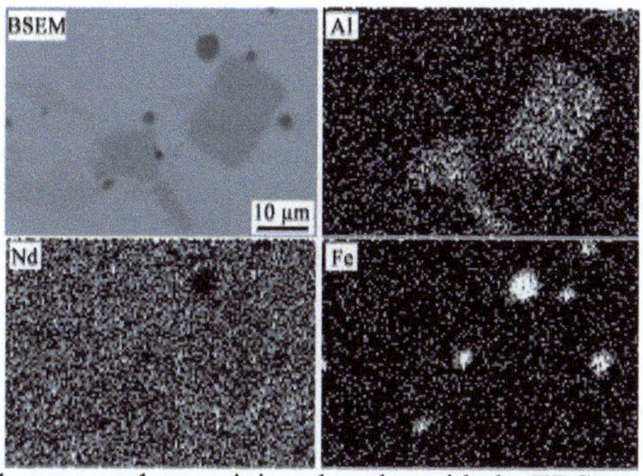

Figure 7. The BSEM of the rare earth containing slag also with the EDS mappings of Al, Nd, and Fe.

3.3 Refinement of the REO containing oxides

The concentration of the oxides and the metal was listed in Table II. At the experimental temperature, Fe_2O_3 can hardly exists in the oxides phase. The concentration of the oxide phase was normalized without considering the appearance of Fe. Normally, the content of Fe measured by ICP is about 0.2% in all slag samples. The concentration of the oxide phase and the metal phase confirms that the rare earth oxides are concentrated in the slag phase and the iron oxides are reduced to the metal phase by the reduction process.

After the reduction process, boron oxides content in the oxide phase is 4.59%. Referring to the calculated results from Figure 4, the carbothermic reduction of B_2O_3 is at about 1650°C. However, some researchers have investigated the production of boron carbides. The reaction becomes thermodynamically feasible above 1400°C.[26] After the reduction process of 4h at 1500°C, the content of B in the metal phase is 0.06%.

Table II. The concentrations of the oxide phase and the metal phase

Exp. No.	Oxide Phase (mass %)					Metal Phase (mass %)					
	Nd_2O_3	Pr_2O_3	La_2O_3	Al_2O_3	B_2O_3	Nd	Pr	La	Al	B	C
1	77.38	11.28	5.35	1.39	4.59	0.08	0.01	<0.01	0.03	0.06	3.45
2	78.55	11.10	5.29	1.36	3.70	0.04	0.01	<0.01	0.02	0.13	3.78
3	80.26	11.53	5.49	1.57	1.15	0.02	<0.01	<0.01	0.01	0.45	2.56

*No.1 Reduction process; No.2 Refining by C_{sat}-iron bath; No.3 Refining by NdFeB added iron bath

For the refinement of the REO containing oxides, the oxides gotten from the reduction process with concentration listed in Table 2 were used in the refinement process. After the refinement by C_{sat}-iron bath, the concentration of the boron oxides in the oxides phase shows a slight decrease, and the content of B in the metal phase is 0.13%. It confirms that the reduction of B_2O_3 by carbon at 1500°C is available. In the refinement process of holding for 4h at 1500°C, 20g rare earth containing slag and 60g pre-melted iron bath were used. Because of the mass ratio of slag and metal decrease, the interface of the slag and metal increase. It helps the reduction of B_2O_3.

Considering the carbon thermal reduction of B_2O_3 is not efficient, the raw NdFeB materials were added to the C_{sat}-iron bath. After this refinement process, the content of B_2O_3 in the oxides phase decrease dramatically, and the content of boron in the metal phase increase apparently. By this method, the purity of the rare earth oxides reached 98.2%. The purities of the rare earth oxides by the reduction process and by the refinement using C_{sat}-iron metal are 94.0 and 94.9%, respectively.

Conclusion

In order to recover the rare earth elements from permanent magnet, the NdFeB materials were oxidized under Air. Following by a high temperature reduction process, the rare earth oxides and the iron based metal phase separated successfully. For high purity of the rare earth oxides, the refinement by C_{sat}-iron bath and NdFeB added C_{sat}-iron bath were used.

High temperature helps in the oxidation process of the NdFeB materials both in the oxidation rate and the decomposition of the magnet. While the oxidation process is an energy consuming process. The experimental temperature of 700°C and holding time of 30min were the optimal conditions for oxidation.

By the reduction process, the rare earth oxides with a purity of 94% were gotten. After the refinement by C_{sat}-iron bath, the impurity of the rare earth oxides improves slightly to 94.9%. However, the impurity of the oxides reached 98.2% by using the refinement of NdFeB added C_{sat}-iron bath. Hence, a new method to obtain rare earth oxides from permanent magnet was achieved. This method is feasible both to the magnet scraps and sludge.

Acknowledgement
This study was financially supported by the National Key Basic Research Program of China (973) (2012CB722805).

References:
[1] S. Doris et al., *Study on Rare Earths and Their Recycling*, Final Report for The Greens/EFA Group in the European Parliament, (2011). 3-6.
[2] H. George C, L. Jinfang, G. Alexander, and M. Melania, "Current Status of Rare Earth Permanent Magnet Research in USA," Proceedings of 19th International Workshop on Rare Earth Permanent Magnets & Their Applications, (2006), 12-22.
[3] B. Koen, J. Peter Tom, B. Bart, G. Tom Van, Y. Yongxiang, W. Allan, and B. Matthias, "Recycling of Rare Earths a Critical Review," Journal of Cleaner Production, 51 (2013), 1-22.
[4] L. Ran, B. Matthias, D. Stefanie, M. Andreas, M. Cornelia, and S. Doris, "Application of Rare Earths in Consumer Electronics and Challenges for Recycling," iEEE International Conference on Consumer Electronics (Berlin Germany), (2011), 286-290.
[5] M. Stefania and R. Marcello, "Rare Earth Elements as Critical Raw Materials: Focus on International Markets and Future Strategies," Resources Policy, 38 (2013), 36-43.
[6] D. Xiaoyue and T.E. Graedel, "Global In-Use Stocks of the Rare Earth Elements: A First Estimate," Environmental Science & Technology, 45 (9) (2011), 4096-4101.
[7] J. Nan, D. Han, M. Yang, M. Cui, and X. Hou, "Recovery of metal values from a mixture of spent lithium-ion batteries and nickel-metal hydride batteries," Hydrometallurgy, 84 (1-2) (2006), 75-80.
[8] S. Nishihama, N. Sakaguchi, T. Hirai, and I. Komasawa, "Extraction and Separation of Rare Earth Metals Using Microcapsules Containing Bis (2-Ethylhexyl) Phosphinic Acid," Hydrometallurgy, 64 (1) (2002), 35-42.
[9] F. Yang, F. Kubota, Y. Baba, N. Kamiya, and M. Goto, "Selective extraction and recovery of rare earth metals from phosphor powders in waste fluorescent lamps using an ionic liquid system," Journal of Hazardous Materials, 254–255 (2013), 79-88.
[10] I. Masahiro, M. Koji, and M. Ken-ichi, "Novel Rare Earth Recovery Process on NdFeB Magnet Scraps by Selective Chlorination using NH_4Cl," Journal of Alloys and Compounds, 477 (1-2) (2009), 484-487.
[11] A.T. Dinsdale, "SGTE data for pure elements," Calphad, 15 (1991), 317-425.
[12] J.M. Le Breton, J. Teillet, P.J. McGuiness, D.S. Edgley, and R. Harris, "The oxidation of Nd-Fe-B permanent magnet at 400°C: a SEM, microhardness and Mössbauer study," Magnetics, IEEE Transactions on, 28 (5) (1992), 2157-2159.
[13] Y. Li, H.E. Evans, I.R. Harris, and I.P. Jones, "The Oxidation of NdFeB Magnets," Oxidation of Metals, 59 (2003), 167-182.
[14] G.W. Warren, G. Gao, and Q. Li, "Corrosion of NdFeB Permanent Magnet Materials," Journal of Applied Physics, 70 (10) (1991), 6609-6611.
[15] M. Nakamoto, K. Kubo, Y. Katayama, T. Tanaka, and T. Yamamoto, "Extraction of Rare Earth Elements as Oxides from a Neodymium Magnetic Sludge," Metal. Mater. Trans B, 43 (3) (2011), 468-476.
[16] T. Saito, H. Sato, S. Ozawa, J. Yu, and T. Motegi, "The Extraction of Nd from Waste NdFeB Alloys by the Glass Slag Method," Journal of Alloys and Compounds, 353 (11) (2003), 189-193.
[17] U. Tetsuya, "Recovery of Rare Earths from Magnet Sludge by $FeCl_2$," Materials Transactions, 43 (1) (2002), 55-62.
[18] T. Osamu, H. Toru, Okabe, and U. Yoshiaki, "Recovery of Neodymium from a Mixture of Magnet Scrap and Other Scrap," Journal of Alloys and Compounds, 408-412 (2006), 387-390.
[19] O. Takeda, T.H. Okabe, Y. Umetsu, "Phase Equilibrium of the System Ag-Fe-Nd, and Nd Extraction from Magnet Scraps using Molten Silver," Journal of Alloys and Compounds, 379 (1-2) (2004), 305-313.

[20] A.M. Martinez, O. Kjos, E. Skybakmoen, A. Solheim, and G.M. Haarberg, "Extraction of Rare Earth Metals from Nd-based Scrap by Electrolysis from Molten Slag, " ECS Transactions, 50 (2012), 453-461.
[21] K. Miura, M. Itoh, and K.-I. Machida, "Extraction and Recovery Characteristics of Fe Element from Nd-Fe-B Sintered Magnet Powder Scrap by Carbonylation," Journal of Alloys and Compounds, 466 (1-2) (2008), 228-232.
[22] D.S. Edgley, J.M. Le Breton, S. Steyaert, F.M. Ahmed, I.R. Harris, and J. Teillet, "Characterisation of High Temperature Oxidation of Nd-Fe-B Magnets," Journal of Magnetism and Magnetic Materials, 173 (1-2) (1997) 29-42.
[23] S.C. Parida, S. Dash, Z. Singh, R. Prasad, K.T. Jacob, and V. Venugopal, "Thermodynamic Studies on $NdFeO_3$(s), Journal of Solid State Chemistry," 164 (1) (2002), 34-41.
[24] A. Roine, Outolumpu HSC Thermochemistry for Windows, Ver.5.1 (Ourokumpu Research Oy, Pori, Finland).
[25] S.K. Thomas and C.J. Buckley, "Metastable States in Alumina-Rich La_2O_3-Al_2O_3 and Nd_2O_3-Al_2O_3," Journal of American Ceramic Society, 88 (1) (2005) 191-195.
[26] G. Goller, C. Toy, A. Tekin, and C.K. Gupta, "The Production of Boron Barbide by Carbothermic Reduction," High Temperature Materials and Processes, 15 (1-2) (1996) 117-122.

Rare Metal Technology 2015
Edited by: Neale R. Neelameggham, Shafiq Alam, Harald Oosterhof, Animesh Jha, David Dreisinger, and Shijie Wang
TMS (The Minerals, Metals & Materials Society), 2015

STUDY ON ELECTROLYSIS FOR NEODYMIUM METAL PRODUCTION

Go-Gi Lee[1], Sung-Koo Jo[1], Chang-Kyu Lee[1], Hong Youl Ryu[2], Jong Hyeon Lee[2]

[1]RIST (Research Institute of Industrial Science and Technology)
32 Hyojadong, Namku, Pohang, Gyeongbuk, Republic of Korea
[2]Graduate School of Energy Science and Technology, Chungnam National University;
79 Daehak-ro, Yuseong-gu, Daejeon 305-764, Republic of Korea

Keywords: Electrowinning, Neodymium, Electrodeposition, Convection

Abstract

This study deals with research on the reduction process of neodymium oxide by molten salt electrowinning. Neodymium fluoride and lithium fluoride electrolyte were used for feed stock of electrolyte and neodymium oxide was added as precursor. Cell current was confirmed that cell potential was applied as constant potential of 4.5V. Argon gas injection into the electrolyte was carried out to confirm convection effect on current efficiency. Applied current was gradually decreased without argon flow, but applied current was kept within a certain range with argon flow in electrolyte.

Introduction

Permanent magnets based on rare earth components have become more and more important in today's technology. Neodymium is an important metal primarily because of its use in Nd–Fe–B magnets. The electrolytic production of neodymium metal is considered as more attractive than the current industrial process that is based on calciothermic reduction of NdF3 [1, 2]. Molten salt electrolysis of neodymium from Nd_2O_3 can be a viable alternative to calciothermic reduction, the advantages of which enables continuous operation and therefore it is more suitable for mass production of the metal than the batch mode of a reduction process. And regeneration process of spent salts is not required unlike calcium reduction process [3]. Several studies on electrolyte were reported in the literature, in various media using alkali or alkali earth haloid melts, mixed chloride–oxide melts [4] and fluorides-oxide melt [5]. Physicochemical properties of the neodymium electrolysis system melts were widely researched. The researched properties contained electrical conductivity [6], dissolution behavior of Nd [7], solubility limit of Nd_2O_3 [8]. Studies of electrochemical property were inadequate for convection effect. In this study molten LiF-NdF3-Nd_2O_3 electrolyte was used as an electrolyte and the cell efficiency in the molten salt electrowinning process was investigated with respect to the convection effect using argon gas flow.

Experimental

A carbon crucible containing the electrolyte is introduced into a refractory Inconel square pillar as shown Figure 1. Neodymium fluoride (20kg), lithium fluoride (30kg) and in some experiments neodymium oxide (7kg) were weighed and placed into the graphite crucible. Anode and cathode were positioned to maximize the spacing between the cell walls, graphite anode and molybdenum cathode. Cathode and anode surface were 1008cm^2. SUS316 tube was inserted into the electrolyte for introducing the small bubble to agitate the electrolyte during electrolysis. Inconel vessel was sealed from atmospheric contamination. Inconel vessel was heated at 1150°C for cell operation using the resistance heat furnace. Voltage was charged with 4.5 volt across the anode and cathode by a DC power supply (AoneTech, Model EX30-700, Korea).

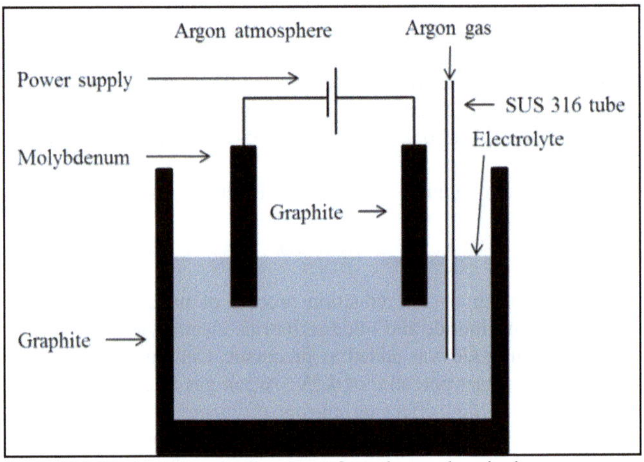

Figure 1. Schematic diagram of an electrochemical reactor.

Result and discussion

Constant voltage of 4.5 V was charged to the cell for 1 hour without argon flow. Initial applied current was decreased from 140A to 80A sharply within short time, but it was found that the applied current was gradually decreased by 50A after 1hour as shown Figure 2.

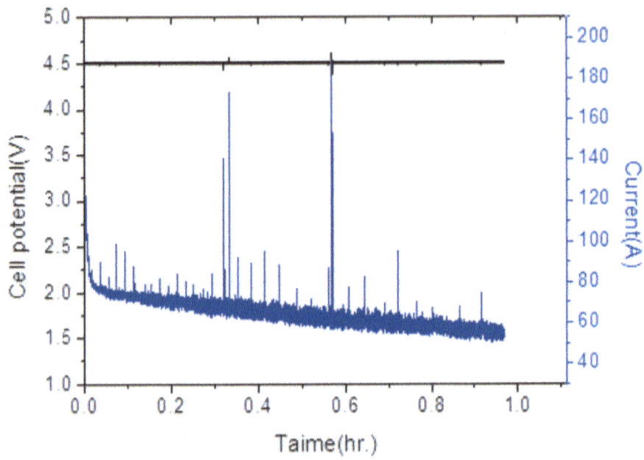

Figure 2. Chronoamperemetry for electrowinning in LiF-NdF3-Nd2O3 system at 1150oC without argon flow.

Constant voltage of 4.5 V was charged to the cell with agitating electrolyte by argon gas injection for 4.3 hour. Initial applied current was decreased from 180A to 120A sharply with more short time compared to no argon injection as shown Figure 3. In this case, it was found that the applied current was maintained with approximate 120A. And also it was observed that applied current was fluctuated within a certain range during all the time.

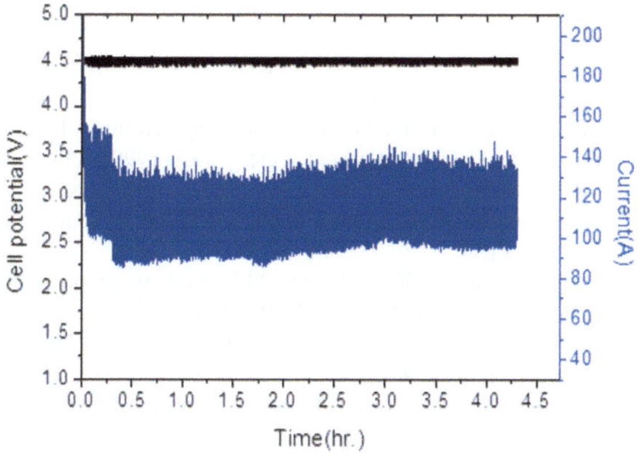

Figure 3. Chronoamperemetry for electrowinning in LiF-NdF_3-Nd_2O_3 system at 1150°C to inject argon gas in electrolyte.

It seems that in the LiF–NdF_3–Nd_2O_3 system the neodymium metal is cathodically produced by the electroreduction of neodymium fluorides while at the anode the oxidation of neodymium oxyfluorides generates oxygen according to the following equations:

$$2[NdF6]^{3-} + 6e^{-} = 2Nd(s) + 12F \quad (1)$$

$$3[NdOF5]^{4-} - 6e^{-} = 3/2O_2(g) + 3Nd^{3+} + 15F^{-} \quad (2)$$

It has been reported in Ref. [9] that oxygen is generated at the anode during the electrolytic production of neodymium from oxide–fluoride melts. The produced oxygen subsequently reacts with the carbon anode and CO and CO_2 gases are produced. Anode efficiency was decreased that CO and CO_2 on anode prevent contact between anode and electrolyte called as bubble curtain effect. Cell performance was improved because of these gases on the anode surface were easily removed by electrolyte flow by argon gas injection. In the future, the study will focus on the mechanism of bubble curtain, and its effect during neodymium electrolysis.

Conclusion

Electrolysis of neodymium was carried out using LiF-NdF_3-Nd_2O_3 electrolyte at 1150°C. Cell performance was improved by convection effect of electrolyte of argon gas injection resulting in easy flotation of CO and CO_2 gases formed on the near surface at anode.

Acknowledgement

This work was supported by the "Energy Efficiency & Resources Core Technology Program" of the Korea Institute of Energy Technology Evaluation and Planning (KETEP), granted financial resource from the Ministry of Trade, Industry & Energy, Republic of Korea.(No. 20122010300041)

References

1. R. A. Sharma and R. N. Seefurth, “Metallothermic Reduction of Nd2O3 with Ca in CaCI2-NaCI Melts,” Journal Electrochemical society, 135 (1988), 66-71.
2. P. T. Velu and R.G. Reddy, “Calciothermic reduction of neodymium fluoride,” The Minerals, Metals & Materials Society, (2005), 1155-1159.
3. G. J. Kipouros and R. A. Sharma, “Electrolytic Regeneration of the Neodymium Oxide Reduction-Spent Salt,” Journal Electrochemical society, 137 (1990), 3333-3338.
4. Y. Castrillejo et al., “Solubilization of rare earth oxides in the eutectic LiCl-KCl mixture at 450 °C and in the equimolar CaCl2-NaCl melt at 550oC,” Journal of Electroanalytical Chemistry, 545 (2003), 141-/157.
5. E. Stefanidaki, C. Hasiotis and C. Kontoyannis, “Electrodeposition of neodymium from LiF–NdF3–Nd2O3 melts,” Electrochimica Acta 46 (2001), 2665–2670.
6. X. W. Hu et al., “Electrical conductivity and Nd solubility of NdF3-LiF-Nd2O3 melts,” Metals & Materials Society, (2007) 79–82.
7. S. G. Chen et al., “Cathodic process of Nd and its dissolution behavior in molten fluoride,” Rare Metals, 13 (1994) 46–49.
8. J. E. Murphy, D. K. Dysinger and M. F. Chambers, “Electrowinning neodymium metal from chloride and oxide-fluoride electrolytes,” Metals & Materials Society, (1995), 1313–1320.
9. A. Kaneko, Y. Yamamoto, C. Okada “Electrochemistry of rare earth fluoride molten salts,” Journal of Alloys and Compounds, 193(1993) 44 - 46

EXPERIMENTAL INVESTIGATION OF RECYCLING RARE EARTH METALS FROM WASTE FLUORESCENT LAMP PHOSPHORS

Patrick M. Eduafo[1], Mark L. Strauss[1], Brajendra Mishra[1]

[1]Colorado School of Mines (Kroll Institute of Extractive Metallurgy);
1500 Illinois Street; Golden, CO, 80401, USA
Waste fluorescent lamp, phosphor powder, rare earth metals, recycle, leaching.

Abstract

Direct acid leaching of rare earth metals (REMs) from waste lamp phosphors is non-selective, leading to co-dissolution of impurities mainly calcium, phosphorus, silicon, iron and zinc. Calcining the phosphor powder at different temperatures followed by leaching with HCl isn't efficient for the extraction of all the chief rare earths especially cerium (Ce), lanthanum (La) and terbium (Tb). Therefore, multi-stage leaching operations are required for the separation of the REMs in phosphor dust. This paper will outline a new process for extracting Ce, La, Tb, Eu and Y from waste fluorescent lamps using a three-stage leaching and separation process. A flowsheet has been developed which shows this process is selective and efficient.

Introduction

Increasing demands for rare earth metals (REMs) due to rapid technological growths in several high-technology applications such as green energy, efficient fuel vehicles, defense related applications, emissive displays and fluorescent lamps, etc. coupled with shortage of supply has forced a paradigm shift towards finding alternative routes for rare earth production and supply. The concentration of rare-earth oxides in lamps with trichromatic phosphors can be as high as 27.9 wt%, but the actual recycled phosphor fractions contain about 10.0 wt% of rare earths oxides [1], which makes them an important secondary resource.

A number of processes have been investigated for recovery of REMs present in phosphor powder. Takahashi's group carried out a series of studies on the hydrometallurgical separation and recovery of rare earths from phosphors in the fluorescent lamp wastes [2], [3], [4]. After optimization of the leaching conditions, 92% of yttrium and 98% of europium were dissolved at sulfuric acid concentration of 1.5 M, temperature of 70˚C, leaching time of 1 h, and pulp concentration of 30 g/L. Wang et al. have conducted leaching experiments on spent trichromatic phosphor mixtures and recovered 96.28% of yttrium with 4 M hydrochloric acid in combination with 4.4 g/L hydrogen peroxide [5]. Rabah proposed a process for the recovery of europium, yttrium and some valuable salts from spent fluorescent lamps by pressure leaching with a H_2SO_4/HNO_3 mixture at 125 ˚C and 5 MPa for 4 hours and dissolved 92.8% of the europium and 96.4% of the yttrium present in the mixture [6]. De Michelis and his co-workers produced yttrium oxalate of 99% purity from synthetic sulphuric acid leach solution by oxalic acid precipitation. Yang et al. [7] developed a two stage leaching process with sulphuric acid in the first stage and nitric acid in the second stage and recovered 99% yttrium and 95% europium from phosphor powder. A lot of emphasis on recovery of REMs from spent fluorescent lamps has been on Y and Eu. This paper proposes a method for extraction of all chief REEs – Ce, La, Tb, Y and Eu.

Experimental

A. Materials

Phosphor powder from the spent fluorescent lamps was provided by Veolia ES Technical Solutions, (IL, USA). All the chemicals used were of analytical grade reagents (Sigma Aldrich, USA). Deionized water was used for preparation of solutions of known concentration and 2% HNO_3 was used for dilution before chemical analysis.

B. Analysis

Elemental composition in the collected phosphor powder was determined by chemical analysis. Concentration of metal ions present in the leach liquor was determined by lithium fusion and inductively coupled plasma-optical emission spectrometer (ICP-OES, Optima 5300 DV Perkin Elmer Co., MA, USA) for quantification.

C. Procedure

Phosphor powder coating of the fluorescent lamps obtained was sieved with a sieve of 325 Tyler mesh to separate any broken glass particles present. Leaching of the powder was carried out with different acid reagents on a hot plate with a magnetic stirring facility in a Pyrex glass beaker. Samples were taken at the end of each leaching experiment and analyzed for REE and impurities in the leachate.

Results and Discussion

A. Phosphor Dust Characterization

Size based separation was done to determine which size fraction will eliminate the most impurities and retain the bulk of the REE. Dry sieving the powder to 44 μm results in an upgrade of the REE content by 95.56% and removal of significant impurities (61.92%). The chemical content of the phosphor powder used for the experiments show some variations, which stems from how the powders were generated by the lamp recycling company. As can be seen in Table 1, the chief rare earths are Y, Eu, Ce, La and Tb, which are constituents of trichromatic lamp phosphors. The main impurities include Ca and P from calcium halophosphate phosphor, Al and Si from alumina (Al_2O_3) and silica (SiO_2) used in the barrier layer between the phosphor layer and the glass tube. Iron is possibly from the crushing and grinding equipment used in processing the end-of-life lamps.

Table 1: Elemental composition of the phosphor powders used in this study.

	Elements (wt%)									
	Y	**Eu**	**Ce**	**La**	**Tb**	**Ca**	**P**	**Fe**	**Al**	**Si**
Powder 1	9.50	0.64	1.48	2.09	0.65	18.50	9.54	0.31	2.72	4.33
Powder 2	9.65	0.67	1.66	2.15	0.69	17.50	8.35	0.32	2.50	6.33
Powder 3	5.46	0.37	1.05	1.27	0.41	21.34	9.95	0.38	2.13	6.61

Particle size and QEMSCAN analysis were also performed for physical characterization of the phosphor powder. The results (data not shown) show that the sieved powders have a size distribution mainly from <10 μm to 40 μm. The main gangue minerals in the powder is quartz/glass, calcite and apatite while the rare earth bearing minerals included fractions of Ce-phosphate as well as Tb and Eu bearing minerals. The Eu/Tb bearing minerals are only weakly associated with gangue minerals such as iron oxides, indicating that Fe may be easily removed in purification. Apatite and quartz have a greater association with Eu/Tb bearing minerals. Therefore, optimizing apatite removal may be most important in selective leaching in the prewash.

B. Direct Acid Leaching

Based on literature review and thermodynamic modeling, it is understood that strong acids can be used to leach rare earths from the waste phosphors directly. A feasibility test was performed using optimized conditions (1.5 M H_2SO_4, temperature of 70°C, leaching time of 1 hour, and pulp concentration of 30 g/L) from Takahashi group as reference in order to identify the best leaching reagent for extracting the REEs from phosphor dust [8]. As evident from Figure 1, HCl gave the highest extraction of europium and yttrium while the extraction of Ce, Tb and La was poor in all the lixiviants used in this test.

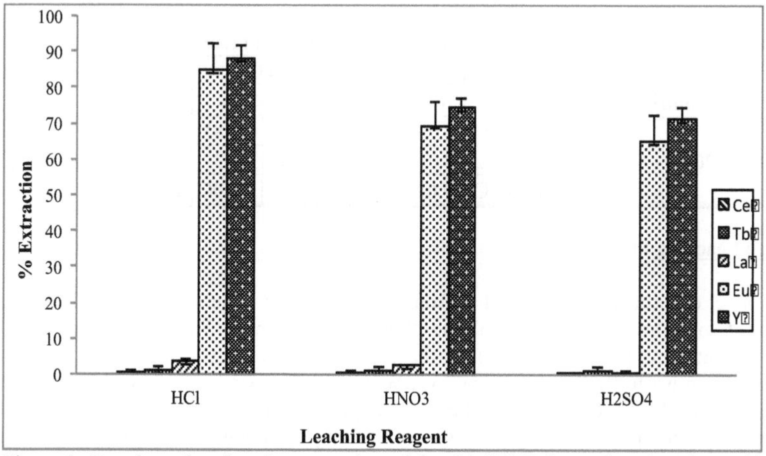

Figure 1: Bar chart showing the extraction of the chief REE from phosphor dust with different leaching reagent.

There is also co-dissolution of other elements such as Al, Ca, P, Fe Si and Zn during leaching (Table 2) and they constitute impurities in the leachate, which have to be removed in downstream purification processes. Leaching with sulfuric acid gives the advantage of significant reduction of Ca extraction into the leach liquor but produces lower REE extraction compared to leaching with hydrochloric acid. The feasibility test shows that direct acid leaching using low concentration acid reagents isn't efficient for the extraction of all chief rare earths in phosphor dust.

Table 2: Extraction of impurities from powder into the acid reagent.

Leaching Conditions	% Extraction				
	Al	Ca	Fe	P	Si
1.5 M HCl, 70°C, 30 g/L, 1 hr, 200 rpm	38.71	98.24	97.61	94.29	10.23
1.5 M H_2SO_4, 70°C, 30 g/L, 1 hr, 200 rpm	33.66	11.43	93.51	92.68	10.31

C. Calcining Powder and Leaching

In the industrial processing of rare earth minerals, high temperature treatments such as calcining and roasting are employed to decompose the minerals to allow easy recovery of the rare earth values [9][10]. Following the thermal treatment approach of rare earth minerals, a study was conducted where the phosphor dust was calcined at various temperatures for 1 hour for decomposition of the rare earth phosphate matrix in the phosphor powder to improve the leachability of terbium, which has high market price. Table 3 shows this approach results in poor dissolution of Ce, La and Tb as well as lower extraction of Eu and Y. The results don't justify using this energy intensive cost process for extracting REMs from waste lamp phosphors.

Table 3: Leaching results after calcining the powder (1 hr) at different temperatures.

Leaching Conditions		% Extraction				
		Ce	La	Tb	Eu	Y
6 M HCl, 30 g/L, 70°C, 1 hr, 200 rpm	550°C	0.79	2.60	1.28	73.85	75.80
	850°C	2.72	5.38	3.09	74.32	74.79
	950°C	9.62	10.46	19.09	76.00	80.31

D. Three-Step Leaching Process

The varying level of extraction of Eu and Y from La, Tb and Ce is quite significant implying varying level of solubility of the individual phosphors used in lamps. The higher solubility of the Eu and Y containing phosphor (YEO) can be exploited to increase overall REE recovery from phosphor from waste fluorescent lamps. A new process for extracting the chief rare earth elements from waste fluorescent lamps has been developed. It employs a three acid leach steps using hydrochloric acid under both mild and strong leaching conditions. The flow sheet for the proposed process is shown in Figure 2.

The wash step is carried out with 0.5 M HCl to remove the main impurities such as Ca, P and Fe in the phosphor dust. The first leaching step with a 1.5 M HCl is done to collect the Eu and Y fraction. The residue is then sent to a second leaching step under a strong oxidizing environment with 4.5 M HCl to extract Ce, Tb and La.

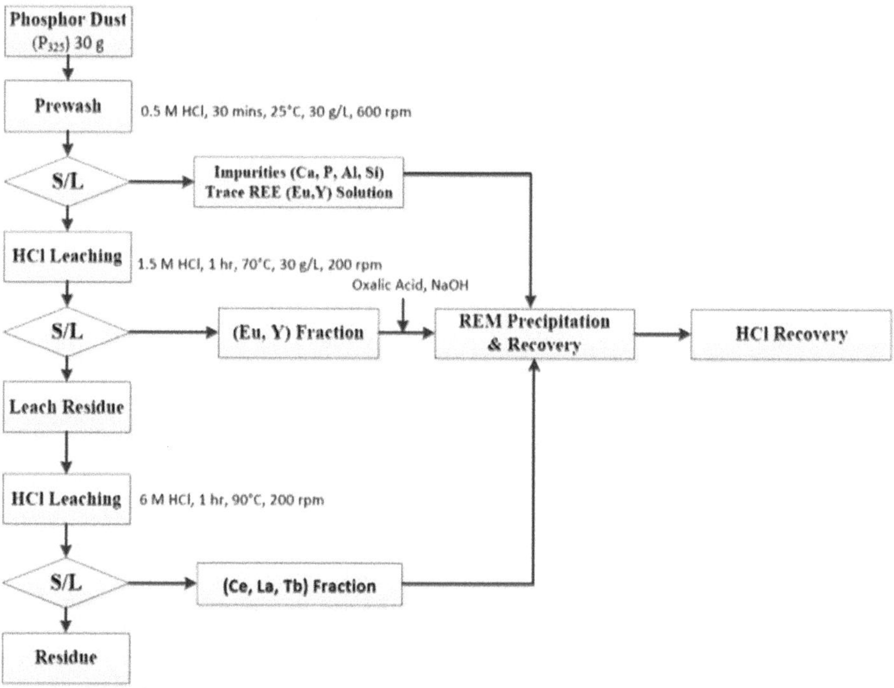

Figure 2: Flowsheet for multistage leach process for REMs extraction from spent fluorescent lamp phosphors.

This method was tried on three different powders and the results are shown in Table 4. There is significant removal of impurities from the leach liquor in the first and second stage leach liquor. Thus this process approaches the problem of impurities in the leachate during downstream processing. The extraction levels of Ce, La and Tb is less than 75% and must be improved but it is important to note that using high concentrations of acid or reagents will lead to an expensive downstream processing to produce the mixed rare earth oxides (REOs).

Table 4: Leaching efficiency of REMs from phosphor for three stages

		% Extraction							
		Ce	Tb	La	Eu	Y	Ca	P	Fe
Powder 1	Wash	BDL	1.40	8.20	6.20	3.90	93.20	89.15	60.46
	1st Leach	2.60	3.00	2.00	86.00	92.30	4.10	4.50	25.40
	2nd Leach	68.60	68.80	75.40	1.40	1.20	≤0.18	4.60	9.40
Powder 2	Wash	0.30	0.40	4.30	5.70	3.60	94.00	87.10	64.60
	1st Leach	1.40	2.40	1.20	84.10	88.60	88.60	88.60	88.60
	2nd Leach	78.90	73.90	78.20	0.90	0.80	≤0.20	5.10	9.10
Powder 3	Wash	1.40	1.70	4.30	7.00	4.80	94.30	83.90	83.90
	1st Leach	3.20	3.60	1.50	90.10	93.30	2.70	2.90	28.20
	2nd Leach	68.60	66.50	71.50	1.30	1.10	≤0.16	8.30	21.10

Leaching conditions: Wash: 0.5 M HCl, 30 g/L, 0.5 hr, 25°C, 200 rpm; 1st Leach: 1.5 M, 1 hr, 70°C, 200 rpm; 2nd Leach: 4.5 M, 1 hr, 90°C, 200 rpm; BDL: below detection limits

The mass flow of the phosphor powder after the sieving and leaching steps in the process as a fraction of the starting powder is shown in Table 5. It can be seen about 20 wt% of the powder remains in the residue after the second leach. The bulk of this final residue is recalcitrant material such as silica (fine glass), iron oxide and the green phosphor, $LaPO_4$:Ce,Tb.

Table 5: Mass balance of the powder during the three-stage process.

	Powder (wt%)		
		Leachate	Residue
Undersize	83.32 ± 1.67		
Wash		64.83 ± 3.24	35.17 ± 0.08
1st Leach		74.98 ± 3.75	25.02 ± 0.05
2nd Leach		80.37 ± 4.02	19.63 ± 0.05

Undersize is the 44 μm phosphor powder obtained after sieving.

The overall process is selective, efficient and economical. However, the leaching conditions for each step should be further optimized to avoid wasting the reagents and improve the susceptibility of using this process in a commercial operation.

CONCLUSIONS

Direct acid leaching of REMs from phosphor dust results in dumping large amount of impurities into the leach solution. This approach is non-selective and also results in poor dissolution of REMs. Calcining the powder at different temperatures followed by leaching with HCl isn't efficient for the extraction of all the chief rare earths especially cerium, lanthanum and terbium. Thus selective multi-stage leaching operations are required for the separation of REEs in phosphor dust.

After dissolution of the REEs, they can be recovered by precipitation with oxalic acid and sodium hydroxide to produce mixed REOs after calcination. Further research is required to improve the dissolution of terbium and also selectively precipitate the REEs from the leach liquors to produce high purity mixed REOs.

REFERENCES

[1] Wang, X.H., Mei, G.J., Zhao, C.L., Lei, Y.G., 2011. Recovery of rare earths from Spent Fluorescent Lamps. 5th International Conference on Bioinformatics and Biomedical Engineering, (iCBBE), Wuhan (China), 10-12 May 2011, pp. 1-4.

[2] Takahashi, T., Tomita, K., Sakuta, Y., Takano, A., Nagano, N., 1996. Separation and recovery of rare earth elements from phosphors in waste fluorescent lamp (part II)—Separation and recovery of rare earth elements by chelate resin. In: Reports of the Hokkaido Industrial Research Institute, No. 295, pp. 37–44.

[3] Takahashi, T., Takano, A., Saito, T., Nagano, N., 1999. Separation and recovery of rare earth elements from phosphors in waste fluorescent lamp (part III) -Separation and recovery of rare earth elements by multistage countercurrent extraction. In: Reports of the Hokkaido Industrial Research Institute, No. 298, pp. 37–47.

[4] Takahashi, T., Takano, A., Saitoh, T., Nagano, N., 2003. Separation and recovery of rare earth phosphor from phosphor sludge in waste fluorescent lamp by multistage countercurrent solvent extraction. In: Reports of the Hokkaido Industrial Research Institute, No. 302, pp. 41–48.

[5] Wang, X.H., Mei, G.J., Zhao, C.L., Lei, Y.G., 2011. Recovery of rare earths from Spent Fluorescent Lamps. 5th International Conference on Bioinformatics and Biomedical Engineering, (iCBBE), Wuhan (China), 10-12 May 2011, pp. 1-4.
[6] Rabah, M.A., 2008. Recyclables recovery of europium and yttrium metals and some salts from spent fluorescent lamps. Waste Manage. 28, 318-325.
[7] F. Yang, F. Kubota, Y. Baba, N.Kamiya, and M. Goto, "Selective extraction and recovery of rare earth metals from phosphor powders in waste fluorescent lamps using an ionic liquid system," Journal of Hazardous Materials, vol. 254-255, pp. 79- 88. 2013.
[8] Takahashi, T., Takano, A., Saitoh, T., Nagano, N., 2003. Separation and recovery of rare earth phosphor from phosphor sludge in waste fluorescent lamp by multistage countercurrent solvent extraction. In: Reports of the Hokkaido Industrial Research Institute, No. 302, pp. 41–48.
[9] Merritt, R.R. (1990) High temperature methods for processing monazite: I Reaction with calcium chloride and calcium carbonate, J. Less Common Metals, 166:197–210.
[10] Merritt, R.R. 1990a. High temperature methods for processing monazite. II. Reaction with sodium carbonate, J. Less Common Metals, 166:211–219.

AUTHOR INDEX
Rare Metal Technology 2015

A
Abdel-Rehim, A. 193
Ahn, J. 157
Allanore, A. 177
Aune, R. 145

B
Bakr, M. 193
Bian, Y. 239

C
Chen, Y. 19
Chorney, M. 169
Chuanhua, L. 45
Cong, W. 231

D
Deliang, Z. 11
Ding, W. 239
Downey, J. 127
Dreisinger, D. 87
Dudley, S. 127, 169

E
Eduafo, P. 253

F
Feng, J. 45
Fitch, R. 87
Forstner, C. 87

G
Gleason, W. 127
Gleason, W. 169
Guo, H. 209
Guo, S. 239

H
Han, L. 135
Huang, H. 127

J
James, R. 127
Jegal, Y. 157
Jiang, L. 209, 239
Jiang, T. 55, 63, 79, 107
Jinhui, P. 45
Jo, S. 249

K
Kaihua, L. 231
Kennedy, M. 145
Kim, H. 117
Kim, J. 117
Kim, M. 117
Kumakura, H. 101
Kumar, J. 117, 157

L
Lambotte, G. 177
Le, T. 11
Lee, C. 249
Lee, G. 249
Lee, J. 117, 249
Li, B. 19, 187
Li, C. 71
Li, H. 71, 209
Li, Q. 55, 63, 79, 107
Liang, L. 231
Libo, Z. 45
Liu, Q. 135
Liu, S. 55
Liu, T. 135
Liu, X. 3
Liu, Y. 217
Liu, Z. 217
Lu, C. 239
Lu, X. 239
Lv, G. 217
Lv, P. 55

M
Mishra, B. 253

N
Nagai, K. 101
Nagai, T. 101
Nakamura, T. 37
Nakanishi, B. 177
Niehoff, T. 11

Q
Qin, W. 3, 225
Qing, Y. 11
Qingdong, M. 231
Qiu, G. 3, 225

R

Ramakrishna, C. 157
Rosenberg, E. 127, 169
Ryu, H. 249

S

Shaohua, J. 45
Shen, M. 19
Sheng, Y. 135
Smith, K. 27
Strauss, M. 253
Sun, T. 145

T

Tang, K. 239
Thriveni, T. 157
Tranell, G. 145
Twidwell, L. 127, 169

V

Verbaan, N. 87
Vidal, E. 27

W

Wallace, G. 127
Wang, B. 135
Wang, J. 3, 225
Wang, M. 187
Wang, Q. 19
Wang, S. 71
Wang, X. 187
Wang, Y. 79

X

Xie, B. 209
Xie, Z. 63
Xu, B. 55, 63, 79, 107

Y

Yan, X. 209
Yanai, S. 101
Yang, T. 135
Yang, X. 71
Yang, Y. 55, 63, 79, 107
Yin, W. 107
Yongguo, L. 11
Young, C. 127, 169
Yurko, J. 27

Z

Zhang, G. 217
Zhang, L. 71
Zhang, T. 217
Zhang, X. 79
Zhang, Y. 217
Zhao, H. 3, 225
Zhongling, W. 11

SUBJECT INDEX
Rare Metal Technology 2015

A

Acid Chloride Leaching....87
Ag....107
Alkaline Pressure Oxidation....55
Alkaline Processing....193
Apatite....145
Au....107

B

Ball-Milling....193
Beryllium....27
Bio-Hydrometallurgy....3
Bolivia....87

C

Calcified Vanadium Slag....209
Carbonaceous Refractory Gold Ores....63
Carbonation....157
Characterisation....169
CMC....79
Comprehensive Utilization....187
Concentration....169
Convection....249
Copper....45, 87
Critical Metals....37
Cyanide Leaching....63

D

Decopperized Anode Slime....107
Directroasting....209
Distribution Factor....217
Double Refractory Gold Ore....55

E

Earths....127
Electrochemical....19
Electrodeposition....249
Electrowinning....249
Extraction....177
Extraction Separation....217
Extractive....27

F

Flotation....55, 63

G

Gallium....87
Gold....87
Gold Leaching....71, 79
Gold Recovery....11

H

Heavy Metals Removal....157

I

Impurity....19
Indium....87
Industrial Oxygen....11
Industrial Practice....3

K

Kinetics....193
Korean Red Mud....157
Kroll Process....27

L

Leaching....127, 135, 253
Leaching Kinetics....11
Leaching Method....107
Lead....87
LiCl-KCl....19

M

Malku Khota....87
Mass Transfer....135
Metallurgy....27
Microfluid....45
Microreactor....45
Modification....169
Molten Oxide Electrolysis....177

N

NaCl....19
Neodymium....249
Non-Cyanide Method....71

P

Pd....107
Permanent Magnet....239
Perovskite-Type Oxide....101
PGM....101
Phosphor Powder....253
Preoxidation....55
Pressure Leaching....217
Pretreatment....107
Process Optimization....231
Product Grade....231

R

Rare Earth 239
Rare Earth Elements 145, 157, 169
Rare Earth Metals 253
Rare Earth Oxide Electrolyte 177
Rare-Earth 135
Recovery 169
Recycle 253
Recycling 37, 101, 177, 239
REE 145
Refractory Gold Concentrate 71
Refractory Ore 11
Removal 19
Resin 169
Resource Efficiency 37
Roasting 63
Rupture 169

S

Silver 87
Simulation 135
Soda Leaching 209
Solvent Extraction 45, 157
Statistical Optimization 127
Stone Coal 187
Sulfide Minerals 79

T

Thiosulfate Consumption 79
Titanium Sponge 231

U

Urban Mining 37

V

Vanadium 187
Vanadium Slag 217

W

Waste Fluorescent Lamp 253
Without Roasting 217
Wolframite 193

Y

Yttrium Recovery 157

Z

Zambia 3
Zinc 87

MIX
Papier aus verantwortungsvollen Quellen
Paper from responsible sources
FSC® C105338

If you have any concerns about our products,
you can contact us on
ProductSafety@springernature.com

In case Publisher is established outside the EU,
the EU authorized representative is:
Springer Nature Customer Service Center GmbH
Europaplatz 3, 69115 Heidelberg, Germany

Printed by Libri Plureos GmbH
in Hamburg, Germany